NUMERICAL ANALYSIS

An intelligent approach to numerical computation

Bill Dalton B.Sc., Ph.D.

Head of Department of Mathematics
Harrow School

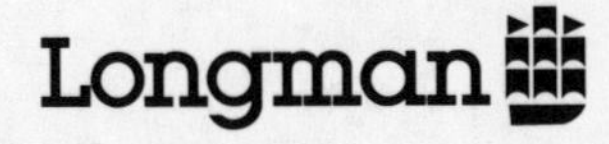

To Dorothy, Michael and Tim for their love, for the help and encouragement they have given me and for being such a tremendous family.

Longman Group UK Limited
Longman House, Burnt Mill, Harlow, Essex CM20 2JE, England and Associated Companies throughout the world.

First published 1991

Set in 10/12 pt Times
Printed in Singapore
Produced by Longman Singapore Publishers Pte Ltd

ISBN 0 582 06419 8

Acknowledgements

We are grateful to the following for permission to reproduce copyright material;

Oxford & Cambridge School Examinations Board for questions, part questions & adapted questions from past examination papers; University of Cambridge Local Examinations Syndicate for questions from past examination papers.

The following abbreviations have been used to indicate sources of questions and Tasks where appropriate:

(C) Cambridge Local Examination Syndicate Syllabus C

(MEI) Mathematics for Education and Industry Project: Oxford and Cambridge School Examinations Board

(SMP) School Mathematics Project set by Cambridge Local Examination Syndicate

(Y&S) Yakowitz and Szidarovsky An introduction to numerical computation 2nd edition Macmillan

We have been unable to trace the copyright holder in *An introduction to numerical computation* (2nd Edition) by S Yakowitz & F Sziderovsky (pub Macmillan Inc, 1989) & would appreciate any information that would enable us to do so.

Cover photo by Roman Tomaschitz. Limit set (computer graphic) plate X.

Contents

Foreword 1

Introduction 3

List of algorithms 23

Chapter 1 Errors 25

1.1 How errors occur 26
1.2 Measurement of error 30
1.3 Error analysis 39
Exercise 1 54

Chapter 2 Polynomials 1 64

2.1 Introduction to polynomials 64
2.2 Polynomial interpolation 68
2.3 The general Lagrange interpolating polynomial 74
2.4 Interpolating polynomials of higher degree 77
2.5 Proof of Theorem 2.1 78
2.6 Errors 80
Exercise 2 85

Chapter 3 Solution of equations 1 94

Introduction 94
3.1 The bisection method 97
3.2 Stopping rules 101
3.3 Convergence 104
3.4 The secant method 106
3.5 Convergence 109
3.6 The fixed point iterative method 109
3.7 Geometrical interpretation of the fixed point iterative method 116
3.8 Errors 120
3.9 Error analysis in the bisection method 120
3.10 Convergence 122

3.11 Error analysis in the secant method 122
3.12 Error analysis in the fixed point iterative method 122
3.13 Using the convergence property of the fixed point iterative method 124
3.14 Acceleration of convergence 128
Exercise 3 131

Chapter 4 Solution of systems of linear equations 141

Introduction 141
4.1 The method of Gaussian elimination for solving a system of equations 143
4.2 Gaussian elimination applied to an augmented matrix 149
4.3 An algorithm for Gaussian elimination 152
4.4 Sum check in Gaussian elimination 153
4.5 Tridiagonal systems of equations 155
4.6 Errors using Gaussian elimination 157
4.7 Pivoting strategy 160
4.8 Ill-conditioned systems 163
Exercise 4 167

Chapter 5 Polynomials 2 174

5.1 Taylor polynomials (centre = 0) 174
5.2 Taylor polynomials (centre ≠ 0) 176
5.3 Taylor's theorem 183
5.4 Horner's method of polynomial evaluation 188
Exercise 5 196

Chapter 6 Solution of equations 2 204

6.1 The Newton-Raphson method 204
6.2 Graphical representation of the Newton-Raphson method 210
6.3 Convergence 213
6.4 Errors and convergence 213
6.5 Finding the real solutions of a polynomial 216
Exercise 6 225

Chapter 7 Numerical differentiation and integration 231

Introduction 231
7.1 Numerical differentiation 232
7.2 To calculate approximately the value of $\frac{dy}{dx}$ 233
7.3 To calculate approximately the value of $\frac{d^2y}{dx^2}$ 236
7.4 The error term in numerical differentiation 237
7.5 Numerical integration 241
7.6 The trapezium rule 242
7.7 Simpson's rule 250
7.8 The error term in numerical integration 258
Exercise 7 268

Chapter 8 Numerical solution of differential equations 282

Introduction 282
8.1 Differential equations 283
8.2 Euler's method 288
8.3 The error in Euler's method 293
8.4 An alternative approximation to $\frac{dy}{dx}$ 295
8.5 Taylor methods 298
8.6 Runge-Kutta methods 304
8.7 Second order differential equations 311
Exercise 8 321

Tasks 335

Appendix 1 352
Appendix 2 353
Appendix 3 355
Bibliography 357
Answers 358
Index 376

Foreword

The aim of this book is to assist the reader to gain a clear understanding of the problems and the principles of numerical analysis.

The book has eight chapters. Each chapter is graded in difficulty. The first part of each chapter is intuitive in character and explains and describes techniques. The second part is more difficult conceptually and algebraically. This latter part tends to deal with the theory of the subject and it is here that most of the proofs in the book will be found.

At the end of each chapter, there are the exercises. Again, these have been divided into two sections. In the first section, the exercises are based largely on the text and follow the order of the chapter. Some of the exercises (particularly the earlier ones) are very simple, but the level of difficulty increases as the exercises progress. Finally, there is a (smaller) collection of harder, miscellaneous questions. Most of these are of a level of difficulty associated with A-level: many are from former A-level papers.

The chapters are not independent.

Chapter 1 'Errors' is fundamental and Sections 1.1 and 1.2 represent an absolute minimum for the required reading from this chapter.

Chapter 2 'Polynomials 1' also contains fundamental material and Section 2.1 should be read by everyone.

Thereafter, the construction of the chapters allows for great versatility in the design of courses.

A first course in numerical analysis might consist of, for example:

Chapter	Sections
1	1.1, 1.2, 1.3a
2	2.1
3	3.1, 3.2, 3.3, 3.6, 3.7
5	5.1
6	6.1, 6.2
7	7.1, 7.2, 7.5, 7.6, 7.7
8	8.1, 8.2, 8.4, 8.5 and hopefully part of 8.6

This course could be extended by including some of:

Chapter	Sections
1	1.3b, 1.3c
3	3.4, 3.5, 3.8, 3.9
4	4.1, 4.2, 4.3
5	5.2, 5.4
6	6.3, 6.5, 6.6
8	8.6

Further extensions might include:

Chapter	Sections
3	3.10, 3.11, 3.12, 3.13, 3.14
4	4.4, 4.6, 4.7
6	6.4
7	7.3, 7.4, 7.8
8	8.3, 8.7

Finally:

Chapter	Sections
1	1.3d
2	2.2 – 2.6
4	4.5, 4.8
5	5.3

A special feature of this book is that the important results in each chapter are highlighted. It is hoped that emphasising the important results will assist in the understanding, will make for quick and easy reference and will be useful for revision purposes.

Numerical analysis is now firmly linked to computers and it is no longer reasonable to expect students to work in numerical analysis without the benefit of a computer. So, we have provided algorithms for all the main techniques. These are written in a form which will, it is hoped, enable students to write programs for their own computers with considerable ease. A full description of the language used to describe the algorithms is given in the 'Introduction'.

At the end of the book there are twenty-one tasks. These are a collection of largely open-ended questions of the type due to be set for the coursework component in the new A-level examinations.

Finally, numerical analysis is fun. Go and explore the modern world of computation. Be prepared to be surprised. Enjoy yourself.

Introduction

In this book we shall be studying numerical analysis. Before getting down to the details, it is essential to have some idea of what numerical analysis is all about and so this Introduction will begin by attempting to answer the question, 'what is numerical analysis?'

The question is easier to ask than to answer. However, one definition is 'numerical analysis is an intelligent approach to numerical calculation'. This definition is certainly reasonable, but it is not perhaps very informative. The important words are 'intelligent' and 'numerical calculation'. Numerical analysis is concerned with how calculations are carried out and as the reader is probably aware there are often several methods of carrying out a given calculation. Hence, we may ask which of the available methods is the best one. It is tempting to write that numerical analysis is the art of using our intelligence to select that best method, but there is a snag. What do we mean by 'best' method? Again, there is no easy answer to this question. The 'best' method could be the method that produces the answer in the shortest possible time – that is, the fastest method. This answer seems promising. When so many calculations are performed on a computer and computer time is expensive, a fast method of calculation must have some appeal. There is another answer, however. The 'best' method is the method that produces the most accurate answer. This answer may come as something of a surprise to some readers. Perhaps they had assumed that the answers to all correctly performed calculations would be accurate. Alas, they are not and it is one of the main purposes of this book to show that, in calculations, errors are often unavoidable. If, then, errors can not be avoided, we may ask: 'Is it possible for the size of the error to be reduced?' or 'Is it possible to calculate a 'bound' which will tell us something about the size of the largest error that can occur?'

Showing how to reduce errors and how to calculate bounds limiting the size of errors will also form a major part of this book.

So let us now repeat the definition:

> Numerical analysis is an intelligent approach to numerical calculation.
>
> Or:
>
> In numerical analysis we use intelligence and mathematical skills to examine the techniques of numerical calculation. We consider the speed and the accuracy of different techniques and use this knowledge to select the method that is 'best' for the particular problem under consideration.

It might be useful to describe here the contents of the remainder of the Introduction.

The next section describes six problems. Each problem is linked to one or more of the chapters that follow. The problems are meant to illustrate the background to the chapters and to give an overview of the kind of difficulties that the techniques described in the chapters were designed to overcome.

The third section describes and discusses algorithms.

The fourth section contains results from pure mathematics that will be needed in later parts of the book. These could easily be omitted on a first reading.

Finally, a list of the notation used in the book is provided.

Six problems

The first problem concerns the way that numbers are stored in a computer's memory. Since there is a reasonable amount of technical detail, this is by far the most challenging of the six problems. While it is not necessary to follow through the working of every line, it is essential that the reader grasp the nature of the fundamental problem: namely that most numbers must, of necessity, be stored inaccurately in a computer's memory. If nothing else, read and absorb the final paragraph of Problem 1.

Problem 1

How does a computer store numbers in its memory?

For this section we need a small amount of binary arithmetic. Any reader not familiar with this subject should read Appendix 1 before proceeding.

Information is passed in a computer by a sequence of electronic impulses. The system is 'on' if an impulse is present and 'off' if no impulse is present. Hence, it is natural to think of recording information in a computer using a number system consisting of just two elements. Such a system is the binary system. Before a computer can 'understand' a number, that number must be written in binary form.

So, for example, when 55.125 is typed into a computer, the denary (base 10) number 55.125 is written by the computer in binary form. Observing that

$$\begin{aligned}55.125 &= 32 + 16 + 4 + 2 + 1 + \tfrac{1}{8}\\ &= 1\times2^5 + 1\times2^4 + 0\times2^3 + 1\times2^2 + 1\times2^1 + 1\times2^0 + 0\times2^{-1} + 0\times2^{-2} + 1\times2^{-3}\end{aligned}$$

it may be seen that the binary form of 55.125_{10} is 110111.001_2 (where the subscripts 10 and 2 indicate the base in which the number is written). The computer then performs all arithmetic and memory operations in terms of binary arithmetic. Because the binary numbers concerned in these operations can be quite long, it is useful to have a unit in which to work.

The basic unit of storage used is the **byte**, which consists of eight binary digits. So, for example, the eight binary digits 00111011 constitute one byte. This byte represents the number

$$\begin{aligned}&0\times2^7 + 0\times2^6 + 1\times2^5 + 1\times2^4 + 1\times2^3 + 0\times2^2 + 1\times2^1 + 1\times2^0\\ &= 32 + 16 + 8 + 2 + 1\\ &= 59_{10}\end{aligned}$$

By considering the smallest and the largest numbers that can be stored in one byte:

00000000 (= 0) and 11111111 (= 255)

we see that 256 different numbers can be stored in a single byte. Of course, we need more than 256 numbers if computing is to be meaningful. To accommodate a large range of numbers, more than one byte is made available for storage.

Most computers presently use a system similar to the following. When a denary number x is typed in, the computer transforms x into binary form. The binary form is then written in the particular form:

$$x = \pm\, m \times 2^e$$

where m lies between $\frac{1}{2}$ and 1 and is called the **mantissa** and e is an integer (positive, negative or zero) and is called the **exponent**.

So, for example, the denary number 7.625 which is 111.101 when written as a bicimal would be entered into the computer in the form

$$.111101 \times 2^3$$

or, more precisely, in the form $.111101 \times 2^{11}$ since the exponent e must also be written in binary form.

.111101 is the mantissa and 3 or 11_2 is the exponent.

More precisely, the mantissa is chosen so that $\frac{1}{2} \leqslant m < 1$. There is no difficulty in putting this into effect: the value of e may be adjusted to ensure that the desired inequality is satisfied. The reason, incidently, for requiring that $m \geqslant \frac{1}{2}$ is that when this inequality holds, the first digit in the binary representation of m must be 1. Since we know that the first digit of m is always 1 there is no need to specify this digit. Instead, we may store in this slot the sign of x.

A number written in the form $\pm m \times 2^e$ is said to be in floating point form. More of this in Chapter 1.

We shall assume that the mantissa has three bytes of storage allocated in the computer for its use and that the exponent has one byte. This provides a reasonable and realistic model of a modern computer. In this case, the mantissa has twenty-four binary digits for its specification (the first of these digits being 1, of course). Hence the smallest number that may be represented in m is:

$$\underbrace{.100 \ldots 000}_{24 \text{ digits}} = 0.5$$

and the largest number is:

$$\underbrace{.111 \ldots 111}_{24 \text{ digits}} = 1 - 2^{-24} \approx 1$$

The exponent e has 1 byte of computer memory at its disposal so e could take any integral value in the range 0 to 255. However, to allow for negative exponents the number that is actually stored in this byte is:

exponent $+128$

So the exponent actually takes values in the range:

$$-128 \leqslant e \leqslant 127$$

In diagrammatic form we have:

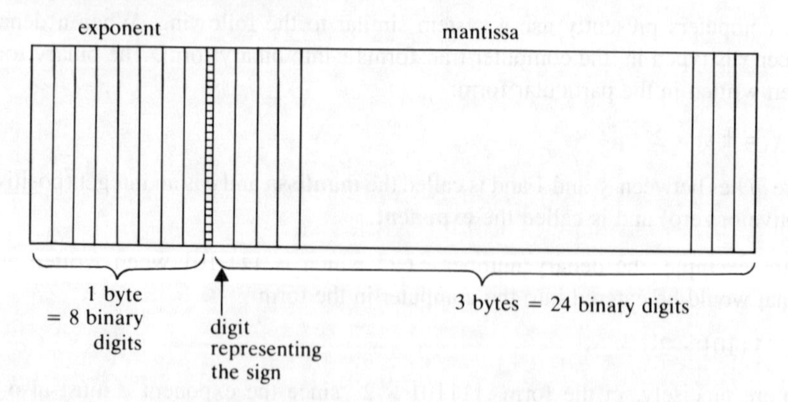

Figure I.1

A number written in this way will be called a **machine number.**

To illustrate how this works in practice, the number

$$1011.1001111001 = .10111001111001 \times 2^4$$

has mantissa $m = .10111001111001$ and exponent $e = 4$.

Since e = 100 when written in binary notation, the number would be stored as a machine number as in the diagram below.

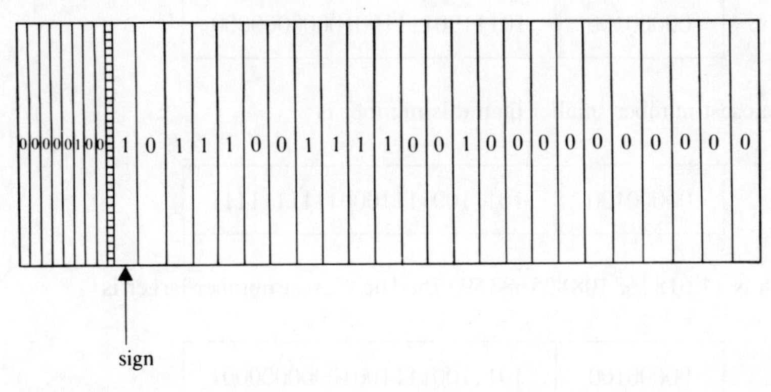

Figure I.2

The reader may verify that the number we have been considering is the denary number 11.6181640625.

It would seem, at first sight that having three bytes or twenty-four binary digits in which to store the mantissa is almost extravagant in its use of memory. Surely, with twenty-four binary digits, this system must be very accurate indeed. However, the first stirrings of doubt may be aroused when we ask a few simple questions.

(a) What are the smallest and the largest numbers that may be stored in our computer?
(b) What are the two numbers closest to 11.6181640625 that may be stored in this computer?
(c) What happens when the computer is required to store an infinite decimal, for example π, $\sqrt{2}$ or $\frac{2}{3}$? What is the greatest error that can occur when a number is stored in our computer's memory?
(d) How many numbers can be stored in our computer with perfect accuracy?

The answers to these questions are set out below.

(a) The smallest number that may be stored is:

$$\underbrace{.100 \ldots 000}_{24 \text{ binary digits}} \times 2^{-128} \approx 1.5 \times 10^{-39}$$

Any number smaller than this will be recorded as zero. And this can have unfortunate consequences involving apparent attempts to 'divide by zero'.

The largest number is:

$$\underbrace{.111 \ldots 111}_{24 \text{ binary digits}} \times 2^{127} \approx 1.7 \times 10^{38}$$

Any number larger than this would cause the computer to 'overflow' and terminate the calculation.

(b) We have seen that the denary number 11.618 164 0625 is stored by our computer as

00000100	101110011110010000000000

The closest number smaller than this number is

00000100	101110011110001111111111

which is 11.618 163 108 825 683 593 75. The closest number larger is

00000100	101110011110010000000001

which is 11.618 164 122 104 644 775 390 625. Hence, the machine number

00000100	10111001111001

must represent not only 11.618 164 0625 but many other numbers as well. By observing the two machine numbers nearest to 11.618 164 0625 we see that

00000100	10111001111001

must represent all the numbers in the interval

[11.618 163 585 662 841 796 875 to 11.618 164 092 302 322 387 695 312 5]

It may be seen then, that a single computer number must represent not one, but many real numbers.

(c) There is only a finite amount of space within a computer's memory to store any number so when a decimal requiring infinitely many decimal places is encountered, the first n decimal places are recorded and the rest ignored. Hence, every infinite decimal is recorded inaccurately in a computer's memory. The error involved in this process is precisely that part of the number that is ignored. If we assume that rounding is used then, since our computer stores twenty-four binary digits, the maximum error in the mantissa will be

$$\underbrace{.000 \ldots 000}_{24 \text{ binary digits}}111111111 \ldots = 2^{-25} \approx 2.98 \times 10^{-8}$$

Hence our infinite decimal will be correct to more than seven but less than eight decimal places. We conclude from this that numbers are stored in our computer's memory to an accuracy of between seven and eight digits.

(d) To answer this question consider first the set of numbers lying between 0.5 and 1. More precisely, consider the set of numbers x where $0.5 \leqslant x < 1$. Clearly, when written as machine numbers in the form $m \times 2^e$, all these numbers will have exponent $e = 0$. m consists of twenty-four binary digits but since the first of these is always taken to be 1, there are twenty-three binary digits in the mantissa which are free to take the values 0 or 1.

Hence there are $2^{23} = 8\,388\,608$ machine numbers lying between 0.5 and 1 that may be stored with perfect accuracy in the computer. This may seem to be a very large number indeed but since there are infinitely many numbers lying between 0.5 and 1 the result is perhaps not quite so impressive as might at first have been thought.

Next, consider the set of numbers lying between 1 and 2. Observe that when a number in this range is written in the form $m \times 2^e$, the exponent $e = 1$. Again, there are 2^{23} machine numbers in this interval so there are 8388608 numbers between 1 and 2 that may be stored with perfect accuracy.

Consider now the set of numbers lying between 2 and 4. When such a number is written in the form $m \times 2^e$, the exponent $e = 2$. Again, there are 2^{23} numbers in the interval [2,4) that may be stored with perfect accuracy.

This may seem to be pointlessly repetitive: there are 8388608 numbers in each interval that may be stored accurately. What is alarming is that while the totality of accurately stored numbers remains constant, the length of the interval is increasing rapidly. To continue:

Interval	e	**Total of numbers in the interval that are stored accurately**
[4,8)	3	$2^{23} = 8388608$
[8,16)	4	$2^{23} = 8388608$
[16,32)	5	$2^{23} = 8388608$
[32,64)	6	$2^{23} = 8388608$
[2048, 4096)	12	$2^{23} = 8388608$
[524 288, 1048576)	20	$2^{23} = 8388608$

Hence, the 'density' of accurate numbers decreases sharply as larger and larger numbers are considered. Or, to state this another way, the same machine number must represent an increasing range of decimal numbers as larger numbers are considered. It becomes easy to see how, for example, subtracting two large decimal numbers that are almost equal can lead to a serious loss in accuracy. (This point is discussed in greater detail in Chapter 1.)

And it is this last point that concerns us particularly. If a computer is performing hundreds (possibly millions) of calculations per second, how will the involuntary errors incurred in most of the calculations affect the final result? Will the errors spiral out of control? Or can the errors somehow be contained to lie within acceptable limits? Might it be possible to calculate maximum values for the limits of the errors so that we have some idea of how accurate the final answer is likely to be? Hopefully, this brief outline of computer arithmetic will have convinced the reader of the need to study 'errors' in greater detail. This we shall do in Chapter 1.

Problem 2

Many people, in the course of their work, come across sets of data. For example:

(1) Investment abroad 1979–81

Amount of investment (1 unit = 1 million pounds)

	1979		**1980**				**1981**	
Quarter	**3**	**4**	**1**	**2**	**3**	**4**	**1**	**2**
Amount	1059	781	764	853	110	472	882	826

(2) Speed of a car at 1 second intervals

Time (s)	0	1	2	3	4	5	6	7	8
Speed ($m s^{-1}$)	0	0.005	0.021	0.09	0.21	0.35	0.73	1.81	4.2

What can we do with such data? At the most basic level, we might simply 'look at it'. This could involve plotting the data as a sequence of points on a graph, as in Figures I.3 and I.4.

(1) Investment abroad 1979–81

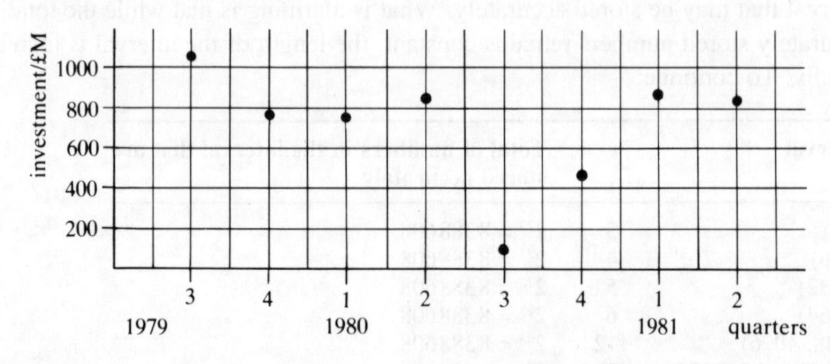

Figure I.3

(2) Speed of a car at 1 second intervals

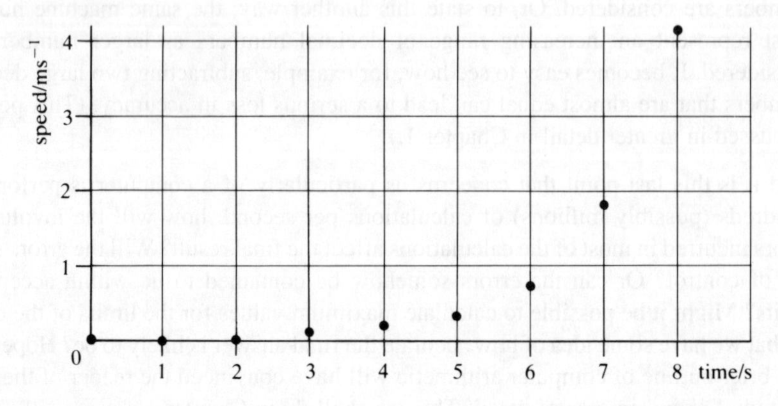

Figure I.4

But what next? Reasonable questions to ask are:

(a) 'What are the values of the data at the in-between points?' In the economics data, we might want to know the value of overseas investment in the month of April 1980, or in the week ending 11 January 1980. What values would have been observed **before** the readings began and **after** the readings had been completed? The latter would, of course, be an example of forecasting.

(b) 'How quickly is the quantity measured by the data increasing or decreasing at any particular moment?'

(c) The physics data is a sequence of velocity readings. 'Is there a way of using the data to estimate the distance travelled in (say) the first 5 seconds?'

(d) 'Is it possible to find an equation that is satisfied by the data?' That is, can we find an equation whose graph passes through each of the data points? Clearly, this would be extremely useful and would go some way towards solving the questions raised in (a), (b) and (c).

Consider the first data set, Investment abroad 1979–81. In Chapter 2 we show how to construct the equation whose graph is shown in the figure below.

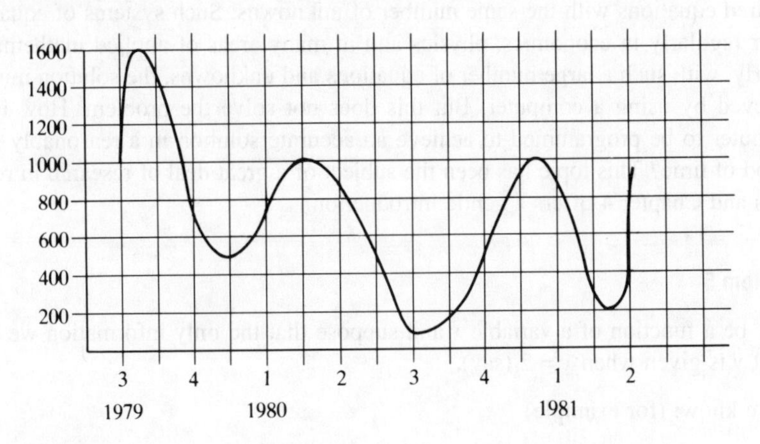

Figure I.5

Questions (b) and (c) form the main content of Chapter 7.

Problem 3

Solve the equation $x^x = 2$.

This is very nearly the simplest equation it is possible to write down and yet, as the reader may verify, it does not seem to be particularly easy to solve. In fact, this equation cannot be solved using the normal techniques of algebra.

Frequently, students who are beginning an A-level course feel puzzled that it should be possible to write down a 'simple' equation that cannot be solved. Equations, they feel, are either easy to solve or hard to solve but they are not impossible to solve. Sadly, it is a fact that very many equations (most equations in fact) cannot be solved using the techniques of algebra. What then, are we to do when we encounter such an equation? One answer is to turn to Chapters 3 and 6. In these chapters, some simple methods of obtaining 'numerical solutions' to such equations are described and investigated. A 'numerical solution' means finding, to a reasonable degree of accuracy, a numerical value (or values) for x that will satisfy the equation. Note the phrase 'to a reasonable degree of accuracy'. If this suggests to the reader that this chapter might also be concerned with errors, then the reader is learning very quickly.

Problem 4

It is hoped that everyone reading this book is familiar with the idea of simultaneous equations and would have no difficulty in solving the system of two equations for two unknowns:

$$3.5x + 2.8y = 8$$
$$98x - 27y = 5$$

In the 'real' world, however, we are more likely to encounter systems of over a hundred equations with the same number of unknowns. Such systems of equations occur regularly in economics, physics and in many areas of applied mathematics. Clearly, with such a large number of equations and unknowns, the solution must be achieved by using a computer. But this does not solve the problem. How is the computer to be programmed to achieve an accurate solution in a reasonably short period of time? This topic has been the subject of a great deal of research in recent years and Chapter 4 offers a gentle introduction.

Problem 5

Let y be a function of a variable x and suppose that the **only** information we have about y is given when $x = 2$ (say).

So we know: (for example)

$$y = 1 \quad \text{when } x = 2$$
$$\frac{dy}{dx} = 0.5 \quad \text{when } x = 2$$
$$\frac{d^2y}{dx^2} = 0.1 \quad \text{when } x = 2$$

and so on.

We would like to know something about y and $\frac{dy}{dx}$ etc. at values of x other than $x = 2$. Is there any way in which we can use the large amount of information available at $x = 2$ to extend our knowledge of y beyond $x = 2$? The answer is 'yes' and the technique is described in Chapter 5.

Observe that the situation here is almost exactly the opposite of that described in Problem 3. There, we had a small amount of information about y (the value) at a large number of values of x. Here, we have a great deal of information about y centred at just one value of x.

The technique described in Chapter 5 may also be used to write a moderately unpleasant algebraic expression in a rather more congenial form.

Problem 6

Perhaps one of the most important topics in the whole of applied mathematics is the study of differential equations. It is difficult to convey adequately the importance to mathematics of differential equations and almost impossible to imagine what applied mathematics would be like without them. And yet we have very little idea of how to solve very many types of differential equations. Indeed, it is not too far-fetched to say that any mathematician worth his salt can write down three unsolvable differential equations before breakfast.

If so many differential equations cannot be solved, how have they become so important? The answer is that differential equations can be solved numerically – which is to say that, instead of finding a general algebraic solution (which is usually impossible), we calculate good approximations to the numerical values of the solution. Thus, instead of writing down an algebraic expression, we write down a sequence of numbers. These numbers represent approximations to the numerical values of the solution. It turns out that this is sufficient for most purposes. We describe in Chapter 8 how this may be achieved.

Algorithms

Numerical analysis fundamentally is concerned with solving problems. A numerical method which can be used to solve a problem is called an **algorithm**. More precisely, an algorithm is a complete and unambiguous set of instructions which may be used to obtain a solution to a mathematical problem.

The chapters that follow, describe the methods and techniques in two ways. First the technique is developed, more or less from first principles, and this is presented in the usual mathematical manner. Secondly, an algorithm is provided for that technique. This will consist of a 'this is what you do to solve the problem' set of instructions. Since it is assumed that most of these techniques will be implemented on a computer, the algorithms are presented in a form that will enable them to be rewritten easily in a computer language (for example, BASIC, FORTRAN or PASCAL or their variants). Program listings are not given, except in one case where BASIC is used. Transporting programs from one machine to another can be time-consuming and frustrating and it is the author's opinion that algorithms written in a computer pseudocode are more useful than program listings.

The computer pseudocode will use the following statements.

1 Repetition

For $i = 1, 2, \dots, n$

this statement means that the indented statements, up to the instruction 'endloop', are to be executed in sequence first with $i = 1$, then with $i = 2$, ... , and finally $i = n$. When all the indented statements have been executed with $i = n$, the program moves on to the next step. The 'endloop' statement just marks the end of the 'for' statement.

Example

```
for x = 0, 1, 2, ... , 9
        print x
        print x²
endloop
```

will produce the following output:

0
0
1
1
2
4
3
9
•
•
•
81

In some cases, we shall not know, before the program is executed, the value of n. In such cases, we shall write

For $i = 1, 2, \ldots$, until satisfied

In this case, the indented statements will be executed in sequence, first with $i = 1$, then with $i = 2$, and so on until some preset condition has been satisfied. Then the program moves on to the next step. The nature of the preset condition, and how it is to be satisfied, will be clear from the text.

2 Choice

Our second computer statement has the form:

```
if 'expression' then
        'first instruction'
otherwise
        'second instruction'
endif
```

The 'expression' is a mathematical expression which must be either true or false. For example, the 'expression' might be '$x = y$' or '$u > 3$'. The 'instructions' might take the form 'put $x = 5$' or 'goto step 20'.

The computer statement operates as follows:

- if 'expression' is true then 'first instruction' is executed and the program then jumps to the step following 'endif',
- if 'expression' is not true then 'second instruction' is executed and the program then jumps to the step following 'endif'.

The ‘endif’ statement simply marks the end of the ‘if’ statement.

Sometimes a simpler statement is used:

```
if ‘expression’ then
    ‘instruction’
endif
```

In this case, the rule is:

- if ‘expression’ is true then ‘instruction’ is executed and the program jumps to the next statement,
- if ‘expression’ is not true then ‘instruction’ is not executed and the program jumps to the next statement,

Example   The algorithm

```
for y = 2, 3, ... , 10
    if y is a prime number then
        print y
    otherwise
        print *
    endif
endloop
```

will have output:

```
2
3
*
5
*
7
*
*
*
```

3 Assignment

The symbol := is used to assign a number to a letter (or to a memory). So we might write

$C := 4$

Then C takes the value 4.

Example   The algorithm:

```
for i = 2, 3, ... , 1000000
    if i is prime then
        p := i
    endif
endloop
print p
```

will print the largest prime number less than 1000000.

The algorithm:

```
for i = 2, 3, ... , 1000000
    if i is prime then
        p := i
    endif
    print p
endloop
```

will print all prime numbers less than 1000000

4 Input and output

The input is the information that is fed into the computer before the program can begin to operate.

The output is the information that we want out of the computer at the end of the calculation. In simple terms, the output represents 'the answer'.

Example

Given a number N: find the sum of all the perfect squares that are less than N.

```
    input : N
    s := 0
    for x = 1, 2, ... , until satisfied
        if x² ≥ N then
            goto  ●
        endif
        s := x² + s
    endloop
●   output : s
```

Note: this is not necessarily the most elegant algorithm for the solution of this problem.

Throughout this book we shall write the algorithms using these four types of statement. It is hoped that the reader will find it easy to convert the algorithms into computer programs. But the observation made in the note above will be relevant in many of the algorithms we present. The aim has been to achieve clarity by limiting the number of algorithmic statements to four. When readers know what has to be done then they can use the resources of the computer language in which they are working to achieve an elegant and an economical program.

Results from pure mathematics

We need a definition and two theorems.

Continuous functions

On several occasions we shall talk about a 'continuous' function. A precise definition of what is meant by 'continuous' requires reasonably advanced mathematics. However, an informal definition will be sufficient for our purposes.

Definition

> A function is continuous in a certain interval if the graph of the function in that interval is a single unbroken line.

So, for example, the functions whose graphs are illustrated below are continuous.

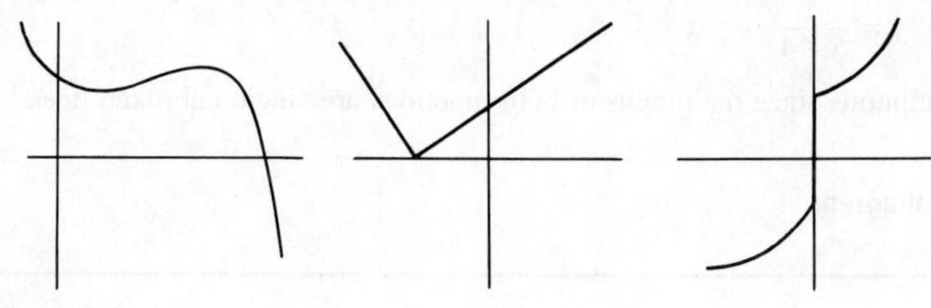

Figure I.6

The only discontinuous functions that we are likely to meet are those functions that are undefined for one or more values of x. Consider for example, the function given by

$$y = \frac{1}{x-1}$$

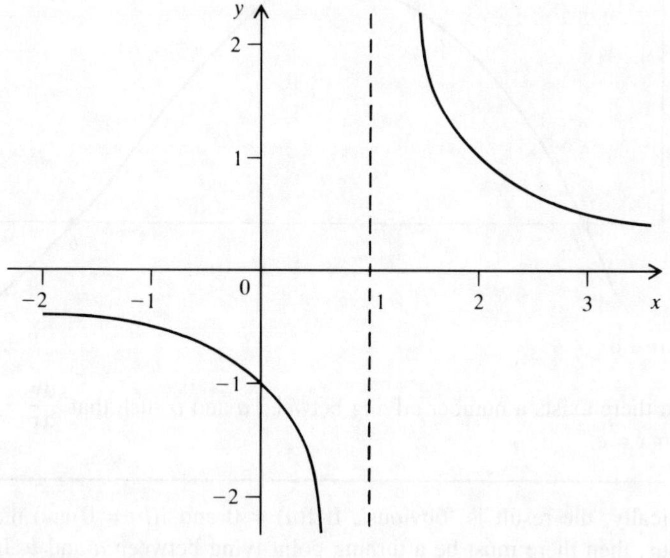

Figure I.7

The function is not defined when $x = 1$ and it is clear that the graph of $y = \frac{1}{x-1}$ is broken into two disjoint sections by the line $x = 1$. We may say, then, that the function $y = \frac{1}{x-1}$ is not continuous and that the discontinuity occurs at $x = 1$.

However, the functions

$$y = \frac{1}{x-1} \qquad x > 1$$

and

$$y = \frac{1}{x-1} \qquad x < 1$$

are continuous since the graphs of both functions are single unbroken lines.

Rolle's theorem

Let $y = f(x)$ be a function which may be differentiated at every point in the interval $[a,b]$. Suppose further that the function crosses the x-axis at both end points of the interval.

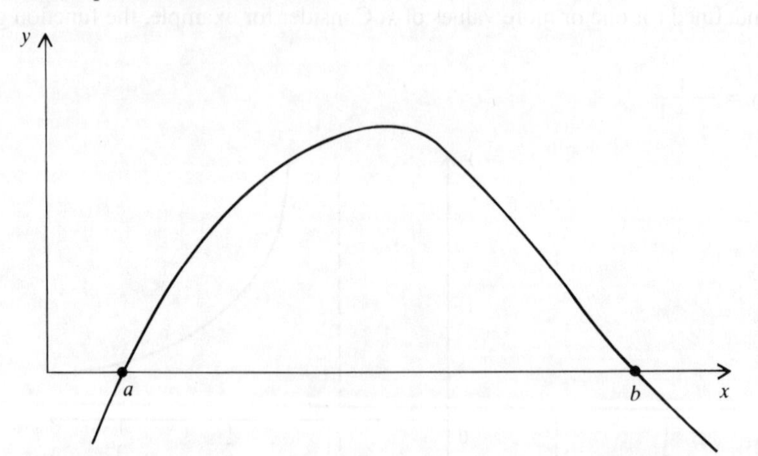

Figure I.8

Then there exists a number c lying between a and b such that $\frac{dy}{dx} = 0$ when $x = c$.

Geometrically, the result is 'obvious'. If $f(a) = 0$ and $f(b) = 0$ and the curve is continuous, then there must be a turning point lying between a and b. In the next figure, there are three possible candidates for the number c: c_1, c_2 and c_3.

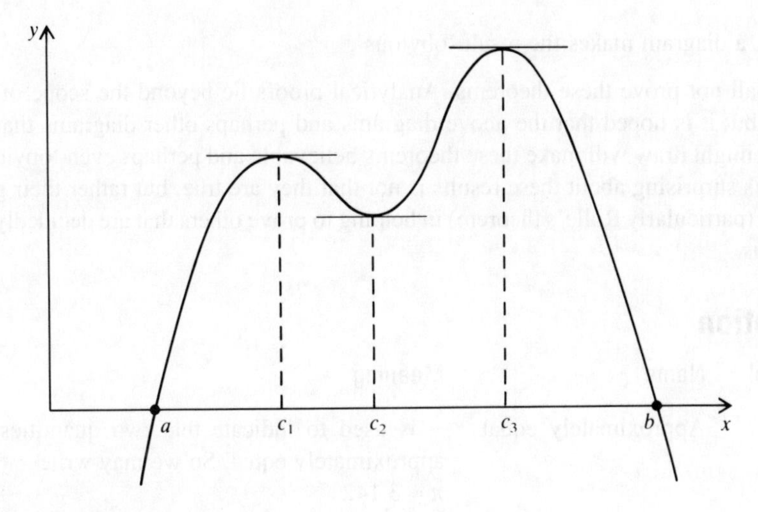

Figure I.9

The intermediate value theorem

> Suppose that $y = f(x)$ is continuous in the interval $[a,b]$. Let f_{min} be the smallest value taken by $f(x)$ in the interval $[a,b]$ and let f_{max} be the largest value taken by $f(x)$ in this interval.
>
> If l is any number lying between f_{min} and f_{max} then there exists a number c lying between a and b such that $f(c) = l$.

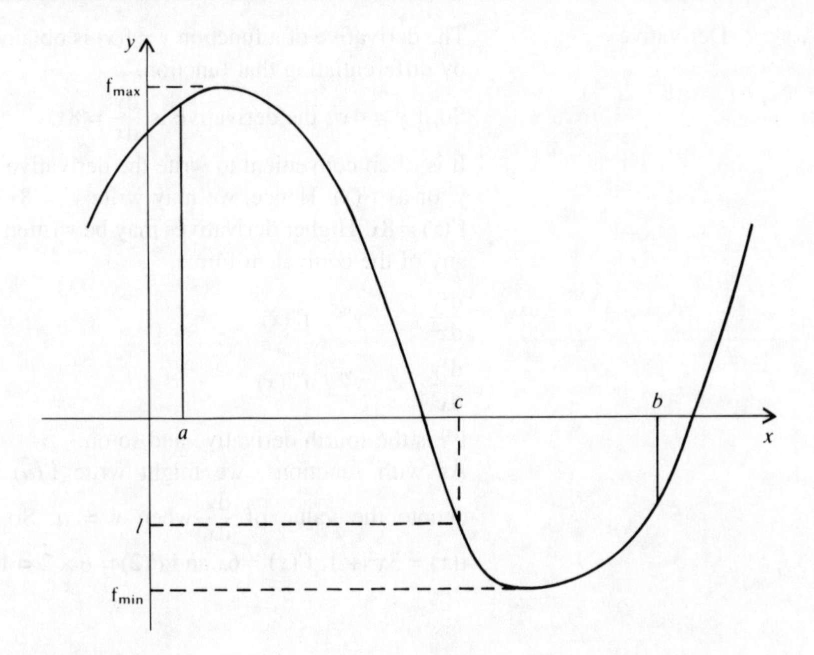

Figure I.10

Again, a diagram makes the result 'obvious'.

We shall not prove these theorems. Analytical proofs lie beyond the scope of this book, but it is hoped that the above diagrams and perhaps other diagrams that the reader might draw will make these theorems believable and perhaps even 'obvious'. What is surprising about these results is not that they are true, but rather their great power (particularly Rolle's theorem) in helping to prove others that are decidedly not obvious.

Notation

Symbol	Name	Meaning
$\approx$	Approximately equal	$\approx$ is used to indicate that two quantities are approximately equal. So we may write $\pi \approx 3.142$ $\sin 45° \approx 0.7071$ diameter of the Earth $\approx$ 12500 km
$y = \mathrm{f}(x)$	Function	$y = \mathrm{f}(x)$ indicates a 'functional' relationship between y and x. So if $y = 3x^2 + 1$, we may write $y = \mathrm{f}(x)$ where $\mathrm{f}(x) = 3x^2 + 1$. The notation is useful if we do not know, or do not wish to specify, the precise functional relationship between x and y. We shall write $\mathrm{f}(a)$ to denote the value of the function $\mathrm{f}(x)$ at $x = a$. So if $\mathrm{f}(x) = 3x^2 + 1$, $\mathrm{f}(2) = 3 \times 2^2 + 1 = 13$.
$\frac{\mathrm{d}y}{\mathrm{d}x}$, y'	Derivative	The derivative of a function $y = \mathrm{f}(x)$ is obtained by differentiating that function.
$\mathrm{f}'(x)$		So if $y = 4x^2$, the derivative is $\frac{\mathrm{d}y}{\mathrm{d}x} = 8x$. It is often convenient to write the derivative as y' or as $\mathrm{f}'(x)$. Hence, we may write $y' = 8x$ or $\mathrm{f}'(x) = 8x$. Higher derivatives may be written in any of the equivalent forms $\frac{\mathrm{d}^2y}{\mathrm{d}x^2}$ y'' $\mathrm{f}''(x)$ $\frac{\mathrm{d}^3y}{\mathrm{d}x^3}$ y''' $\mathrm{f}'''(x)$ $\mathrm{f}^{(4)}$ is the fourth derivative and so on. As with functions, we might write $\mathrm{f}'(a)$ to denote the value of $\frac{\mathrm{d}y}{\mathrm{d}x}$ when $x = a$. So if $\mathrm{f}(x) = 3x^2 + 1$, $\mathrm{f}'(x) = 6x$ and $\mathrm{f}'(2) = 6 \times 2 = 12$.

Symbol	Name	Meaning
$\lvert x \rvert$	Modulus	The modulus of a number is the positive value of that number. So $\lvert -4 \rvert = 4$, $\lvert 3.14 \rvert = 3.14$ and $\lvert -0.0021 \rvert = 0.0021$. In computer language, abs(x) is often used to calculate the modulus of x.

Two important applications of the modulus are

(a) Find x if $\lvert x \rvert = 6$.

From the definition, we have $x = +6$ or $x = -6$.

(b) Find the range of values satisfied by x if $\lvert x \rvert < 4$.

Again, from the definition, it may be seen that $-4 < x < 4$.
In general,

if $\quad \lvert x \rvert \leqslant c \quad$ then $\quad -c \leqslant x \leqslant +c$

Symbol	Name	Meaning
$[a, b]$	Intervals	We specify intervals of real numbers in three forms: $[a, b]$ means the set of x such that $a \leqslant x \leqslant b$ $(a, b]$ means the set of x such that $a < x \leqslant b$ $[a, b)$ means the set of x such that $a \leqslant x < b$.
1(0.2)2		Taking values from 1 to 2 in steps of 0.2, i.e. 1 1.2 1.4 1.6 1.8 2.

List of algorithms

Algorithm	Name	Page
2.1	Neville's algorithm	77
3.1	The bisection method	100, 102
3.2	The secant method	109
3.3	The fixed point iterative method	112
3.4	Aitken's process (accelerated convergence)	130
4.1	Gaussian elimination	152
4.2	Tridiagonal systems	156
4.3	Gaussian elimination with a pivoting strategy	162
5.1	Horner's method	195, 217
6.1	The Newton-Raphson method	205
6.2	To obtain all real solutions of a polynomial	219
7.1	The trapezium rule	250
7.2	Simpson's rule	258
8.1	Euler's method	290
8.2	The simple Runge-Kutta method (modified Euler method)	305
8.3	Heun's method	307
8.4	The Runge-Kutta method of order four	309

For second order differential equations:

Algorithm	Name	Page
8.5	Euler's method	313
8.6	The simple Runge-Kutta method	314
8.7	The Runge-Kutta method of order four	319

Chapter 1

Errors

If nonmathematicians are asked what an 'error' is they would probably give replies something like 'An error is a mistake; it is doing something incorrectly; it is a flaw or a malfunction.'

If this were the definition of error that we were going to adopt, then this chapter might take the following form:

> **Chapter 1**
> errors = mistakes.
> don't make them.
> end of Chapter 1.

We shall, however, use a more specialised definition of error.

Definition

Let x represent some quantity and let x^* denote an approximation to x. Then the error in using x^* as an approximation to x is given by the difference $x - x^*$, that is:

error = exact value − approximate value

or, in symbols:

$$\text{error} = x - x^*$$

This chapter will explain how errors arise in numerical calculations and how the size of these errors may be calculated. We observe that if calculations are performed on numbers which are subject to errors then the results of these calculations will also be subject to (possibly larger) errors. We shall also show how estimates for the size of these (possibly larger) errors may be calculated. In this way, it is hoped that the proliferation of errors throughout a calculation may be controlled.

1.1 How errors occur

The errors that we shall consider in this chapter arise because human beings – and computers – can write decimals to only a finite number of decimal places. We saw in the Introduction how errors can occur when numbers are stored in the memory of a computer. But errors arise also in everyday arithmetic.

For example, to write $\frac{1}{3}$ as a decimal, we should have to write = 0.33333 ... where the 3s continue to infinity. But no human being has the time to write a decimal containing infinitely many numbers. So, to be useful, the decimal representation of $\frac{1}{3}$ must somehow be shortened. In fact, we must find a way of writing an infinite decimal as a decimal of finite length.

A lazy person might write $\frac{1}{3} = 0.3$. A rather more conscientious person might write $\frac{1}{3} = 0.333$. The computer described in the Introduction would write $\frac{1}{3} = 0.333\,333\,33$.

We know that all these representations of $\frac{1}{3}$ are inaccurate just as we know that some degree of inaccuracy is inevitable. But some of these representations are more inaccurate than others and, to help us judge the level of inaccuracy, we calculate the error that occurs when a decimal is shortened and written to a finite number of decimal places.

In the examples above, the error involved in taking $\frac{1}{3}$ to be 0.3 is:

$$0.333\,333\,333\,... - 0.3 = 0.033\,333\,333\,...$$

The error involved in taking $\frac{1}{3}$ to be 0.333 is:

$$0.333\,333\,333\,... - 0.333 = 0.000\,333\,333\,...$$

The error incurred by our computer is:

$$0.333\,333\,333\,... - 0.333\,333\,33 = 0.000\,000\,003\,333\,3\,...$$

Clearly, the more digits that we take in the finite decimal that is to represent $\frac{1}{3}$, the smaller the error and the more accurate the ensuing calculations are going to be.

Suppose now that we wished similarly to consider the fraction $1\frac{2}{3}$. Written as a decimal

$$1\tfrac{2}{3} = 1.666\,666\,66\,...$$

with infinitely many 6s. To write this as a decimal of finite length, we could write 1.6666 using just four places after the decimal point. The error would then be:

$$1.666\,666\,666\,... \quad - 1.6666 = 0.000\,066\,666\,...$$

If, however, we had written the decimal representation of $1\frac{2}{3}$ as 1.6667 (using again four places after the decimal point) the error would now be:

$$1.666\,666\,666\,... \quad - 1.6667 = -\,0.000\,033\,333\,...$$

and the error in this case is one half of that immediately above.

It does appear then that even when we take the same number of digits in a decimal representation, there are better and worse ways of obtaining a finite decimal representation of an infinite decimal.

It is hoped that the above discussion will have persuaded the reader of the need for a clear understanding of how decimals may be shortened. With this in mind, we introduce some notation.

Definition

The decimal number x is written in floating point form if x is written as:

$$x = 0.d_1d_2d_3 \ldots d_n \times 10^e$$

where $d_1 \neq 0$ and e is an integer which may be positive, negative or zero.

Examples

(a) 113.2591 is written in floating point form as $0.113\,259\,1 \times 10^3$.
(b) 0.003 529 1 is written in floating point form as $0.352\,91 \times 10^{-2}$.
(c) $-0.000\,000\,23$ is written in floating point form as -0.23×10^{-6}.
(d) 0.3333 is written in floating point form as 0.3333×10^0.

The decimal part $.d_1d_2d_3 \ldots d_n$ is called the **mantissa** and if n digits are used to represent the mantissa we say that x is written in n decimal digit (floating point) form. The number e is called the **exponent**.

Example 1.1

Write the numbers (a) 439.0057, (b) 0.000 005 39 in floating point form. In each case, write down the mantissa and the exponent.

(a) $439.0057 = 0.439\,005\,7 \times 10^3$ in floating point form. The mantissa is the seven decimal digit number 0.439 005 7 and the exponent is 3.
(b) $0.000\,005\,39 = 0.539 \times 10^{-5}$ in floating point form. The mantissa is the three decimal digit number 0.539 and the exponent is -5.

The reason for requiring that $d_1 \neq 0$ is to ensure that the floating point representation of a number is unique. If d_1 could take the value zero, all the following would be valid floating point representations of 113.2591

$0.113\,259\,1 \times 10^3$
$0.001\,132\,591 \times 10^5$
$0.000\,000\,000\,001\,132\,591 \times 10^{14}$

By insisting that $d_1 \neq 0$, we ensure that the floating point representation is unique.

Using the idea of floating point form, we now describe two ways of writing a decimal in finite digit form. The two methods are called chopping and rounding.

In chopping, the decimal is literally ‘chopped off’ after the agreed number of digits. If we were to use chopping to write a decimal to three decimal digits in floating point form, we should write down the first three digits of the mantissa and ignore the rest.

Example 1.2

Using chopping to write $x = 1398.56$ as a three digit floating point number. Find the error in writing x in this way.

$$1398.56 = 0.139\,856 \times 10^4 \text{ in floating point form}$$
$$= 0.139 \times 10^4 \text{ after chopping to three decimal digits.}$$

The error is

$$\begin{array}{rl} & 0.139856 \times 10^4 \\ & \underline{0.139 \quad\;\; \times 10^4} - \\ & 0.000856 \times 10^4 \\ = & 0.856 \times \quad 10^{4-3} \\ = & 0.856 \times \quad 10^1 \end{array}$$

Technically, if x^* is the number obtained by chopping x to n decimal digits, then x^* is the n digit number lying in the interval $[0,x]$ which is closest to x.

When rounding a decimal x to (say) three decimal digits, we choose the three digit number that is closest to x. In practice, this means that we examine the fourth digit of the mantissa; if this digit is 5 or greater, the third digit of the mantissa is increased by 1; if, however, the fourth digit is less than 5, the third digit is left unchanged. When this procedure has been accomplished, the first three digits of the mantissa are taken as the rounded three digit representation of the decimal. More generally, if a decimal is to be rounded to k decimal digits, the $(k+1)$th digit of the mantissa is examined and if it is found that this digit is 5 or greater then the kth digit is increased by 1. Otherwise, the kth digit is left unchanged. The rounded decimal is then taken to be the first k digits of the mantissa with, of course, the necessary power of 10.

Example 1.3

Use rounding to write the number $y = 0.002\,817\,52$ in four decimal digit form. Calculate the error in writing y in this form.

$0.002\,817\,52 = 0.281\,752 \times 10^{-2}$ in floating point form. Counting to the fifth decimal digit reveals a 5 in that position. In this case, the fourth decimal digit must be increased by 1. This causes the 7 to be replaced by an 8. Writing down the first four digits gives the required four digit form to be 0.2818×10^{-2}

The error is $.281\,752 \times 10^{-2} - 0.2818 \times 10^{-2}$ which is negative.

We calculate

$$\begin{array}{rl} & 0.281\,800 \times 10^{-2} \\ & \underline{0.281\,752 \times 10^{-2}} - \\ & 0.000\,048 \times 10^{-2} \\ = & 0.48 \times 10^{-2-4} \\ = & 0.48 \times 10^{-6} \end{array}$$

The error, then is -0.48×10^{-6}

Notation

To indicate that a number has been chopped or rounded to a certain number of decimal digits, we shall make remarks such as, 'The number is written correct to n decimal digits (using rounding).' Or, 'The number has been written in chopped n decimal digit form.'

If it is obvious that the numbers are all decimals, we may drop the word 'decimal' and write 'n digit form' or 'n digit numbers'. Also, if it is not specified which of chopping or rounding is being used, we shall assume that numbers are being rounded.

Further, if the approximation x arises because the decimal x has been chopped or rounded, the error $= x - x^*$ is given the special name of **roundoff error**. Hence roundoff error is the (frequently unavoidable) error that is introduced when a decimal is shortened by chopping or rounding.

The question may well be asked: 'Which of the two methods, chopping or rounding should be used?' It is clear that rounding provides the smaller roundoff error and so there seems to be a strong case for choosing this method. However, on a computer, rounding is a (relatively) time consuming and so expensive operation. Since the differences in the roundoff error in the two systems are small at the accuracy to which a computer works, most computers use a chopping technique. This example illustrates nicely the conflict that often arises in numerical analysis: is the increase in accuracy that is achieved by a more complicated technique really worth the time and expense that is required to achieve it? There are no easy answers to this question.

If the reader is by now beginning to think that we are making a fuss about nothing, consider the following simple expression:

$$\frac{1}{3}\left(\frac{2}{11} - \frac{1}{6}\right)$$

The exact value of this expression is

$$\frac{1}{3}\left(\frac{12 - 11}{66}\right) = \frac{1}{3} \times \frac{1}{66} = \frac{1}{198}$$

Written as a chopped eight digit decimal, in floating point form this becomes

$$0.505\,050\,50 \times 10^{-2}$$

Now evaluate this expression working in chopped two digit arithmetic:

$$\begin{aligned} 0.33(0.18 - 0.16) &= 0.33 \times 0.02 \\ &= 0.0066 \\ &= 0.66 \times 10^{-2} \end{aligned}$$

Working now in rounded two digit arithmetic produces:

$$\begin{aligned} 0.33(0.18 - 0.17) &= 0.33 \times 0.01 \\ &= 0.0033 \\ &= 0.33 \times 10^{-2} \end{aligned}$$

It may be seen that neither of these answers has even one digit accuracy. Hence, it may be seen that when decimals are written in shortened form, very serious errors can arise even in trivially short calculations. It may be argued, of course, that writing decimals in two digit form was inviting serious inaccuracy and that a minimum of three decimal digits should always be taken. This may be good advice, but as we shall see later in the chapter, it does not solve the problem. Even when working in three or four decimal digit arithmetic, it is disturbingly easy to produce serious inaccuracy in simple calculations.

The whole point of this example is to show how, even in simple calculations, errors can proliferate as the calculation continues. The next step is to increase our understanding of how errors can accumulate in the course of a calculation. To help in this we introduce two ways of measuring errors: the **absolute error** and the **relative error.**

1.2 Measurement of error

As before, we shall write x to denote the exact value of a decimal and x^* to denote an approximation to x.

Definition

We define the error in using x^* as an approximation to x by

$$\text{error} = x - x^*$$

and the absolute error by

$$\text{absolute error} = \left| x - x^* \right|$$

Note: the notation used here, $\left| x - x^* \right|$ is defined in the notation section at the end of the Introduction.

The absolute error is simply the (positive) magnitude of the error. Often, we are not interested in whether the error is positive or negative: it is the size of the error that is important and this is exactly what is measured by the absolute error.

Example 1.4

The solution to a problem is given as $x = 3.436$. It is known that the absolute error in the solution is at most 0.01. Find the interval within which the exact value of the solution must lie.

$$x^* = 3.436$$

and if x is the exact value of the solution then

$$\left| x - x^* \right| < 0.01$$

so $\qquad \left| x - 3.436 \right| < 0.01$

From the definition of absolute value

$$-0.01 < x - 3.436 < 0.01$$

Therefore $3.436 - 0.01 < x < 3.436 + 0.01$
and $3.426 < x < 3.446$

Since, in almost all calculations, some decimals will appear in chopped or rounded form, it would clearly be useful to have some general results which measured roundoff error. These are most conveniently presented in terms of absolute error or, in this case, absolute roundoff error. These important results form our first illustration of the use of absolute error. To introduce the ideas, we begin with a simple example.

Example 1.5

Let $x = 28.619\,788$. Calculate the roundoff error if chopping is used to write x correct to three decimal digits.

$x = 0.286\,197\,88 \times 10^2$ in floating point form
$= 0.286 \times 10^2$ after chopping

Roundoff error $= x - x^*$ and we calculate

$$\begin{array}{r} 0.286\,197\,88 \times 10^2 \\ 0.286 \qquad \times 10^2 - \\ \hline 0.000\,197\,88 \times 10^2 \end{array}$$

$$\text{Hence } |x - x^*| = 0.000\,197\,88 \times 10^2 = 0.197\,88 \times 10^{2-3} = 0.197\,88 \times 10^{-1}$$

We use this method to obtain a more general result.

Theorem 1a

Let y be any number and suppose that the exponent of y is e. That is:

$y = 0.d_1d_2d_3 \ldots \times 10^e$

If y^* is the chopped three digit form of y, then

$|y - y^*| < 10^{e-3}$

Further, if y^* is the chopped k digit form of y then

$|y - y^*| < 10^{e-k}$

A proof of Theorem 1a is presented in Appendix 2. Example 1.5 illustrates the theorem.

This type of analysis is useful if we want to keep track of the errors that occur when the numbers in a calculation are chopped to a certain number of decimal digits. The problem when numbers are rounded is similar and, again, a simple numerical example is considered first.

Example 1.6

Let $x = 0.004\,585\,29$. Find the rounding error if x is rounded to three decimal digits.

$$\begin{aligned} x &= 0.458\,529 \times 10^{-2} \text{ in floating point form} \\ x^* &= 0.459 \times 10^{-2} \text{ after rounding} \\ \text{Roundoff error} &= 0.458\,529 \times 10^{-2} - 0.459 \times 10^{-2} \\ &= -0.000\,471 \times 10^{-2} \end{aligned}$$

Hence
$$\begin{aligned} |x - x^*| &= 0.000\,471 \times 10^{-2} \\ &= 0.471 \times 10^{-2-3} \\ &= 0.471 \times 10^{-5} \end{aligned}$$

We may again produce a general result:

Theorem 1b

Let y be as described in Theorem 1a. If y^* is the rounded three digit form of y then

$$|y - y^*| < 0.5 \times 10^{e-3}$$

Further, if y^* is the rounded k digit form of y then

$$|y - y^*| < 0.5 \times 10^{e-k}$$

A proof of Theorem 1b is also presented in Appendix 2.

Example 1.7

The number w is given as 0.015 74 and this is known to be correct to four decimal digits. Unfortunately, it is not known whether chopping or rounding has been used to give 0.015 74. Calculate the smallest interval within which w must lie.

Write
$$\begin{aligned} w^* &= 0.01574 \\ &= 0.1574 \times 10^{-1} \text{ in floating point form} \end{aligned}$$

Hence the exponent of w is -1.

Case 1 Chopping was used.

From Theorem 1a

$$\begin{aligned} |w - w^*| &< 10^{e-4} \\ &= 10^{-1-4} \\ &= 10^{-5} \end{aligned}$$

Since $w - w^*$ is always positive (or zero) when chopping is used, we have

$$w - w^* < 10^{-5}$$
$$w < w^* + 10^{-5}$$
$$w < 0.015\,74 + 10^{-5}$$

Hence, $w < 0.015\,75$

The smallest interval that must contain w is

$$[0.015\,74,\ 0.015\,75]$$

Note that from the definition of chopping we could have written immediately $w < 0.015\,749\,999\,9$ ...
Hence the smallest interval that must contain w is

$$[0.015\,74,\ 0.015\,749\,999\ ...]$$

Since 0.015 749 999 ... and 0.015 75 represent the same number, the two methods are (thankfully) in agreement.

Case 2 Rounding was used.

From Theorem 1b,

$$|w - w^*| < 0.5 \times 10^{e-4}$$
$$= 0.5 \times 10^{-1-4}$$
$$= 0.5 \times 10^{-5}$$

Hence, from the definition of modulus:

$$-0.5 \times 10^{-5} < w - w^* < 0.5 \times 10^{-5}$$

and $w^* - 0.5 \times 10^{-5} < w < w^* + 0.5 \times 10^{-5}$

$$0.015\,74 - 0.5 \times 10^{-5} < w < 0.015\,74 + 0.5 \times 10^{-5}$$
$$0.015\,735 < w < 0.015\,745$$

This result could have been deduced immediately from the definition of rounding. It is, however, always gratifying to see that results obtained by the application of theorems agree with simple common sense. We shall use the idea contained in this result in Section 1.3d.

To illustrate some of the difficulties associated with the use of absolute error, consider the following two totally dissimilar examples.

(a) The value of the expression $\frac{1}{3}(\frac{2}{11} - \frac{1}{6})$ when calculated in two digit arithmetic (using chopping) was 0.66×10^{-2}. Hence, the error is

$$0.505\,050\,50 \times 10^{-2} - 0.66 \times 10^{-2}$$
$$= -0.154\,949\,50 \times 10^{-2} \text{ (8 decimal digits)}$$

and the absolute error is $0.154\,949\,50 \times 10^{-2}$.

Similarly, the error when using two digit arithmetic (and rounding) was

$$0.505\,050\,50 \times 10^{-2} - 0.33 \times 10^{-2}$$
$$= 0.175\,050\,50 \times 10^{-2} \text{ (8 decimal digits)}$$

and the absolute error is $0.175\,050\,50 \times 10^{-2}$.

Both absolute errors are reasonably small numbers so perhaps the approximations obtained by using two digit arithmetic are not as bad as we first thought. Is this reasoning correct?

(b) The engineers digging the cross Channel tunnel meet their counterparts who are digging from the other side of the Channel. The two parts of the tunnel, it is found, are 50cm away from being perfectly in line. The absolute error is 50cm and the chief engineer gets a knighthood.

A heart surgeon placing a valve in a diseased heart places the valve 50cm away from where it ought to be. The absolute error is 50cm and the surgeon gets the sack.

In both cases, the absolute error was 50cm. So errors of equal magnitude earned one person great praise and the other person public disgrace. And rightly so. The point of this little fable is that in some situations, knowledge of the absolute error is insufficient to allow us to form an opinion on the value of the approximation x^*. In such a situation, it is often better to measure the error relative to the exact value x, so that we are measuring the magnitude of the error in relation to the size of the numbers we are working with. We are thus led to the following definition.

Definition

With x and x^* as defined above, the **relative error** in using x^* as an approximation to x is defined as:

$$\text{Relative error} = \frac{x - x^*}{x}$$

Example 1.8

Calculate the relative error in evaluating $\frac{1}{3}(\frac{2}{11} - \frac{1}{6})$ in two digit arithmetic.

The relative error in performing the calculation in two digit arithmetic (chopping) is

$$\text{relative error} = \frac{-0.154\,949\,50 \times 10^{-2}}{0.505\,050\,50 \times 10^{-2}} = -0.306\,800\,01$$

and the relative error in performing the calculation in two digit arithmetic (rounding) is

$$\text{relative error} = \frac{0.175\,050\,50 \times 10^{-2}}{0.505\,050\,50 \times 10^{-2}} = 0.346\,599\,99$$

As expected, the relative errors are large, indicating that we were right to be concerned about the loss in accuracy caused by working in two digit arithmetic. The comment at the end of Example (a) at the top of this page is therefore incorrect.

The following example shows how the idea of relative error can be used with equations.

Example 1.9

Solve the equation $\frac{2}{3}(x-1) - \frac{3}{7}x = -\frac{67}{100}$.

What would be the solution if all calculations were carried out in three digit arithmetic (using chopping)?

Find (a) the absolute error and (b) the relative error that arise from the use of three digit arithmetic (using chopping).

To solve the equation:

$$21 \times \frac{2}{3}(x-1) - 21 \times \frac{3}{7}x = -\frac{67}{100} \times 21$$

giving

$$\begin{aligned} 14x - 14 - 9x &= -\frac{1407}{100} \\ 5x &= -0.07 \\ x &= -0.014 \end{aligned}$$

Working now in three decimal digit arithmetic (with chopping)

$$\begin{aligned} 0.666\,(x-1) - 0.428x &= -0.67 \\ 0.238x - 0.666 &= -0.67 \\ x &= -\frac{0.004}{0.238} \\ &= -0.0168 \end{aligned}$$

The absolute error is

$$\begin{aligned} &\left| -0.014 - (-0.0168) \right| \\ &= \left| -0.014 + 0.0168 \right| \\ &= 0.0028 \end{aligned}$$

The relative error is

$$\frac{(-0.014 - (-0.0168))}{-0.014} = \frac{0.0028}{-0.014} = -0.2$$

Here again, we have a small absolute error and a much larger relative error, indicating that the error is large in relation to the magnitude of the exact solution. Hence, the solution obtained by three digit arithmetic cannot, in this case, be regarded as highly accurate.

In Theorems 1a and 1b, bounds on the errors arising from chopping or rounding a number to a fixed number of decimal digits were calculated. Using those results, bounds on the relative errors that arise in these circumstances may similarly be calculated. In this case, the result for the rounding operation will be proved, leaving 'chopping' as an exercise.

Theorem 1c

As before, let $y = 0.d_1d_2d_3 \ldots \times 10^e$. If y^* is the rounded three decimal digit form of y then the relative error

$\dfrac{y - y^*}{y}$ introduced by this process satisfies

$$-0.5 \times 10^{-3+1} < \frac{y - y^*}{y} < 0.5 \times 10^{-3+1}$$

More generally, if y^* is the rounded k decimal digit form of y then the relative error satisfies

$$-0.5 \times 10^{-k+1} < \frac{y - y^*}{y} < 0.5 \times 10^{-k+1}$$

Proof

From Theorem 1b, $|y - y^*| < 0.5 \times 10^{e-3}$

Dividing both sides by the positive number $|y|$ gives

$$\frac{|y - y^*|}{|y|} < \frac{0.5 \times 10^{e-3}}{|y|}$$

For the left-hand side:

$$\frac{|y - y^*|}{|y|} = \left|\frac{y - y^*}{y}\right|$$

For the right-hand side: $y = 0.d_1d_2d_3 \ldots \times 10^e$ and by definition $d_1 \geqslant 1$.

Hence $y \geqslant 0.1 \times 10^e$ and the right-hand side satisfies

$$\frac{0.5 \times 10^{e-3}}{|y|} \leqslant \frac{0.5 \times 10^{e-3}}{0.1 \times 10^e} = 0.5 \times 10^{-3+1}$$

Putting all this together gives:

$$\left|\frac{y - y^*}{y}\right| < 0.5 \times 10^{-3+1}$$

From the definition of modulus we have the required inequality

$$-0.5 \times 10^{-3+1} < \frac{y - y^*}{y} < 0.5 \times 10^{-3+1}$$

The proof for the general case is similar.

For chopping, we have:

Theorem 1d

With y as described in Theorem 1c:

If y^* is the chopped three decimal digit form of y, the relative error introduced by this process satisfies

$$-10^{-3+1} < \frac{y - y^*}{y} < 10^{-3+1}$$

More generally; if y^* is the chopped k decimal digit form of y, the relative error satisfies

$$-10^{-k+1} < \frac{y - y^*}{y} < 10^{-k+1}$$

The proof is left as an exercise for the reader.

Example 1.10

It is given that $x = 6.3$, but it is known that this value is correct only to two decimal digits (using rounding). Find the extreme values between which the relative error must lie.

From Theorem 1c, since rounding to two decimal digits has been used,

$$-0.5 \times 10^{-2+1} < \text{relative error} < 0.5 \times 10^{-2+1}$$

Therefore $\quad -0.05 < \text{relative error} < 0.05$

Example 1.11

The solution to a problem is given as $u = 35.62$. It is known that the relative error in the solution is at most 2%. Find, to six decimal digits, the range of values within which the exact value of the solution must lie.

The relative error is at most 0.02

Hence $\quad -0.02 < \dfrac{u - u^*}{u} < 0.02$

If $\dfrac{u - u^*}{u} < 0.02$,

then $\quad u - u^* < 0.02u$ (since $u > 0$)

and $\quad u(1 - 0.02) < u^*$

giving $\quad u < \dfrac{u^*}{1 - 0.02} = \dfrac{35.62}{0.98}$

$$= 36.346\,938\,78$$

If $-0.02 < \dfrac{u - u^*}{u}$

then $\quad u - u^* > -0.02u$

and $\quad u(1 + 0.02) > u^*$

giving $\quad u > \dfrac{u^*}{1 + 0.02} = \dfrac{35.62}{1.02}$

$\quad = 34.921\,568\,63$

Hence $\quad 34.9215 < u < 36.3469 \quad$ (to six decimal digits)

The results of Theorems 1c and 1d will be used again in Section 1.3d. In the meantime, we summarise the results so far.

Summary

The absolute error measures the magnitude of the error.

The relative error measures the magnitude of the error in relation to the size of the exact value. Often the relative error will be a particularly useful indicator of the value of an approximation when the numbers concerned are either very large in magnitude, or close to zero.

One point to bear in mind is that often the exact value of a particular quantity is not known exactly even though we do have some idea of the size of the error involved. If the error is small, then x should not be very different from x^* and we may estimate the relative error by calculating:

$$\frac{x - x^*}{x^*} = \frac{\text{error}}{x^*}$$

So far, we have discussed in some detail how rounding errors occur and how errors may be measured. But consider that, in a modern computer, upwards of a million calculations are made every second and many programs run for many minutes (some, indeed, run for over an hour). If each number involved was subject to even a small error, what would happen when those millions of small errors acted on each other? Would the errors multiply, proliferate and spiral out of control? Certainly, in the early days of computers, it was thought that the errors would be uncontrollable and that any answer given by a computer would be unreliable. No one, it was thought, would know whether the errors had remained within reasonable bounds or whether they had become large enough to significantly affect the results of the calculation. In fact, it is now known that it is possible to keep track of the errors within a calculation. Using the techniques of 'error analysis', the errors in a particular calculation can be shown to lie within calculable bounds. Of course, such calculations are, except in very simple cases, extremely complicated. But we shall examine some of the basic ideas of error analysis and perhaps give the reader a feel for what is a most remarkable subject.

1.3 Error analysis

1.3a Interval arithmetic

We begin by considering what is probably the simplest type of error analysis: **interval arithmetic**. To see how this works, suppose that we want to evaluate $\frac{5.29}{3.42}$. A calculator would give the answer:

$$\frac{5.29}{3.42} = 1.546\,783\,626$$

and this appears to be acceptable.

But suppose now that 5.29 and 3.42 are known to be accurate only to the three decimal digits given. Suppose, in fact, that 5.29 and 3.42 have arisen from rounding to three decimal digits. Since 5.29 and 3.42 are known to have only two digit accuracy, we may ask: 'Is it sensible to write the quotient $\frac{5.29}{3.42}$ to ten decimal digits?'

The answer must be: 'Of course not.' Such a procedure gives an impression of ten digit accuracy that the calculation does not possess.

But how accurate is the calculation? If 5.29 and 3.42 are each accurate to two decimal digits, must it follow that $\frac{5.29}{3.42}$ is also accurate to two decimal digits and, more importantly, how do we find out? To answer these questions, we proceed as follows.

Since 5.29 is the result of rounding to three decimal digits, the exact value of the number that 5.29 represents must lie between 5.285 and 5.295. We shall express this by writing [5.285,5.295] to represent this unknown number. In this way, a number is being represented by an interval: hence the name, interval arithmetic.

Similarly, the exact value of the number represented by 3.42 must lie between 3.415 and 3.425. This number is represented by [3.415,3.425].

Now we see that the quotient could be as small as

$$\frac{5.285}{3.425} = 1.543\,065\,693$$

(dividing the smallest possible value of the numerator by the largest possible value of the denominator) or as large as

$$\frac{5.295}{3.415} = 1.550\,512\,445$$

(dividing the largest possible value of the numerator by the smallest possible value of the denominator). The exact value of the quotient must, then, lie between 1.543 065 693 and 1.550 512 445 and so may be written [1.543 065 693, 1.550 512 445]. We see that the original answer 1.546 783 626 is accurate only to two decimal digits.

The technique of interval arithmetic consists of the following steps:

1 Calculate the smallest and the largest possible values that can be taken by the numbers appearing in the expression.

2 Replace each number in the expression by the relevant interval and calculate the smallest and the largest values that the expression can take.

3 Form the interval consisting of all numbers lying between the two extreme values in **2**.

Then the exact value of the calculation must lie within this interval. Some examples will make the method clear.

Example 1.12

(a) Calculate the value of $\dfrac{1.4 + 0.36}{5.2 \times (12 - 0.99)}$

(b) If it is known that each of the numbers in this expression is the result of rounding to two decimal digits, calculate the smallest interval within which the exact value of the expression must lie. What is the accuracy of the original answer?

(a)
$$\frac{1.4 + 0.36}{5.2\,(12 - 0.99)} = \frac{1.76}{57.252}$$
$$= 0.030\,741\,284$$
$$= 0.307\,412\,84 \times 10^{-1}$$

(b) Since each number is the result of two digit rounding, we may write:

$$1.4 = [1.35, 1.45] \qquad 0.36 = [0.355, 0.365]$$
$$5.2 = [5.15, 5.25] \qquad 12 = [11.5, 12.5]$$
$$0.99 = [0.985, 0.995]$$

Replacing each number by the corresponding interval gives:

$$\frac{[1.35, 1.45] + [0.355, 0.365]}{[5.15, 5.25] \times ([11.5, 12.5] - [0.985, 0.995])}$$

For the numerator:

$$[1.35,1.45] + [0.355,0.365] = [1.35 + 0.355, 1.45 + 0.365]$$
$$= [1.705, 1.815]$$

For the denominator:

$$[11.5,12.5] - [0.985,0.995] = [11.5 - 0.995, 12.5 - 0.985]$$
$$= [10.505, 11.515]$$

and
$$[5.15,5.25] \times [10.505,11.515] = [5.15 \times 10.505, 5.25 \times 11.515]$$
$$= [54.100\,75, 60.453\,75]$$

Hence the expression becomes $\dfrac{[1.705, 1.815]}{[54.100\,75, 60.453\,75]}$

Again, the smallest possible value is:

$$\frac{\text{smallest value numerator}}{\text{largest value denominator}} = \frac{1.705}{60.453\,75}$$

and the largest possible value is

$$\frac{\text{largest value numerator}}{\text{smallest value denominator}} = \frac{1.815}{54.10075}$$

and the exact answer lies in the interval

$$[0.028\,203\,378, 0.033\,548\,518]$$

Hence we cannot be certain of the accuracy of any of the digits in the mantissa of the calculated answer in (a).

Example 1.13

In an experiment, time (in seconds) is measured correct to the nearest tenth of a second. A quantity w is known to be correct to four decimal digits (after chopping) and a constant k is known to be correct to two decimal digits (after chopping). It is necessary to evaluate

$$x = k\sin(wt) \qquad \text{(measuring angles in degrees)}$$

when $t = 10.2$, $w = 3.141$ and $k = 0.35$.

Find the maximum error in the calculation.

From the information:

$$t = [10.15, 10.25]$$
$$w = [3.141, 3.142] \quad \text{(since } w \text{ could have been as large as } 3.1419999\ \ldots)$$
$$k = [0.35, 0.36]$$

Hence

$$wt = [3.141 \times 10.15, 3.142 \times 10.25] = [31.881\,15, 32.2055]$$

Since, for y in the range $0 \leqslant y \leqslant 90$, $\sin y$ is an increasing function of y, we have:

$$\text{Minimum value of } \sin wt = \sin 31.881\,15 = 0.528\,159\,00$$
$$\text{Maximum value of } \sin wt = \sin 32.2055 = 0.532\,957\,50$$

and

$$k\sin wt = [0.35, 0.36] \times [0.528\,159\,00, 0.532\,957\,00] = [0.184\,855\,65, 0.191\,864]$$

If x is evaluated at $t = 10.2$, $w = 3.141$, $k = 0.35$ we obtain

$$x = 0.35\sin(3.141 \times 10.2) = 0.185\,669\,59$$

Hence the maximum error in the calculated value is

$$0.191\,864 - 0.185\,669\,59 = 0.619\,441 \times 10^{-2}$$

The great advantages of interval arithmetic are that it is easy to understand and that is will always work. One disadvantage is that in a long and complicated calculation, the arithmetic can become well nigh intractable. What is needed is an algebraic technique that can be used with greater generality and which can be used to establish general results. Such a technique is introduced in Section 1.3d . We now provide a third example of interval arithmetic. This example gives some bad news.

Example 1.14

Calculate the value of $E = \dfrac{15.21}{9.861 - 9.859}$

It is known that each of the three numbers in this expression has been rounded to four decimal digits. Calculate the bounds within which the true value of E must lie.

Using a calculator $E = \dfrac{15.21}{9.861 - 9.859} = 7605$

15.21 represents a number which lies in the interval

[15.205, 15.215]

9.861 represents a number which lies in the interval

[9.8605, 9.8615]

9.859 represents a number which lies in the interval

[9.8585, 9.8595]

$$\begin{aligned}\text{Hence } E &= \frac{[15.205, 15.215]}{[9.8605, 9.8615] - [9.8585, 9.8595]} \\ &= \frac{[15.205, 15.215]}{[9.8605 - 9.8595, 9.8615 - 9.8585]} \\ &= \frac{[15.205, 15.215]}{[0.001, 0.003]}\end{aligned}$$

Minimum value of $E = \dfrac{15.205}{0.003} = 5068.3333$

Maximum value of $E = \dfrac{15.215}{0.001} = 15\,215$

Hence the true value of E must lie in the interval

[5068.3333, 15215]

This is – to put it mildly – a disturbing result. A simple calculation involving numbers written to four digit accuracy has not only lost all accuracy, but has produced a possible absolute error of 15215 – 7605 = 7610 and a possible relative error of $\dfrac{15215 - 7605}{15215} = 0.500\,16$. A calculation such as this, occurring perhaps as part of a larger calculation in a computer program, would cause havoc and produce a completely unreliable solution. What has gone wrong?

The answer is simple and is a well known phenomenon in numerical analysis. The gross inaccuracy has arisen because, when two numbers – almost equal in value – are subtracted, there is a loss of accuracy. When the result of this subtraction is used in a further calculation, the loss of accuracy is compounded. And if, in particular, the result of the subtraction is used as the denominator in an expression, the further loss of accuracy is particularly acute.

In Example 1.14, the denominator of E requires the subtraction of two numbers almost equal in value,

$$9.861 - 9.859 = 0.002$$

Because the numbers 9.861 and 9.859 have been rounded to four decimal digits, we know that the first three digits of each number are accurate. The fourth digits, however, are not reliable as these digits may have been altered in the rounding process. But the only nonzero number in the difference, 0.002, came from subtracting the two unreliable digits. Hence all accuracy has been lost in this subtraction. When this now totally unreliable number is used in a further calculation, it is not surprising that the result is a very inaccurate answer.

This process – of loosing accuracy through the subtraction of almost equal numbers and then using the suspect difference as a denominator – is alarmingly common and has been given the name of **subtractive cancellation.** This process is definitely to be avoided.

The next example illustrates further the difficulties that can befall the unwary.

Example 1.15

The four numbers x, y, z and w are given by:

$$x = 8.6329157 \qquad y = 8.6328398 \qquad z = 0.0000131239$$
$$w = 976.259$$

Working entirely in five digit arithmetic (using chopping) calculate

(a) $x - y$ (b) $\dfrac{(x - y)}{z}$ (c) $(x - y) \times w$ (d) $\dfrac{w}{(x - y)}$ and comment on the accuracy of your answers.

Using five digit arithmetic with chopping gives

$$x^* = 0.86329 \times 10^1 \qquad y^* = 0.86328 \times 10^1$$
$$z^* = 0.13123 \times 10^{-4} \qquad w^* = 0.97625 \times 10^3$$

The table below gives the values, in five digit arithmetic, of the calculations (both exact and approximate), the absolute errors and the relative errors.

Expression	**Exact values working with x, y, z, w**	**Approx values working with x^*, y^*, z^*, w^***	**Absolute error**	**Relative error**
(a) $x - y$	0.759×10^{-4}	0.1×10^{-3}	0.241×10^{-4}	0.31752
(b) $\dfrac{(x - y)}{z}$	0.57833×10^1	0.76202×10^1	0.18369×10^1	0.31762
(c) $(x - y) \times w$	0.74098×10^{-1}	0.97625×10^{-1}	0.23527×10^{-1}	0.31751
(d) $\dfrac{w}{(x - y)}$	0.12862×10^8	0.97625×10^7	0.30995×10^7	0.24098

Because x and y have been chopped to five decimal digits, all the digits in x^* and y^* are accurate. Nevertheless, x^* and y^* are very close together in value and accuracy has been lost in the subtraction. This is reflected in the large relative error that occurs in each of the calculations. In (a) and, to a certain extent, in (c) the absolute error is not large, emphasising that a small absolute error can go hand in hand with a large relative error. In (d), the situation is partially reversed: the absolute error is enormous while the relative error (while still large) is the smallest of the four.

The following section illustrates a further instance of subtractive cancellation and some ways of dealing with the problem of loss of accuracy.

1.3b Reducing the error

It is often possible to reduce the error that arises from the subtraction of 'almost equal' numbers, once the student has become aware of the problem. Suppose, for example, that at some point in a calculation or in a computer program, it is required to calculate

$$\frac{A}{1-\cos x}$$

This expression is acceptable so long as $x \neq 360n°$ for some integer n. But what happens if x is 'almost' $360n°$ for some n?

Suppose that we are working in six digit arithmetic and that $x = 360.1°$. Then, working in six digit arithmetic

$$\cos x = 0.999\,998$$
$$1-\cos x = 0.000\,002$$

Now, $\cos x$ has five digit accuracy while the only nonzero digit in $1-\cos x$ has been calculated using the unreliable digit in $\cos x$. Hence all accuracy has been lost in the subtraction. Going on to calculate $\dfrac{A}{1-\cos x}$ will lead to subtractive cancellation and the loss of accuracy may well become even worse.

The situation can be improved by observing that

$$\begin{aligned} 1-\cos x &= \frac{(1-\cos x)(1+\cos x)}{1+\cos x} \\ &= \frac{1-\cos^2 x}{1+\cos x} \\ &= \frac{\sin^2 x}{1+\cos x} \end{aligned}$$

Hence, it is possible to write $1-\cos x$ in a form that does not involve the subtraction of 'almost equal' quantities.

As a second example, consider the quadratic equation

$$100x^2 - 10011x + 10.011 = 0$$

The standard method of solving this equation would be to use the formula:

$$x = \frac{-b \pm \sqrt{b^2 - 4ac}}{2a}$$

after, perhaps, dividing by 100.

Hence, if $x^2 - 100.11x + 0.10011 = 0$

$$x = \frac{100.11 \pm \sqrt{100.11^2 - 4 \times (0.10011)}}{2}$$

Working in five digit arithmetic (and rounding)

$$x = \frac{100.11 \pm \sqrt{10022 - 0.40044}}{2}$$

$$= \frac{100.11 \pm \sqrt{10022}}{2}$$

$$= \begin{cases} \frac{200.22}{2} = 100.11 \\ \text{and } 0 \end{cases}$$

In fact, the solutions rounded to 5 decimal digits are 100.11 and 0.00100.

So the method has worked well for the larger solution, but not at all well for the smaller solution. If we had gone on to use these solutions as denominators of other expressions, the alleged solution $x = 0$ would have caused major problems.

What has happened is that we have subtracted two 'almost equal' numbers (numbers that were in fact exactly equal in five digit arithmetic) and suffered a corresponding loss (total!) of accuracy.

How can we prevent this form from happening? One way would be to rewrite the expression for the solution of a quadratic equation to prevent the subtraction of numbers that were 'almost equal'.

The problem lies, in this case, with the negative sign being assigned to the square root.

i.e. with $\frac{-b - \sqrt{b^2 - 4ac}}{2a}$

We multiply numerator and denominator by $-b + \sqrt{b^2 - 4ac}$

$$\frac{(-b - \sqrt{b^2 - 4ac})(-b + \sqrt{b^2 - 4ac})}{2a(-b + \sqrt{b^2 - 4ac})}$$

$$= \frac{(-b)^2 - (b^2 - 4ac)}{2a(-b + \sqrt{b^2 - 4ac})}$$

$$= \frac{4ac}{2a(-b + \sqrt{b^2 - 4ac})}$$

$$= \frac{2c}{-b + \sqrt{b^2 - 4ac})}$$

Using this expression with $a = 1$, $b = -100.11$, $c = 0.10011$ gives

$$x = \frac{2 \times 0.10011}{100.11 + \sqrt{10022}} \quad \text{(in five digit arithmetic)}$$

$$= \frac{0.20022}{200.22}$$

$$= 0.001$$

which is the exact value of the solution rounded to 5 decimal digits.

It must be pointed out that this alternative method of calculating the value of a 'small' solution of a quadratic equation will not always come up with exactly the correct answer. It will, however, in such cases always come up with an answer that is more accurate than that provided by the usual formula.

For those who have studied the properties of quadratic equations, there is an easier way of computing this more accurate value of the smaller root. This method is based on the well known fact that if α and β represent the roots of the quadratic equation $ax^2 + bx + c = 0$ then

$$\alpha\beta = \frac{c}{a}$$

We then let α take the value of the larger root ($\alpha = 100.11$) and observe that the other root β is given by

$$\beta = \frac{c}{\alpha a} = \frac{0.10011}{100.11 \times 1} = 0.001$$

1.3c Calculus in error analysis

The final two sections of this chapter will indicate how more powerful mathematical techniques may be used in the study of errors. To introduce the first of these, suppose that we are considering a certain function of x : $f(x) = e^{5x}$, say. This function is known precisely, but what will happen if the value of x, at which $f(x)$ is to be evaluated, is known only approximately? If x^* represents the approximation to x, how is $f(x^*)$ related to $f(x)$? Or, more precisely, what is the error in using $f(x^*)$ as an approximation to $f(x)$?

To make progress with this problem, write (as usual)

$$x = x^* + e$$

Then the required error

$$E = f(x) - f(x^*)$$

or

$$E = f(x^* + e) - f(x^*)$$

From elementary calculus,

$$\frac{f(x^* + e) - f(x^*)}{e} \approx f'(x^*)$$

giving $\quad f(x^* + e) - f(x^*) \approx ef'(x^*)$

Hence $\quad E \approx ef'(x^*)$

Example 1.16

Let $f(x) = e^{5x}$. It is given that x takes the value 10 when written correct to the nearest integer. Find the greatest possible error in taking f(10) to represent $f(x)$.

Write $\quad x = 10 + e$ where $|e| \leqslant 0.5$

Then
$$\begin{aligned} E &= f(x) - f(x^*) \\ &\approx ef'(x^*) \\ &\leqslant 0.5 \times 5e^{5x^*} \\ &= 0.5 \times 5e^{5\times 10} \\ &= 1.296 \times 10^{22} \end{aligned}$$

and the greatest error is seen, in this case, to be enormous.

This result is presented as a theorem.

Theorem 1e

Let $f(x)$ be a given function of x. if x^* is an approximation to x and $x = x^* + e$, then

$$f(x) - f(x^*) \approx ef'(x^*)$$

Example 1.17

The diameter of a sphere is measured and is given as 4.5 cm. It is thought that the exact value of the diameter lies between 4.4 cm and 4.6 cm. Find the greatest error in the volume of the sphere. ($V = \frac{4}{3}\pi r^3$)

Assume that $\pi = 3.141\,592\,7$ and ignore the error in this value of π.

Write $V = f(r) = \frac{4}{3}\pi r^3$. Then $f'(r) = 4\pi r^2$

The parameter of interest is the radius. Clearly, $r^* = 2.25$ and the greatest error in the value of r^* is 0.05.

The greatest error in the volume is given by

$$\begin{aligned} f(r) - f(r^*) &\approx ef'(r^*) \\ &\approx 0.05 \times 4\pi(2.25)^2 \\ &\approx 3.180\,862\,609 \end{aligned}$$

1.3d A general approach to error analysis

Now consider a more general approach to error analysis. Let x^* and y^* represent approximate values of x and y. The approximations may have arisen through chopping or rounding to a certain number of decimal digits (as in a computer calculation) or because x^* and y^* represent inaccurate observations on x and y. If x^* and y^* are used in place of x and y in a calculation, then the result of that calculation will be in error. The aim of this section is to show how the error in the calculation may be related to the errors in x^* and y^*. Since all calculations are based on the four operations $\times$, $\div$, $+$, $-$, these operations will be considered in detail.

Notation

The following notation will be used.

$$x - x^* = e_x \quad \text{and} \quad y - y^* = e_y$$

where e_x represents the error in taking x^* as an approximation to x, and similarly with e_y.

Write $$\frac{x - x^*}{x} = \zeta_x \quad \text{and} \quad \frac{y - y^*}{y} = \zeta_y$$

where ζ_x, represents the relative error in taking x^* as an approximation to x, and similarly with ζ_y.

We begin by considering addition and subtraction.

Addition

Clearly $$x = x^* + e_x \quad \text{and} \quad y = y^* + e_y$$

Hence $$\begin{aligned} x + y &= x^* + e_x + y^* + e_y \\ &= x^* + y^* + e_x + e_y \end{aligned}$$

The error in taking $x^* + y^*$ to represent $x + y$ is

$$(x + y) - (x^* + y^*) = e_x + e_y$$

The absolute error is

$$\begin{aligned} \left|(x + y) - (x^* + y^*)\right| &= \left|e_x + e_y\right| \\ &\leqslant \left|e_x\right| + \left|e_y\right| \end{aligned}$$

by the triangle inequality.

This gives the important rule:

> The absolute error in using $x^* + y^*$ to approximate $x + y$ is less than or equal to the sum of the absolute errors caused by using x^* as an approximation to x and y^* as an approximation to y.

Subtraction

In the same way,

$$\begin{aligned} x - y &= x^* + e_x - (y^* + e_y) \\ &= x^* - y^* + (e_x - e_y) \end{aligned}$$

The error in using $x^* - y^*$ as an approximation to $x - y$ is

$$(x - y) - (x^* - y^*) = e_x - e_y$$

and the absolute error is

$$\begin{aligned} \left| (x - y) - (x^* - y^*) \right| &= \left| e_x - e_y \right| \\ &\leqslant \left| e_x \right| + \left| e_y \right| \end{aligned}$$

by the triangle inequality.

This gives:

> The absolute error in using $x^* - y^*$ as an approximation to $x - y$ is less than or equal to the **sum** of the absolute errors caused by using x^* as an approximation to x and y^* as an approximation to y.

Example 1.18

If, in the expression 54.381 – 21.2, both numbers are correct to the accuracy shown, calculate the maximum error that could occur in performing the calculation.

It is not known whether rounding or chopping has been used to write the numbers in the form shown. But it is clear that 54.381 has been written correct to five decimal digits and 21.2 to three decimal digits.

If rounding has been used then, from Theorem 1b, the absolute error $\left| e_1 \right|$ in 54.381 satisfies

$$\left| e_1 \right| \leqslant 0.5 \times 10^{2-5} = 0.5 \times 10^{-3}$$

and the absolute error $\left| e_2 \right|$ in 21.2 satisfies

$$\left| e_2 \right| \leqslant 0.5 \times 10^{2-3} = 0.5 \times 10^{-1}$$

From the above, the absolute error $\left| e \right|$ in the subtraction 54.381 – 21.2 satisfies

$$\left| e \right| \leqslant 0.5 \times 10^{-3} + 0.5 \times 10^{-1} = 0.505 \times 10^{-1}$$

Hence the exact value of the number represented by 54.381 – 21.2 must lie between

$$33.181 - 0.505 \times 10^{-1} \text{ and } 33.181 + 0.505 \times 10^{-1}$$

If chopping has been used, then

$$|e_1| \leqslant 10^{2-5} = 10^{-3} \qquad \text{and} \qquad |e_2| < 10^{2-3} = 10^{-1}$$

Hence $\quad |e| \leqslant 10^{-3} + 10^{-1} = 0.101.$

The exact value of the number must lie between 33.181 – 0.101 and 33.181 + 0.101.

The greatest error arises from chopping and is ± 0.101

When considering multiplication and division, it is convenient to work with the relative error. We have then

$$\frac{x - x^*}{x} = \zeta_x$$

This gives $\quad x - x^* = x\zeta_x$

$$x^* = x - x\zeta_x$$

and $\quad x^* = x\,(1 - \zeta_x)$

However, it is possible that the relative error is negative. In this case, we could write

relative error $= -\zeta_x$

and the above algebra will produce

$$x^* = x\,(1 + \zeta_x)$$

It is usual to write this important equation in this form.

$$\boxed{x^* = x\,(1 + \zeta_x)}$$

Similarly, we write $y^* = y\,(1 + \zeta_y)$

With this notation, we examine:

Multiplication

$$\begin{aligned} x^*y^* &= x(x + \zeta_x)y(1 + \zeta_y) \\ &= xy(1 + \zeta_x)(1 + \zeta_y) \\ &= xy(1 + \zeta_x + \zeta_y + \zeta_x\zeta_y) \end{aligned}$$

If ζ_x and ζ_y are reasonably small, then $\zeta_x\zeta_y$ will be very much smaller than either ζ_x or ζ_y and hence, will be ignored.

We may then write

$$x^*y^* \approx xy(1 + \zeta_x + \zeta_y)$$

Hence using the approximations x^* and y^* will cause xy to take the value $xy(1 + \zeta)$ where $\zeta \approx \zeta_x + \zeta_y$.

The relative error ζ satisfies

$$|\zeta| = |\zeta_x + \zeta_y| \leqslant |\zeta_x| + |\zeta_y|$$

and we obtain the general result

$$|\text{Relative error in } x^*y^*| \leqslant |\text{Relative error in } x^*| + |\text{Relative error in } y^*|$$

or

> The modulus of the relative error in a product is less than or equal to the sum of the relative errors of the factors.

Division

$$x^* \div y^* = \frac{x(1+\zeta_x)}{y(1+\zeta_y)}$$

$$= \frac{x}{y}(1+\zeta_x)(1+\zeta_y)^{-1}$$

$$= \frac{x}{y}(1+\zeta_x)(1-\zeta_y+\zeta_y^2-\ldots) \qquad \text{by the binomial theorem}$$

$$= \frac{x}{y}(1+\zeta_x-\zeta_y-\zeta_x\zeta_y+\zeta_y^2-\ldots)$$

Again, neglecting products of errors such as $\zeta_x\zeta_y$, ζ_y^2

gives

$$x^* \div y^* \approx \frac{x}{y}(1+\zeta_x-\zeta_y)$$

Hence using the approximations x^* and y^* will cause $\frac{x}{y}$ to take the value $\frac{x}{y}(1+\zeta)$ where $\zeta \approx \zeta_x - \zeta_y$.

Since ζ_x and ζ_y may be of opposite sign, we must allow

$$|\zeta| \leqslant |\zeta_x| + |\zeta_y|$$

and we have the same kind of error structure as for multiplication.

$$|\text{Relative error in } x^* \div y^*| \leqslant |\text{Relative error in } x^*| + |\text{Relative error in } y^*|$$

and in general:

> The modulus of the relative error in a quotient is less than or equal to the sum of the relative errors of the quotients.

Example 1.19

$x = 56.31$, $y = 0.4218$, $z = 215.6$ and each number has been rounded to four decimal digits. Calculate (a) the greatest relative error (b) the greatest absolute error that can occur in performing the calculation $\frac{x^2y}{z}$.

(a) Let $\zeta_1, \zeta_2, \zeta_3$, represent the relative errors in writing x, y and z in the above form. Then

$$x^* = x(1 + \zeta_1) \qquad y^* = y(1 + \zeta_2) \qquad z = z^*(1 + \zeta_3)$$

Hence $$\frac{(x^*)^2 y^*}{z^*} = \frac{[x(1 + \zeta_1)]^2\, y(1 + \zeta_2)}{z(1 + \zeta_3)}$$

$$= \frac{x^2 y}{z}(1 + \zeta_1)^2 (1 + \zeta_2)(1 + \zeta_3)^{-1}$$

Expanding $(1 + \zeta_1)^2$ and $(1 + \zeta_3)^{-1}$ using the binomial theorem gives

$$\frac{(x^*)^2 y^*}{z^*} = \frac{x^2 y}{z}(1 + 2\zeta_1 + \zeta_1^2)(1 + \zeta_2)(1 - \zeta_3 + \zeta_3^2 - \zeta_3^3 \ldots)$$

Multiplying out the brackets and, as before, ignoring all products of errors gives

$$\frac{(x^*)^2 y^*}{z^*} \approx \frac{x^2 y}{z}(1 + 2\zeta_1 + \zeta_2 - \zeta_3)$$

$$= \frac{x^2 y}{z}(1 + \zeta) \qquad \text{where} \qquad \zeta \approx 2\zeta_1 + \zeta_2 - \zeta_3$$

Since all decimals have been rounded to four decimal digits,

$$|\zeta_i| \leqslant 0.5 \times 10^{-4+1} \qquad i = 1, 2, 3$$

Since $$|\zeta| \approx |2\zeta_1 + \zeta_2 - \zeta_3| \leqslant 2|\zeta_1| + |\zeta_2| + |\zeta_3|$$

we have $$|\zeta| \leqslant 2 \times 0.5 \times 10^{-3} + 0.5 \times 10^{-3} + 0.5 \times 10^{-3}$$

$$= 2 \times 10^{-3}$$

Hence the greatest value that can be taken by the relative error is 2×10^{-3}.

Observe that this result could have been obtained by a straight appeal to the results on page 51.

But by writing $\dfrac{(x^*)^2 y^*}{z^*} = \dfrac{x^2 y}{z}(1 + \zeta)$ we see more clearly how the approximation $\dfrac{(x^*)^2 y^*}{z^*}$ is related to the exact value $\dfrac{x^2 y}{z}$.

Using techniques such as these, error analysis may be applied in much more complicated situations.

(b) To estimate the maximum absolute error observe that

$$|\text{absolute error}| = |\text{exact value}| \times |\text{relative error}|$$

$$\approx \left| \frac{56.31 \times 0.4218}{215.6} \right| \times |2 \times 10^{-3}|$$

$$= 0.110\,164\,93 \times 2 \times 10^{-3}$$
$$= 0.220\,329\,86 \times 10^{-3}$$

Hence we take $0.220\,329\,86 \times 10^{-3}$ to be our estimate of the maximum absolute error.

Under certain circumstances, we can use these ideas to say more about addition and to prove a result described above about subtraction. Consider first addition.

Addition

$$\begin{aligned} x^* + y^* &= x(1 + \zeta_x) + y(1 + \zeta_y) \\ &= x + y + x\zeta_x + y\zeta_y \\ &= (x + y)\left(1 + \frac{x\zeta_x + y\zeta_y}{x + y}\right) \end{aligned}$$

Hence the relative error in $x^* + y^*$ is

$$\zeta = \frac{x\zeta_x + y\zeta_y}{x + y}$$

We now consider a special case.

Suppose that $\quad -\rho \leqslant \zeta_x \leqslant \rho \quad$ and $\quad -\rho \leqslant \zeta_y \leqslant \rho \quad \ldots (1)$

This would occur, for example, if the errors in both x and y arose from rounding (or chopping) to k decimal digits.

Now, if x and y are both positive, then

$$\frac{x(-\rho) + y(-\rho)}{x + y} < \zeta < \frac{x\rho + y\rho}{x + y}$$

and $\quad -\rho < \zeta < \rho$

Similarly, if x and y are both negative, then again

$$-\rho < \zeta < \rho$$

and we see that if x and y have the same sign, then, ζ the relative error in using $x^* + y^*$ to approximate $x + y$, satisfies the same pair of inequalities as ζ_x and ζ_y. Hence, under the conditions of (1), the addition of two numbers of the same sign does not cause any further loss in accuracy.

What happens if x and y are of opposite sign?

Subtraction

We shall consider only the situation where x and y are similar in magnitude. In this case, $x + y$ will be an instance of the subtraction of numbers that are almost equal. Thus $x + y$ will be close to zero and consequentially

$$\zeta = \frac{x\zeta_x + y\zeta_y}{x + y}$$

may well be very large indeed. We have thus shown algebraically that the subtraction of almost equal numbers can lead to a significant loss of accuracy.

It is time to move on to something new.

Exercise 1

Section 1.1

1 Write the numbers

(a) 543.291 (b) 0.006 789 1 (c) 29.003

(d) −0.068 39 (e) 0.516 321 (f) −0.000 01

in floating point form. In each case, write down the mantissa and the exponent of the number.

2 Which of the following numbers, when written in decimal form, require infinitely many digits to specify their value exactly.

(a) $\frac{1}{128}$ (b) $\frac{1}{9}$ (c) $\sqrt{3}$

(d) $\frac{3}{32}$ (e) $\pi - 3.142$ (f) $\frac{(16.7 - 16.2)}{12.3}$

3 Use chopping to write the decimals below in floating point form with a three digit mantissa.

(a) $0.478\,132 \times 10^{0}$ (b) $0.152\,943 \times 10^{-3}$ (c) $0.999\,999 \times 10^{-1}$

(d) $0.181\,818\,181 \times 10^{4}$ (e) $0.615\,961 \times 10^{7}$

4 Use rounding to write the decimals below in floating point form with a four digit mantissa.

(a) $0.681\,523 \times 10^{-1}$ (b) $0.759\,159\,8 \times 10^{4}$ (c) $0.817\,77 \times 10^{0}$

(d) $0.952\,22 \times 10^{-2}$ (e) $0.531\,57 \times 10^{-3}$

5 Write the numbers given below in finite decimal form using chopping or rounding and the number of digits given in the question.

(a) 538.421 rounding, three digits.

(b) $\frac{\pi}{4}$ chopping, two digits.

(c) 0.000 541 3 chopping, three digits.

(d) $\sqrt{5}$ rounding, four digits.

(e) −0.000 015 4 chopping, two digits.

(f) 25.6998 rounding, five digits.

(g) $\frac{16.31}{2.37}$ rounding, four digits.

6 Write the numbers given below in floating point form using an eight digit mantissa. State whether you are using rounding or chopping.

(a) $8\frac{8}{9}$ (b) $\sqrt{117}$ (c) $\frac{65}{7003}$ (d) $\sqrt[3]{5}$

(e) $\frac{1}{51.3}$ (f) $\frac{3}{5.2 - 4.89}$ (g) $\frac{\sqrt{21}}{\sqrt{2} - 1}$

In each case, state the exponent of the floating point number.

7 If 54.6514 is chopped to three decimal digits, what is the roundoff error?

8 If 0.091 628 is rounded to two decimal digits, what is the roundoff error?

9 Write 4.932 98 correct to four decimal digits using chopping. What is the roundoff error?

Section 1.2

10 Describe the difference between absolute error and relative error. Provide one example to illustrate when absolute error would be a better measure of the error and one example to illustrate when the relative error would be the better measure.

11 If 693 257 is rounded to two decimal digits, what is
(a) the absolute error?
(b) the relative error?

12 If 0.005 998 is rounded to three decimal digits, what is
(a) the absolute error?
(b) the relative error?

13 Write the number 675.973 in four decimal digit form using chopping. What is
(a) the absolute error?
(b) the relative error?

14 Calculate (i) the roundoff error, (ii) the absolute error, (iii) the relative error, when the numbers given below are shortened to the number of digits given in the question.
(a) 678.492 rounded, four decimal digits.
(b) 0.058 143 2 chopped, two decimal digits.
(c) 7.9817 rounded, three decimal digits.
(d) 19.892 chopped, four decimal digits.
(e) 0.000 001 798 rounded, three decimal digits.
(f) 0.9352×10^{-2} chopped, two decimal digits.
(g) $0.876\,32 \times 10^{5}$ rounded, four decimal digits.

15 If $x = \frac{1}{3}$, $y = \frac{3}{7}$, $z = \frac{5}{21}$, evaluate $\frac{(x \times y^2)}{z}$ as accurately as possible.

Use chopping to write each of x, y, z in floating point form with a three digit mantissa. Now perform the calculation again using the chopped numbers and working entirely in three digit arithmetic and using chopping. Calculate (a) the absolute error and (b) the relative error.

16 $x = 17.594$ $y = 1.3529$ $z = 110.89$

(a) Calculate the value of $E = \frac{x + y}{z + y}$ as accurately as possible.
(b) Round each of x, y, z to two decimal digits. Using the rounded forms of x, y, and z and working entirely in two digit arithmetic (using rounding) calculate the value of E.
(c) Calculate the absolute error and relative error in your answers. Comment on your results.

17 If $y = \dfrac{5}{(5 - x^2)}$, calculate the value of y when $x = 2.23607$.

Now calculate the value of y working entirely in two digit arithmetic (using chopping). Calculate the absolute error and the relative error in your answers and comment on your results.

18 If $x = 5.9843129$, $y = 29.683971$, $z = 0.0034848269$, calculate the value of $w = \dfrac{(y - 4x)}{z}$.

(a) as accurately as possible,
(b) by performing all calculations in three digit arithmetic, using rounding.
(c) Calculate the absolute error and the relative error in your answers.

19 Solve the equation

$$\tfrac{1}{3}(x + 1) - 1\tfrac{2}{3}(x - 1) = \tfrac{5}{7}$$

(a) exactly,
(b) working in chopped three digit arithmetic.

Find the absolute error and the relative error in your answers.

20 (a) Solve the equation $\tfrac{2}{7}(1 - x) + \tfrac{2}{3}(x - \tfrac{1}{3}) = 1$.
Now solve this equation by performing all calculations in rounded two decimal digit arithmetic. Comment on the accuracy of your results.
(b) Repeat the question using the equation

$$\tfrac{4}{9}(x - 1) + \tfrac{5}{11}(3 - x) = \tfrac{2}{3}$$

21 Solve

$$\frac{\tfrac{2}{3}}{x + 1.785} - \frac{1\tfrac{5}{7}}{x - 0.4380} = 0$$

(a) exactly,
(b) by working entirely in chopped two digit arithmetic.
(c) Comment on the accuracy of your answer to (b).

22 Solve the simultaneous equations

$$\tfrac{1}{3}x + \tfrac{2}{7}y = -\tfrac{5}{6}$$
$$\tfrac{1}{9}x - \tfrac{3}{11}y = \tfrac{2}{3}$$

(a) exactly,
(b) by performing all calculations in chopped three digit arithmetic.
(c) Calculate the absolute error and the relative error for both x and y.

23 Solve the simultaneous equations

$$\begin{aligned} 3x + 4y &= -2 \\ 2x + 2.6667y &= 3 \end{aligned}$$

(a) exactly,
(b) by performing all calculations in three decimal digit arithmetic, using rounding.
(c) Comment on the accuracy in your answers.

24 Solve the equations

$$\begin{aligned} 5x - 5.5556y &= 3 \\ 4.5x - 5y &= -4 \end{aligned}$$

(a) as accurately as possible,
(b) by working entirely in four digit arithmetic using rounding.
(c) Comment on the accuracy of your results.

25 Solve the equation

$$3x^2 - 1.732\,05x - 8.5359 = 0$$

(a) as accurately as possible,
(b) working entirely in two digit arithmetic (rounded).
(c) Calculate the absolute error and the relative error in your answer to (b). Comment.

26 Solve the equation

$$2.02x^2 + 4.8991x - 32.8773 = 0$$

(a) as accurately as possible,
(b) working in chopped two digit arithmetic.
(c) Calculate the absolute error and the relative error in your answer. Comment.

27 It is known that $u = 0.005\,48$ is correct to three decimal digits, but it is not known whether chopping or rounding has been used to give this value. Calculate the smallest interval within which the exact value of u must lie.

28 The value of x is given approximately as $x = 16.58$. The relative error in this approximation is less than 0.01. Find an interval within which the exact value of x must lie.

29 If an answer is given as 0.0354 and it is known that the relative error is at most 1% find to three decimal digits the range of values within which the exact value of the solution must lie.

30 If an answer is given as 58.29 and it is known that the relative error is at most 1.8%, find to three decimal digits the range of values within which the exact value of the answer must lie.

Section 1.3a

31 In the expression $\dfrac{5.43 - 3.21}{6.97}$

each of the numbers is written correct to three decimal digits using rounding. Find the smallest interval within which the exact value of the expression must lie and calculate the greatest possible error.

32 If $w = \dfrac{5.316}{28.73 - 14.29}$

with each number written correct to four decimal digits (chopped), find the smallest interval within which the exact value of w must lie. Calculate also an estimate of the greatest value of the relative error.

33 If $z = \dfrac{6.1 + 8.8}{0.3 - 0.5}$
with all numbers correct to the accuracy written (using chopping), find the smallest interval within which z must lie.

34 It is given that $a = \dfrac{(0.321)^2}{0.14 + 0.28}$
where each number is correct to the number of decimal digits it contains (using rounding). Find the smallest interval within which the exact value of a must lie. Calculate an estimate of the largest value of the relative error.

35 It is given that $x = 1.5$, $y = 0.76$, $z = 28$ and it is known that each of x, y, and z is correct to two decimal digits (using chopping). In each of the expressions below, calculate the smallest interval within which the exact value of each expression must lie.

(a) $x - y$ (b) $\dfrac{x}{y}$ (c) $y + xz$ (d) x^2 (e) y^2z

(f) $\dfrac{(x + y)^2}{z}$ (g) $\dfrac{y}{(x - z)}$ (h) $\dfrac{xz^2}{y^3}$ (i) $\dfrac{\sqrt{x + y}}{(z + y)^2}$

36 If x = 32° and y = 43°, each to the nearest integer, find the range of values which could be taken by

(a) $\sin(x + y)$ (b) $\tan(3x)$ (c) $\cos(y - x)$ (d) $\sin x + \cos y$

37 The area of a circle is πr^2. If the radius is given as 3.1 cm (to two decimal digits using chopping) and $\pi = 3.142$ (four decimal digits using rounding), calculate the smallest interval within which the exact value of the area must lie. Repeat the question for the circumference ($2\pi r$).

38 If $s = ut + \frac{1}{2}at^2$ and $u = 55\,\text{ms}^{-1}$
$a = 9.8\,\text{ms}^{-2}$
$t = 12\,\text{s}$
all rounded to two decimal digits, find the smallest interval within which s must lie.

39 $y = a\cos(wt)°$ and $a = 2$, $w = 3$, $t = 4$ (all chopped to one digit accuracy). Calculate the smallest interval within which the exact value of y must lie.

40 In the triangle: $A = 56.1°$
$B = 71.8°$
both written to three digit accuracy using rounding. Find the smallest interval within which the exact value of C must lie.

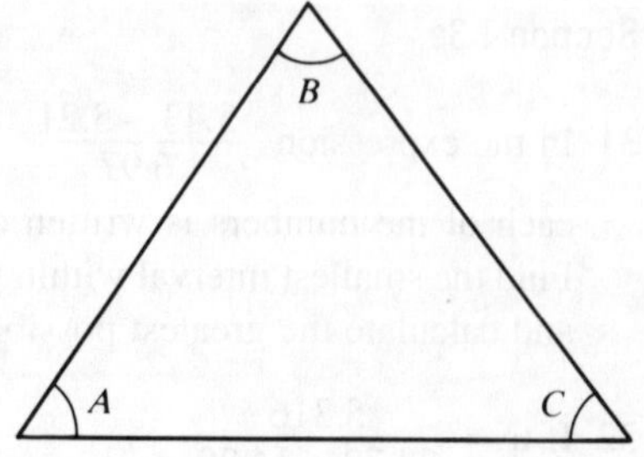

41 The time t of oscillation of a simple pendulum is given by

$$t = 2\pi\sqrt{\frac{l}{g}}$$

where $\pi = 3.142$ (to four decimal digits using rounding),
$g = 9.8\,\text{ms}^{-2}$ (to two decimal digits using rounding),
$l = 0.8\,\text{m}$ (to one decimal digit using rounding).

Use interval arithmetic to calculate an interval within which the exact value of the time of oscillation must lie.

42 In the equation $4.1x^2 - 3.8x - 2.3 = 0$, the coefficients 4.1, –3.8, –2.3 are written correct to two decimal digits using rounding. If the formula

$$x = \frac{-b \pm \sqrt{b^2 - 4ac}}{2a}$$

is used to calculate the two values of the solution, calculate the greatest errors in the solution.

Section 1.3b

43 For each of the expressions below, describe why problems could arise if the expression is evaluated at a certain value of x. In each case, say which value of x is likely to cause a problem and rearrange the expression so as to eliminate the potential difficulty.

(a) $1 - \cos^2 x$

(b) $\dfrac{\sin x - \cos x}{\tan x - 1}$

(c) $\ln(10 + x) - \ln(10)$

(d) $2 - \sqrt{x}$

(e) $x - \sqrt{x^2 - 0.1}$

44 Explain the problems that can arise in evaluating

$$\frac{1}{\sqrt{x+1} - \sqrt{x}}$$

when x is large. Evaluate this expression as accurately as possible when $x = 50000$.

45 Solve the equation

$$x^2 + 0.4002 \times 10^0 x + 0.8 \times 10^{-4} = 0$$

as accurately as possible. Explain why problems arise if the usual formula is used. (Assume that the coefficients 1, 0.4002×10^0, 0.8×10^{-4} are exact.)

46 Obtain as accurately as possible the solutions of the equations

(a) $0.5x^2 + 4x - 0.0003 = 0$

(b) $1.2x^2 - 5.2x + 0.00004 = 0$

(c) $-0.4x^2 + 3.6x + 0.00085 = 0$

47 For which values of x is the expression

$$\frac{4}{x - \sqrt{x^2 - 1}}$$

likely to produce an inaccurate solution? For such a value of x, how could the expression be evaluated with greater accuracy? Evaluate this expression as accurately as possible when $x = 10000$.

Section 1.3c

48 If $u = 3.14$ is known to have been rounded to three decimal digits, calculate approximately the error in evaluating $u^{\frac{1}{8}}$.

49 If $y = 29.13$ is written correct to four decimal digits (chopping) calculate approximately the error in evaluating $y^{\frac{3}{4}}$.

50 If $y = \sin 3x - 2x$ and $x = 0.42 \pm 0.01$ radians, find approximately the error when y is evaluated at $x = 0.42$.

51 It is given that $w = \ln(x^2)$. If $x = 5.1$ (when written correct to two decimal digits using chopping) find approximately the error in evaluating w at $x = 5.1$.

52 If $y = (1 + x^2)^{\frac{1}{2}}$ and $x = 10$ when written correct to the nearest integer, find approximately the error in evaluating y at $x = 10$.

53 $A = \pi r^2$ and $r = 4.5 \pm 0.05$ cm. If $\pi = 3.1415927$, find the error that can occur in A by taking r = 4.5 cm. (Assume that π is not subject to error.)
If now, r is known to be exactly 4 cm calculate the error that can occur by taking $\pi = 3.142$.

54 Find the error that may occur in z if

$$z = \cos^3(2x + 1)$$

and it is known that when x is rounded to two decimal digits, then x takes the value $x = 0.59$ radians.

55 If $v = \frac{p}{q} + \frac{q}{p}$ and $p = 5$, calculate approximately the error that can occur when $q = 6$ (when rounded to the nearest integer).

Section 1.3d

56 Let x^* approximate x with absolute error e_1 and let y^* approximate y with absolute error e_2.

(i) Prove that the absolute error in using $x^* + y^*$ to approximate $x + y$ is at most $e_1 + e_2$.

(ii) Find the maximum absolute error in using $x^* - y^*$ to approximate $x - y$.

If $u = 43.8$ is written correct to three decimal digits using rounding and $v = 5.156$ is written correct to four decimal digits using rounding, calculate the maximum error that could arise in evaluating

(a) $u + v$

(b) $u - v$

(c) $4u - 5v$

57 Using the notation of Question 56: prove that the relative error in using x^*y^* to approximate xy is less than or equal to the sum of the relative errors in x^* and in y^*. State and prove a similar result for $\frac{x}{y}$.
If $x = 0.48$ and $y = 10.2$ and the relative error in both numbers is less that 0.15, calculate the greatest relative error that can occur in calculating
(a) xy (b) $3x^2y$ (c) $7(xy)^2$

58 If $x = 28.45$ with relative error 0.02,
$y = 16.40$ with relative error 0.018,
$z = 5.281$ with relative error 0.10,
calculate the relative errors in the expressions

(a) xy (b) x^2y^2 (c) $\frac{xz^2}{y^2}$ (d) $\frac{yz}{x^3}$

59 If $p = 0.241 \times 10^2$, $q = 0.59 \times 10^{-1}$, $r = 0.6413 \times 10^1$
and all three numbers are correct to the accuracy to which they are written (using rounding), estimate the relative error and the greatest absolute error in calculating

(a) pq^2 (b) $\frac{pq}{r}$ (c) $\frac{qr}{p^2}$ (d) $\frac{pr^3}{q^2}$

Miscellaneous

60 Three quantities, x, y and z, are measured and found to be 1.61, 2.23 and 4.66 respectively, correct to two places of decimals. Find x^2, $(y - z)^2$ and $x^2 - (y - z)^2$ to two places of decimals and in each case calculate the maximum possible error involved.

(MEI)

61 Positive numbers N_1 and N_2 are each rounded off to two decimal places to give n_1 and n_2 respectively. Estimate, in terms of n_1 and n_2, the maximum possible error in using n_1n_2 as an approximation for N_1N_2.
Two numbers, rounded off to two decimal places, are 4.63 and 7.35. Calculate the product of the two numbers, giving the answer to as many significant figures as are known to be accurate.

(MEI)

62 A small sphere of radius r and density ρ falling through a liquid of density ρ_0 and viscosity η reaches a terminal speed υ given by

$$\upsilon = \frac{2}{9} gr^2 \left(\frac{\rho - \rho_0}{\eta} \right)$$

where g is the local gravitational acceleration.
For a small aluminium sphere falling through glycerine $r = 0.200 \pm 0.007$ cm, $\rho = 2.70 \pm 0.05\,\text{g cm}^{-3}$, $\rho_0 = 1.26 \pm 0.03\,\text{g cm}^{-3}$ and $\eta = 9.45 \pm 0.50$ P ($\text{g cm}^{-1}\,\text{s}^{-1}$). Find the terminal speed of the sphere and estimate the maximum possible error in this figure if $g = 981\,\text{cm s}^{-2}$.

(MEI)

63 A charge of q coulombs is uniformly distributed throughout a nonconductive spherical volume with radius r. Let x be a distance of a point from the centre. Then the potential at this point is given by

$$V = \frac{q(r^2 - x^2)}{8\pi e r^3}$$

where e is the permittivity constant. Suppose that nominal values

$$q^* = 6.52,\ r^* = 3.15,\ x^* = 1.12,\ \pi^* = 3.14,\ e^* = 8.85 \times 10^{-12}$$

are given and assume that all numbers are accurate to the number of figures given. That is, the values have been rounded to three digits. Bound the associated error in computing V.

(Y & S)

64 The area of a triangle is

$$A = \frac{ab\sin(\theta)}{2}$$

where a and b are the lengths of two sides and θ is the angle of intersection (in radians). Assume that a^*, b^*, and θ^* are approximations of a, b, and θ, respectively, and that $a^* = b^* = 10\,\text{cm}$ and $\theta^* = \pi/4$. If

$$\delta(a^*) = \delta(b^*) = 0.01\,\text{cm}$$

and

$$\delta(\theta^*) = 0.01\,\text{rad}$$

bound the error in the estimate A^*.

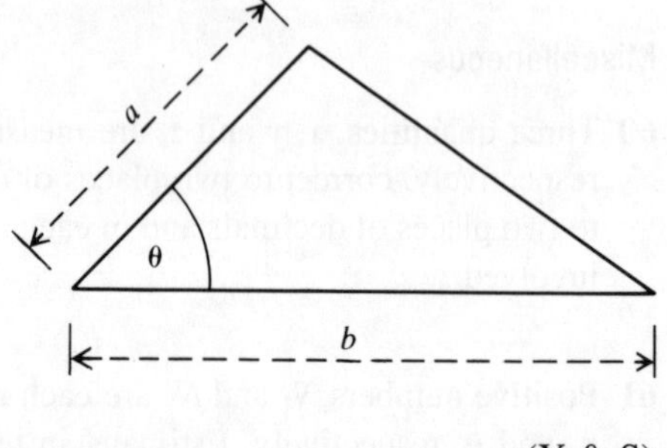

(Y & S)

65 Consider the following resistive circuit:

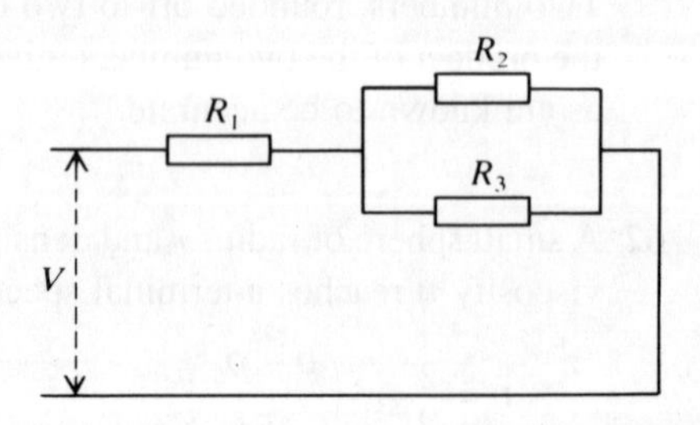

The formula for the net resistance is

$$R = R_1 + \frac{R_2 \times R_3}{R_2 + R_3}$$

Suppose that the nominal values are $R_j^* = j$ ohm for $j = 1, 2, 3$, and $\delta(R_j) = 0.1$. Bound the range of possible values for R about

$$R^* = R_1^* + \frac{R_2^* \times R_3^*}{R_2^* + R_3^*}$$

(Y & S)

66 The time of swing T of a pendulum of length l is given by $T = 2\pi\sqrt{\frac{l}{g'}}$, where $g' = g\left(\frac{r}{r+h}\right)^2$. Find the maximum possible percentage error in T due to small errors of a, b, c and d per cent in h, l, g and r respectively.

(MEI)

67 In a diatomic molecule, the potential energy $U(x)$ resulting from the force fields between two atoms a distance x apart is given by

$$U(x) = \frac{a}{x^{12}} - \frac{b}{x^6}$$

where a and b are positive but not perfectly known. Let a^* and b^* denote approximations of a and b. Presume that e is a given positive error tolerance, and error bounds are denoted by

$$\delta(a^*) = \delta(b^*) = \delta$$

Find what the common value of δ must be so that $\delta(U(x)) = e$. Sketch $\delta(U(x))$ as a function of separation x.

(Y & S)

68 Let (x_0, y_0), (x_1, y_1) be two points on a straight line. Two expressions will give the point at which this line cuts the x-axis.

(i) $x = \dfrac{x_0 y_1 - x_1 y_0}{y_1 - y_0}$ (ii) $x = x_0 - \dfrac{(x_1 - x_0)y_0}{y_1 - y_0}$

(a) Show that both expressions are correct.

(b) If $(x_0, y_0) = (1.31, 3.24)$ and $(x_1, y_1) = (1.93, 4.76)$ use three digit rounding arithmetic to calculate x using both expressions. State which is the better expression and why.

69 If $x = 493.5$, $y = 145.7$ and $z = 28.47$, calculate, working entirely in four digit rounded arithmetic, the values of

(a) $x(y + z)$ and

(b) $xy + xz$.

Which well known law of algebra has been shown not to hold when calculation is performed in four digit arithmetic?

70 (a) Consider the expressions

(i) $(x + y) + z$ and (ii) $x + (y + z)$

Find numbers x, y and z such that when (i) and (ii) are evaluated in six digit arithmetic (rounded), different answers result.

(b) Find another algebraic law that is **not** satisfied when the operations are performed in finite digit arithmetic.

71 An equation that gives, at time t, the size of a population P in terms of the initial population size P_0 and a constant c is:

$$P = P_0 e^{ct}$$

Suppose that the greatest possible error in P_0 is E and the greatest possible error in determining the constant c is δ. Calculate, in terms of E and δ the greatest error in the value of P.

Chapter 2

Polynomials 1

Polynomials form one of the main building blocks of numerical analysis. They also occupy a fundamental position in many other areas of pure and applied mathematics. This chapter introduces the reader to polynomials and describes some of the reasons for their popularity as a tool of mathematics. It will show how to find a polynomial that will 'fit' exactly a given set of data. In this way, the reader will discover an equation whose graph will pass through each individual point of a set of data. In a slightly more advanced section, an expression for the error involved in this technique will then be derived.

First, though, 'What is a polynomial?'

2.1 Introduction to polynomials

The following familiar expressions are all examples of polynomials:

(a) $3x^2 - 2x + 1$ (b) $-6y + 2$ (c) $4a^3 - a$ (d) $8x^4$

The highest power in a polynomial determines the degree of the polynomial. So

(a) is a polynomial of degree two.
(b) is a polynomial of degree one.
(c) is a polynomial of degree three.
(d) is a polynomial of degree four.

Note that it does not matter which letter is used in a polynomial: it is usual, however, to use x and we shall follow this convention from now on.

Definition

A polynomial of degree n (in x) is an expression of the form

$$a_n x^n + a_{n-1}x^{n-1} + a_{n-2}x^{n-2} + \dots + a_1 x + a_0$$

where n is an integer, a_n, a_{n-1}, ... a_0 are numbers and $a_n \neq 0$.

So, for example, $-8x^4 + 2x^3 - x^2 + 2x - 1$ is a polynomial of degree four, $3x^5 - 2x^2 + x$ is a polynomial of degree five.

The numbers $a_n, a_{n-1}, \ldots a_0$ are called the coefficients of the polynomial. In particular, a_n is called the coefficient of x^n, a_{n-1} is called the coefficient of x^{n-1} and so on. a_0 is often called the constant term.

Example 2.1

Which of the following are polynomials? For each expression that is a polynomial, write down the degree of the polynomial and the coefficient of the highest power of x present.

(a) $5x^2 + 8x^{-1}$ (b) $7x^4 - 3x$ (c) $2x - 4x^5$
(d) $6x^3 + 2x^2 - x + y$ (e) $4x^{16} - 8x^4 - \frac{1}{8}$

(a) Not a polynomial: the term $8x^{-1} = \dfrac{8}{x}$ is not allowed.
(b) Polynomial of degree four: coefficient of $x^4 = 7$
(c) Polynomial of degree five. This expression may be written $-4x^5 + 2x$. Coefficient of $x^5 = -4$
(d) Not a polynomial as written: the y term is not allowed. If, however, it is known that y is a polynomial in x (e.g. $y = 2x - 4$) then this expression is a polynomial.
(e) Polynomial of degree 16: coefficient of $x^{16} = 4$.

We will be concerned with the addition, subtraction and multiplication of polynomials and will now look at how these operations are performed.

Addition and subtraction of polynomials

These operations are performed by adding or subtracting the coefficients of like powers of x: some examples should make the method clear.

Example 2.2

Find $(4x^3 - 3x^2 + 2x - 5) + (2x^3 + 5x^2 - 5x)$

Grouping like powers of x together gives

$$4x^3 + 2x^3 - 3x^2 + 5x^2 + 2x - 5x - 5$$

Combining the occurrences of like powers of x (which is equivalent to adding their coefficients) gives

$$\begin{aligned} &(4 + 2)x^3 + (-3 + 5)x^2 + (2 - 5)x - 5 \\ &= 6x^3 + 2x^2 - 3x - 5 \end{aligned}$$

Example 2.3

Find $(5x^3 + 2x^2 - x + 3) - (2x^3 + 6x^2 + 2x - 1)$

Again, grouping like powers of x together:

$$5x^3 - 2x^3 + 2x^2 - 6x^2 - x - 2x + 3 - (-1)$$

Combining the occurrences of like powers of x

$$\begin{aligned}&(5-2)x^3 + (2-6)x^2 + (-1-2)x + (3-(-1))\\ &= 3x^3 - 4x^2 - 3x + 4\end{aligned}$$

It may be seen that to achieve this expression, the coefficients of like powers of x have been subtracted.

Multiplication of polynomials

Two polynomials are multiplied together by multiplying each term in the first polynomial by each term in the second polynomial (or vice versa). This represents a simple extention of the idea of multiplying out brackets in elementary algebra and again, an example should make the process clear.

Example 2.4

Find $(3x^3 + 2x - 1)(-5x^2 + 7)$

We calculate

$$\begin{aligned}&(3x^3) \times (-5x^2) + (3x^3) \times (7) + (2x) \times (-5x^2) + (2x) \times (7)\\ &\quad + (-1) \times (-5x^2) + (-1) \times (7)\\ &= -15x^5 + 21x^3 - 10x^3 + 14x + 5x^2 - 7\end{aligned}$$

Collecting terms together

$$= -15x^5 + 11x^3 + 5x^2 + 14x - 7$$

Note that when two polynomials are multiplied together, a polynomial is produced whose degree is equal to the sum of the degrees of the factor polynomials. We saw above that multiplying a polynomial of degree three by a polynomial of degree two produced a polynomial of degree five.

The importance of polynomials stems from two facts. The first is that their structure makes them remarkably easy to handle. Differentiating or integrating a polynomial is a very straight-forward operation. We shall see in Chapter 5 that calculating the value of a polynomial for a particular value of x can also be made into a straight-forward and efficient operation. Also, as can be seen from the definition, a polynomial is constructed from x using only addition, subtraction and multiplication. Since these operations are fundamental to a computer, it is clear that polynomials, unlike some functions, are ideal for use on a computer.

The second fact is a remarkable result due to Weierstrass. This result is concerned with continuous functions. (See 'Results from pure mathematics' in the Introduction.)

Weierstrass showed that any continuous function can be approximated as closely as we wish by a polynomial. This is indeed a remarkable result. More precisely, it says that given a continuous function f(x) where x lies in an interval $[a,b]$ it is possible to find a polynomial p(x) so that $|f(x) - p(x)|$ is as small as we want it to be for x in $[a,b]$. Since most functions in common use are continuous in most of their domain, this result of Weierstrass is a very powerful result indeed and shows polynomials to be objects worthy of further study.

Weierstrass's result will not be proved.[†] Instead, we shall consider a result with more immediate practical applications.

To illustrate: suppose that the salesmen in a certain company decided to adopt a new selling technique. Their sales this year were £560 000 and they wanted to observe the improvement in sales after 1 year, 2 years, 3 years and 4 years. The figures are given in the table.

Year	0	1	2	3	4
Sales in £100 000	5.6	6.1	4.9	8.3	9.5

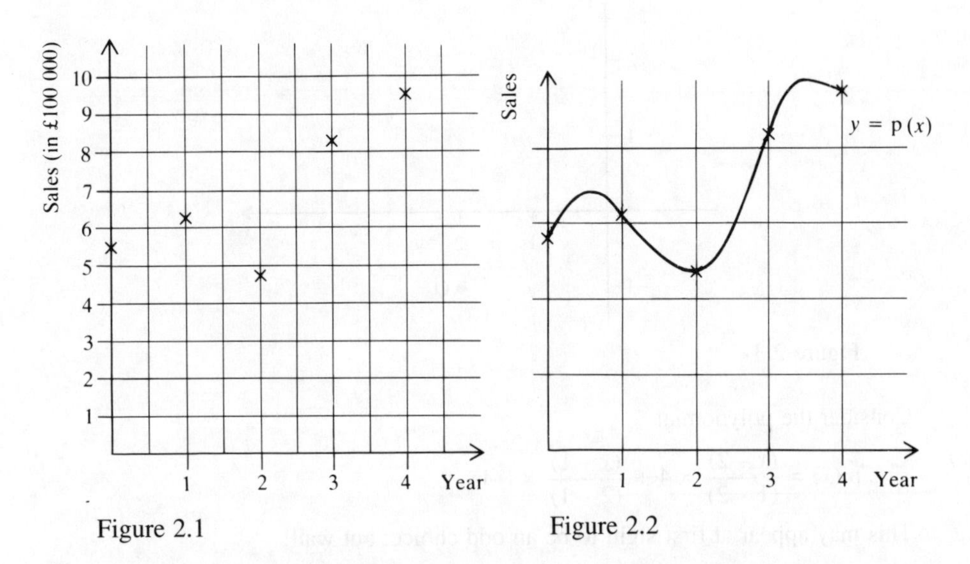

Figure 2.1

Figure 2.2

In the next part of the chapter we shall see how to find a polynomial p(x) whose graph will pass through each of the five points on the sales graph (Figure 2.1) as shown in Figure 2.2. This process in called **polynomial interpolation**.

There is another situation in which polynomial interpolation is of use. This is when a function $y = f(x)$ is known but it is required to replace the (possibly rather

† A proof may be found in Ralston and Rabinowitz *A first course in numerical analysis* 2nd edition. McGraw Hill.

complicated) function f(x) with a simple function. After choosing a number of points lying on the curve $y = \mathrm{f}(x)$, we may use polynomial interpolation to construct a polynomial p(x) which will pass through each of the chosen points. In this way, the polymial p(x) could be thought of as an approximation to the function f(x). This application of polynomial interpolation is very useful if we want to differentiate or integrate an unpleasant function f(x). We shall see how this may be achieved in Chapter 7.

2.2 Polynomial interpolation

We begin with some easy, almost trivial, sets of data and develop the argument step by step.

Find a polynomial whose graph will pass through the points P(1,4) and Q(2,–1) in Figure 2.3.

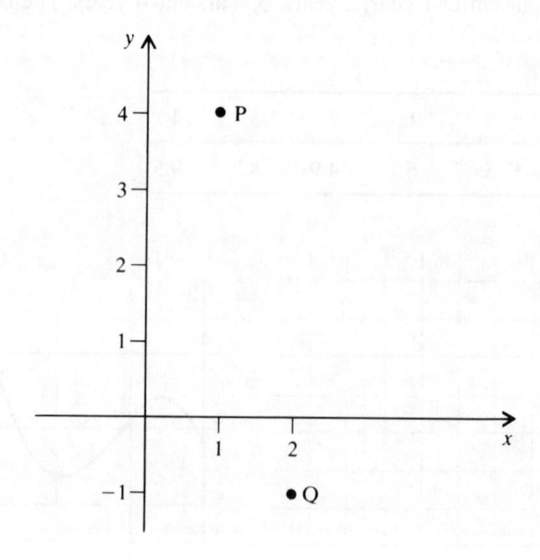

Figure 2.3

Consider the polynomial

$$\mathrm{p}_1(x) = \frac{(x-2)}{(1-2)} \times 4 + \frac{(x-1)}{(2-1)} \times (-1)$$

This may appear at first sight to be an odd choice: but wait!

First, it is clear that $\mathrm{p}_1(x)$ is a polynomial in x. The right-hand side may be written as

$$\frac{(x-2)}{(-1)} \times 4 + \frac{(x-1)}{(1)} \times (-1)$$

$$= \frac{(4x-8)}{-1} + \frac{-x+1}{1}$$

$$= -5x + 9$$

So $\mathrm{p}_1(x)$ is a polynomial of degree one.

Second, the points $x = 1$, $y = 4$ and $x = 2$, $y = -1$ lie on the graph of $y = p_1(x)$.

When $x = 1$,
$$p_1(x) = \frac{(1-2)}{(-1)} \times 4 + \frac{(1-1)}{(1)} \times (-1)$$
$$= \frac{(-1) \times 4}{(-1)} + 0$$
$$= 4$$

When $x = 2$,
$$p_1(x) = \frac{(2-2)}{(-1)} \times 4 + \frac{(2-1)}{(1)} \times (-1)$$
$$= 0 + \frac{1}{1} \times (-1)$$
$$= -1$$

Hence,
$$p_1(x) = \frac{(x-2)}{(1-2)} \times 4 + \frac{(x-1)}{(2-1)} \times (-1)$$

is a polynomial of degree one which passes through both points (1,4) and (2,–1) in the data set. The graph of $y = p_1(x)$ is, of course, the straight line which joins the points (1,4) and (2,–1), Figure 2.4.

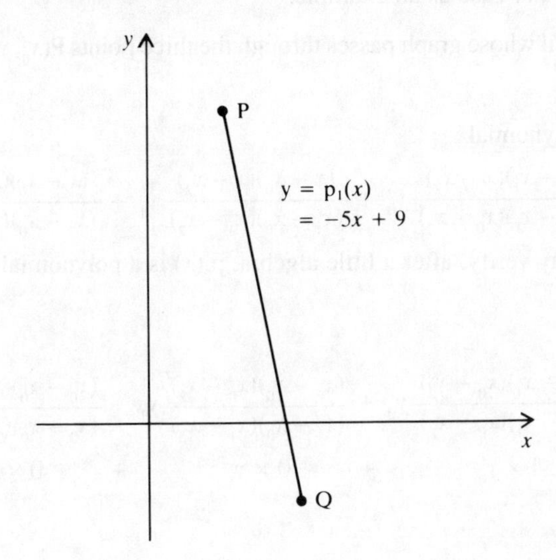

Figure 2.4

To try to unscramble some of this, we repeat this argument with the two general points $P(x_0, y_0)$ and $Q(x_1, y_1)$.

Consider the polynomial

$$p_1(x) = \frac{(x - x_1)}{(x_0 - x_1)} y_0 + \frac{(x - x_0)}{(x_1 - x_0)} y_1$$

As above, $p_1(x)$ is a polynomial of degree one. It may now be seen rather more clearly, however, that

$$\text{when } x = x_0 \qquad p_1(x) = \frac{(x_0 - x_1)}{(x_0 - x_1)} y_0 + \frac{(x_0 - x_0)}{(x_1 - x_0)} y_1 = 1 \times y_0 + 0 \times y_1 = y_0$$

$$\text{when } x = x_1 \qquad p_1(x) = \frac{(x_1 - x_1)}{(x_0 - x_1)} y_0 + \frac{(x_1 - x_0)}{(x_1 - x_0)} y_1 = 0 \times y_0 + 1 \times y_1 = y_1$$

Hence,

$$p_1(x) = \frac{(x - x_1)}{(x_0 - x_1)} y_0 + \frac{(x - x_0)}{(x_1 - x_0)} y_1$$

is a polynomial of degree one whose graph passes through the points (x_0, y_0) and (x_1, y_1).

What happens if there are three points? We begin this time with the general result, and take a particular case as an example.

Find a polynomial whose graph passes through the three points $P(x_0, y_0)$, $Q(x_1, y_1)$ and $R(x_2, y_2)$.

Consider the polynomial

$$p_2(x) = \frac{(x - x_1)(x - x_2)}{(x_0 - x_1)(x_0 - x_2)} y_0 + \frac{(x - x_0)(x - x_2)}{(x_1 - x_0)(x_1 - x_2)} y_1 + \frac{(x - x_0)(x - x_1)}{(x_2 - x_0)(x_2 - x_1)} y_2$$

As the reader may verify, after a little algebra, $p_2(x)$ is a polynomial of degree two. Also,

when $x = x_0$

$$p_2(x) = \frac{(x_0 - x_1)(x_0 - x_2)}{(x_0 - x_1)(x_0 - x_2)} y_0 + \frac{(x_0 - x_0)(x_0 - x_2)}{(x_1 - x_0)(x_1 - x_2)} y_1 + \frac{(x_0 - x_0)(x_0 - x_1)}{(x_2 - x_0)(x_2 - x_1)} y_2 = 1 \times y_0 + 0 \times y_1 + 0 \times y_2 = y_0$$

when $x = x_1$

$$p_2(x) = \frac{(x_1 - x_1)(x_1 - x_2)}{(x_0 - x_1)(x_0 - x_2)} y_0 + \frac{(x_1 - x_0)(x_1 - x_2)}{(x_1 - x_0)(x_1 - x_2)} y_1 + \frac{(x_1 - x_0)(x_1 - x_1)}{(x_2 - x_0)(x_2 - x_1)} y_2 = 0 \times y_0 + 1 \times y_1 + 0 \times y_2 = y_1$$

Similarly,

when $x = x_2 \qquad p_2(x) = y_2$

Hence, $p_2(x)$ is a polynomial of degree two whose graph passes through each of the three points $P(x_0, y_0)$, $Q(x_1, y_1)$ and $R(x_2, y_2)$.

From now on, we shall drop all mention of 'graph' and write about 'a polynomial which passes through three points'. Strictly speaking this is not correct, but the meaning is perfectly clear and the phrase is in common use.

Example 2.5

Find a polynomial of degree two which passes through the three points A(2,4), B(5,1) and C(6,7).

Here, $x_0 = 2, \quad x_1 = 5 \quad$ and $\quad x_2 = 6$
$y_0 = 4, \quad y_1 = 1 \quad$ and $\quad y_2 = 7$

Hence,

$$\mathrm{p}_2(x) = \frac{(x-5)(x-6)}{(2-5)(2-6)} \times 4 + \frac{(x-2)(x-6)}{(5-2)(5-6)} \times 1 + \frac{(x-2)(x-5)}{(6-2)(6-5)} \times 7$$

$$= \frac{x^2 - 11x + 30}{(-3)(-4)} \times 4 + \frac{x^2 - 8x + 12}{(3)(-1)} \times 1 + \frac{x^2 - 7x + 10}{(4)(1)} \times 7$$

$$= \frac{4x^2 - 44x + 120 - 4x^2 + 32x - 48 + 21x^2 - 147x + 210}{12}$$

$$= \frac{21x^2 - 159x + 282}{12} = \frac{7}{4}x^2 - \frac{53}{4}x + \frac{141}{6}$$

The graph of $y = \frac{7}{4}x^2 - \frac{53}{4}x + \frac{141}{6}$ is shown in Figure 2.5.

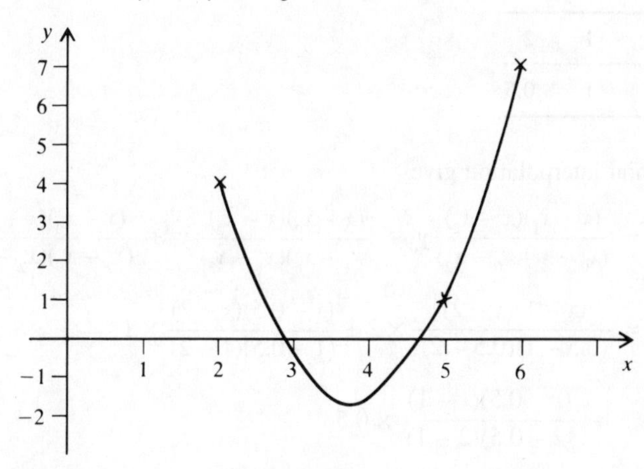

Figure 2.5

The polynomial that we have found to pass through each of the given points is called the **Lagrange interpolating polynomial**. Two observations should be made:

(a) Note the apparent relationship between the number of data points and the degree of the interpolating polynomial. When two points were given, the

Lagrange interpolating polynomial was of degree one. When three points were given, the interpolating polynomial was of degree two. We might suspect that if $n + 1$ points were given, then the degree of the interpolating polynomial would be n; and so it turns out to be.

(b) It may be proved – but we shall not do so here – that the polynomial described above is unique. That is, given (say) three points, there is one and only one polynomial of degree two that passes through each of the three points. And this polynomial is the Lagrange interpolating polynomial.

A second example illustrates how the Lagrange interpolating polynomial may be used to approximate a given function.

Example 2.6

Find a polynomial of degree two, $p_2(x)$, that interpolates the function $f(x) = \frac{1}{x}$ at the points where $x_0 = 0.5$, $x_1 = 1$ and $x_2 = 2$. Draw the graph of $y = \frac{1}{x}$ for values of x ranging from 0.1 to 3 and on the same axes draw the graph of $y = p_2(x)$. Comment on the fit of the interpolating polynomial

(a) over the interval [0.5,2],

(b) outside the interval [0.5,2].

The table of values for the function $y = \frac{1}{x}$ is

x	0.5	1	2
y	2	1	0.5

Polynomial interpolation gives:

$$p_2(x) = \frac{(x - x_1)(x - x_2)}{(x_0 - x_1)(x_0 - x_2)}y_0 + \frac{(x - x_0)(x - x_2)}{(x_1 - x_0)(x_1 - x_2)}y_1 + \frac{(x - x_0)(x - x_1)}{(x_2 - x_0)(x_2 - x_1)}y_2$$

$$p_2(x) = \frac{(x - 1)(x - 2)}{(0.5 - 1)(0.5 - 2)} \times 2 + \frac{(x - 0.5)(x - 2)}{(1 - 0.5)(1 - 2)} \times 1$$

$$+ \frac{(x - 0.5)(x - 1)}{(2 - 0.5)(2 - 1)} \times 0.5$$

$$= (x^2 - 3x + 2) \times 2\tfrac{2}{3} + (x^2 - 2.5x + 1) \times (-2) + (x^2 - 1.5x + 0.5) \times \tfrac{1}{3}$$

$$= x^2 - 3.5x + 3.5$$

The graphs $y = \frac{1}{x}$ (dotted line) and $y = x^2 - 3.5x + 3.5$ (solid line) are shown in Figure 2.6.

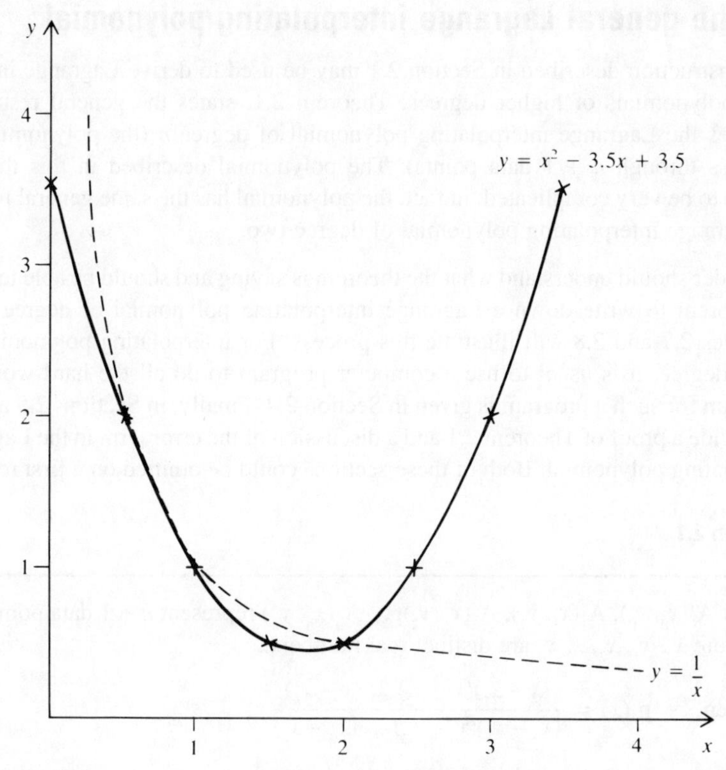

Figure 2.6

(a) The fit is seen to be quite good over the interpolating interval [0.5,2].

(b) Outside this interval, however, it is a different story.

(i) $0 < x < 0.5$

As $x \to 0$ so $\frac{1}{x} \to \infty$ and $p_2(x) \to 3.5$. Hence the difference $f(x) - p_2(x)$ can be very large indeed. For example, when $x = 0.01$, $f(x) - p_2(x) = 100 - 3.4651 = 96.5349$.

(ii) $2 < x$

It is clear that in this case, the graphs are diverging and to use $p_2(x)$ to approximate values of $f(x)$ would be asking for trouble.

It is important to realise that although the interpolating polynomial agrees with the given function at a number of fixed points, the interpolating polynomial may not generally be 'close' to the function it is interpolating. This is particularly noticable for points lying outside the interval of interpolation. In Example 2.6, the approximation was very poor indeed outside the interval [0.5,2].

2.3 The general Lagrange interpolating polynomial

The construction described in Section 2.1 may be used to derive Lagrange interpolating polynomials of higher degrees. Theorem 2.1, states the general result and describes the Lagrange interpolating polynomial of degree n (the polynomial that will pass through $n + 1$ data points). The polynomial described in this theorem appears to be very complicated; in fact, the polynomial has the same general form as the Lagrange interpolating polynomial of degree two.

The reader should understand what the theorem is saying and should be able to apply the theorem to write down a Lagrange interpolating polynomial of degree three. Examples 2.7 and 2.8 will illustrate this process. For interpolating polynomials of higher degree, it is usual to use a computer program to do all the hard work. An algorithm for such a program is given in Section 2.4. Finally, in Sections 2.5 and 2.6 we provide a proof of Theorem 2.1 and a discussion of the error term in the Lagrange interpolating polynomial. Both of these sections could be omitted on a first reading.

Theorem 2.1

Let $A_0(x_0, y_0), A_1(x_1, y_1), A_2(x_2, y_2), \dots A_n(x_n, y_n)$ represent $n + 1$ data points where $x_0, x_1, x_2, \dots x_n$ are distinct real numbers.

Then
$$\begin{aligned} p_n(x) = {} & \frac{(x - x_1)(x - x_2)\dots(x - x_n)}{(x_0 - x_1)(x_0 - x_2)\dots(x_0 - x_n)} y_0 \\ & + \frac{(x - x_0)(x - x_2)\dots(x - x_n)}{(x_1 - x_0)(x_1 - x_2)\dots(x_1 - x_n)} y_1 \\ & + \dots \\ & + \frac{(x - x_0)(x - x_1)\dots(x - x_{n-1})}{(x_n - x_0)(x_n - x_1)\dots(x_n - x_{n-1})} y_n \end{aligned}$$

is a polynomial of degree at most n that passes through each of the $n + 1$ data points.

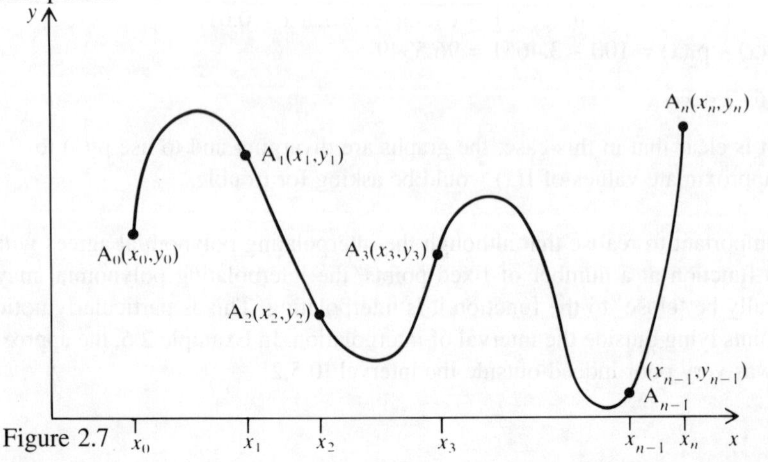

Figure 2.7

Example 2.7

Construct a polynomial $p_3(x)$ of degree three to interpolate the function $f(x) = e^{x^2}$ at the points given by $x_0 = 1$, $x_1 = 1.5$, $x_2 = 2$ and $x_3 = 2.5$. Use your polynomial to estimate the value of e^{x^2} when:

(a) $x = 1.2$ (b) $x = 1.6$ and (c) $x = 2.4$

Calculate the exact value of e^{x^2} at these points and hence write down the errors in your estimates. Comment on your results.

Write $y = e^{x^2}$; then the four points of interpolation are:

x	$x_0 = 1$	$x_1 = 1.5$	$x_2 = 2$	$x_3 = 2.5$
y	$y_0 = 2.7183$	$y_1 = 9.4877$	$y_2 = 54.598$	$y_3 = 518.01$

(all decimals rounded to five digits.)

From the above:

$$
\begin{aligned}
p_3(x) = {} & \frac{(x-1.5)(x-2)(x-2.5)}{(1-1.5)(1-2)(1-2.5)} \times 2.7183 \\
& + \frac{(x-1)(x-2)(x-2.5)}{(1.5-1)(1.5-2)(1.5-2.5)} \times 9.4877 \\
& + \frac{(x-1)(x-1.5)(x-2.5)}{(2-1)(2-1.5)(2-2.5)} \times 54.598 \\
& + \frac{(x-1)(x-1.5)(x-2)}{(2.5-1)(2.5-1.5)(2.5-2)} \times 518.01 \\
= {} & -3.6244\ (x^3 - 6x^2 + 11.75x - 7.5) \\
& + 37.951\ (x^3 - 5.5x^2 + 9.5x - 5) \\
& - 218.39\ (x^3 - 5x^2 + 7.75x - 3.75) \\
& + 690.68\ (x^3 - 4.5x^2 + 6.5x - 3) \\
= {} & 506.62x^3 - 2203.1x^2 + 3114.8x - 1415.6
\end{aligned}
$$

Using $p_3(x)$, we have:

(a) When $x = 1.2$, $p_3(x) = 25.135$ and $f(x) = 4.2207$.
(b) When $x = 1.6$, $p_3(x) = 3.2595$ and $f(x) = 12.936$.
(c) When $x = 2.4$, $p_3(x) = 373.58$ and $f(x) = 317.35$.

Hence the errors are:

When $x = 1.2$, the absolute error $= \left| f(1.2) - p_3(1.2) \right| = 20.914$.

When $x = 1.6$, the absolute error $= \left| f(1.6) - p_3(1.6) \right| = 9.6765$.

When $x = 2.4$, the absolute error $= \left| f(2.4) - p_3(2.4) \right| = 56.23$.

We see then that the errors are large, and we might conclude that, in this case, the interpolating polynomial does not provide a good approximation to e^{x^2}.

The next example shows how polynomial interpolation may be used to derive further information from a set of data.

Example 2.8

The data given below shows the speed in metres per second over successive half second intervals of a car accelerating away from a fixed point.

Time (s)	0	0.5	1	1.5
Speed (ms^{-1})	41	42.5	44.3	47.8

Find a polynomial of degree three that will pass through each of the four points. Using your polynomial, estimate
(a) the speed of the car after 0.75s,
(b) the acceleration after 1s,
(c) the distance travelled in the first second.

Let the required polynomial be known as $p_3(x)$.
As before, we write:

$$\begin{aligned} p_3(x) &= \frac{(x-0.5)(x-1)(x-1.5)}{(0-0.5)(0-1)(0-1.5)} \times 41 \\ &\quad + \frac{(x-0)(x-1)(x-1.5)}{(0.5-0)(0.5-1)(0.5-1.5)} \times 42.5 \\ &\quad + \frac{(x-0)(x-0.5)(x-1.5)}{(1-0)(1-0.5)(1-1.5)} \times 44.3 \\ &\quad + \frac{(x-0)(x-0.5)(x-1)}{(1.5-0)(1.5-0.5)(1.5-1)} \times 47.8 \\ &= -54.667\,(x^3 - 3x^2 + 2.75x - 0.75) + 170\,(x^3 - 2.5x^2 + 1.5x) \\ &\quad + 177.2\,(x^3 - 2x^2 + 0.75x) + 63.733\,(x^3 - 1.5x^2 + 0.5x) \\ &= 1.866x^3 - 2.1985x^2 + 3.6323x + 41 \end{aligned}$$

(a) To estimate the speed after 0.75s, calculate $p_3(0.75)$

$$\begin{aligned} p_3(0.75) &= 1.866 \times (0.75)^3 - 2.1985 \times (0.75)^2 + 3.6323 \times (0.75) + 41 \\ &= 43.275 \end{aligned}$$

(b) The data gives the speed y at time x. Since acceleration is given by $\frac{dy}{dx}$, we estimate the acceleration by calculating

$$\frac{d}{dx}\, p_3(x) = 3 \times 1.866x^2 - 2 \times 2.1985x + 3.6323$$

Hence, the acceleration when $x = 1$ is approximately

$$\begin{aligned} &3 \times 1.866 \times 1 - 2 \times 2.1985 \times 1 + 3.6323 \\ &= 4.8333\,ms^{-2} \end{aligned}$$

(c) The distance travelled in the first second may be approximated by $\int_0^1 p_3(x)dx$. This is:

$$\int_0^1 (1.866x^3 - 2.1985x^2 + 3.6323x + 41)\,dx$$

$$= \left[1.866\frac{x^4}{4} - 2.1985\frac{x^3}{3} + 3.6323\frac{x^2}{2} + 41x\right]_0^1$$

$$= \left(\frac{1.866}{4} - \frac{2.1985}{3} + \frac{3.6323}{2} + 41\right) - 0$$

$$= 42.550\,\text{m}$$

2.4 Interpolating polynomials of higher degree

The examples considered so far have involved interpolating polynomials of degree three or lower. Clearly, if polynomial interpolation is to be effective, we should be able to consider polynomials of degree higher than three. Equally clearly, such a procedure is going to be extremely time consuming without the use of a computer. Therefore, an algorithm is now provided from which a computer program may be written which will enable the values of a Lagrange interpolating polynomial to be calculated.

This algorithm is known as Neville's algorithm. A justification of this algorithm is outlined in Task U.

Algorithm 2.1 Neville's algorithm

To evaluate, at $x = t$, the polynomial $p_n(x)$ which passes through each of the $n + 1$ data points $(x_0, y_0), (x_1, y_1), (x_2, y_2), \ldots, (x_n, y_n)$.

input $x_0, x_1, x_2, \ldots, x_n$
$y_0, y_1, y_2, \ldots, y_n$
(store $y_0, y_1, y_2, \ldots, y_n$ in the first column of an $(n + 1) \times (n + 1)$ array Q)
for $i = 0, 1, \ldots, n$
$Q_{i0} := y_i$
endloop
for $i = 1, 2, \ldots, n$
for $j = 1, 2, \ldots, i$

$$Q_{ij} := \frac{(t - x_{i-j})\,Q_{i\,j-1} - (t - x_i)\,Q_{i-1\,j-1}}{x_i - x_{i-j}}$$

endloop
endloop
output $Q_{nn} = p_n(t)$

2.5 Proof of Theorem 2.1

The proof is broken down into the following steps which show that:

Step 1 The expression for $p_n(x)$ can be written in a more convenient and more comprehensible form.

Step 2 $p_n(x)$ passes through each of the $n + 1$ points $A_0, A_1, A_2, \dots, A_n$.

Step 3 $p_n(x)$ is a polynomial of degree at most n.

Step 1

Recall that the polynomial $p_n(x)$ is proposed, where:

$$
\begin{aligned}
p_n(x) = {} & \frac{(x-x_1)(x-x_2)\dots(x-x_n)}{(x_0-x_1)(x_0-x_2)\dots(x_0-x_n)}y_0 \\
& + \frac{(x-x_0)(x-x_2)(x-x_3)\dots(x-x_n)}{(x_1-x_0)(x_1-x_2)(x_1-x_3)\dots(x_1-x_n)}y_1 \\
& + \dots \\
& + \frac{(x-x_0)(x-x_1)(x-x_2)\dots(x-x_{n-1})}{(x_n-x_0)(x_n-x_1)(x_n-x_2)\dots(x_n-x_{n-1})}y_n
\end{aligned}
$$

This may look horrific, but confidence should be restored somewhat when it is seen that there is a clear pattern in the expressions. To help see this, write $L_0(x)$ to denote the coefficient of y_0

$$\text{so} \qquad L_0(x) = \frac{(x-x_1)(x-x_2)(x-x_3)\dots(x-x_n)}{(x_0-x_1)(x_0-x_2)(x_0-x_3)\dots(x_0-x_n)}$$

Write $L_1(x)$ to denote the coefficient of y_1

$$\text{so} \qquad L_1(x) = \frac{(x-x_0)(x-x_2)(x-x_3)\dots(x-x_n)}{(x_1-x_0)(x_1-x_2)(x_1-x_3)\dots(x_1-x_n)}$$

and, generally, write $L_i(x)$ to denote the coefficient of y_i

$$\text{so} \qquad L_i(x) = \frac{(x-x_0)(x-x_1)\dots(x-x_{i-1})(x-x_{i+1})\dots(x-x_n)}{(x_i-x_0)(x_i-x_1)\dots(x_i-x_{i-1})(x_i-x_{i+1})\dots(x_i-x_n)}$$

Using this notation, we have

$$p_n(x) = L_0(x)y_0 + L_1(x)y_1 + L_2(x)y_2 \dots + L_n(x)y_n$$

Observe that in $L_0(x)$, the numerator consists of the n factors $(x - x_1), (x - x_2), \dots, (x - x_n)$ but does not include $(x - x_0)$. The denominator may be obtained by substituting $x = x_0$ in the expression for the numerator.

In $L_1(x)$, the numerator consists of the n factors $(x - x_0)$ $(x - x_2), (x - x_3), \dots, (x - x_n)$ but does not include $(x - x_1)$. The denominator is just the numerator with x replaced by x_1.

In general, in $L_i(x)$, the numerator consists of the n factors $(x - x_0), (x - x_1), \dots, (x - x_{i-1}), (x - x_{i+1}), \dots, (x - x_n)$ but does not include $(x - x_i)$. The denominator is the numerator with x replaced by x_i.

Hence, although the expression for $p_n(x)$ may look complicated, the pattern in the terms is really quite simple.

Observe also that in each of the expressions $L_0(x), L_1(x), \ldots, L_n(x)$, the denominator is a number, while the numerator is the product of n polynomials of degree one. From the remarks at the beginning of this chapter, we may conclude that each of $L_0(x), L_1(x), \ldots, L_n(x)$ is a polynomial of degree n.

Step 2

The real worth of this construction becomes apparent when we calculate the values of $L_0(x), L_1(x), \ldots, L_n(x)$ for each of $x = x_0, x = x_1, \ldots, x = x_n$. To begin, we calculate the value of each of $L_0(x), L_1(x), \ldots, L_n(x)$ when $x = x_0$. Using functional notation we write these values as $L_0(x_0), L_1(x_0), \ldots, L_n(x_0)$, where

$$L_0(x_0) = \frac{(x_0 - x_1)(x_0 - x_2) \ldots (x_0 - x_n)}{(x_0 - x_1)(x_0 - x_2) \ldots (x_0 - x_n)} = 1$$

$$L_1(x_0) = \frac{(x_0 - x_0)(x_0 - x_2) \ldots (x_0 - x_n)}{(x_1 - x_0)(x_1 - x_2) \ldots (x_1 - x_n)}$$

$$= \frac{0 \times (x_0 - x_2) \ldots (x_0 - x_n)}{(x_1 - x_0)(x_1 - x_2) \ldots (x_1 - x_n)} = 0$$

$$L_i(x_0) = \frac{(x_0 - x_0)(x_0 - x_1) \ldots (x_0 - x_{i-1})(x_0 - x_{i+1}) \ldots (x_0 - x_n)}{(x_i - x_0)(x_i - x_1) \ldots (x_i - x_{i-1})(x_i - x_{i+1}) \ldots (x_i - x_n)} = 0$$

Hence, $L_0(x_0) = 1, \quad L_1(x_0) = 0, \quad L_2(x_0) = 0, \quad \ldots, \quad L_n(x_0) = 0$

Now put $x = x_1$.

$$L_0(x_1) = \frac{(x_1 - x_1)(x_1 - x_2) \ldots (x_1 - x_n)}{(x_0 - x_1)(x_0 - x_2) \ldots (x_0 - x_n)} = 0$$

$$L_1(x_1) = \frac{(x_1 - x_0)(x_1 - x_2) \ldots (x_1 - x_n)}{(x_1 - x_0)(x_1 - x_2) \ldots (x_1 - x_n)} = 1$$

$$L_i(x_1) = \frac{(x_1 - x_0)(x_1 - x_1) \ldots (x_1 - x_{i-1})(x_1 - x_{i+1}) \ldots (x_1 - x_n)}{(x_i - x_0)(x_i - x_1) \ldots (x_i - x_{i-1})(x_i - x_{i+1}) \ldots (x_i - x_n)} = 0$$

Hence, $L_0(x_1) = 0, \quad L_1(x_1) = 1, \quad L_2(x_1) = 0, \quad \ldots, \quad L_n(x_1) = 0$

Similarly, the reader may verify (see Exercise 38 at the end of this chapter) that when $x = x_2$,

$$L_0(x_2) = 0, \quad L_1(x_2) = 0, \quad L_2(x_2) = 1, \qquad L_3(x_2) = 0, \quad \ldots \quad L_n(x_2) = 0$$

and further, that

$$L_i(x_i) = 1 \quad \text{for all} \quad 0 \leqslant i \leqslant n$$
$$L_i(x_j) = 0 \quad i \neq j$$

Our task is now almost complete.

In $\mathrm{p}_n(x) = \mathrm{L}_0(x)y_0 + \mathrm{L}_1(x)y_1 + \mathrm{L}_2(x)y_2 \dots + \mathrm{L}_n(x)y_n$

put $x = x_0$, so

$$\mathrm{p}_n(x_0) = \mathrm{L}_0(x_0)y_0 + \mathrm{L}_1(x_0)y_1 + \mathrm{L}_2(x_0)y_2 \dots + \mathrm{L}_n(x_0)y_n$$

From the above discussion,

$$\begin{aligned}\mathrm{p}_n(x_0) &= 1 \times y_0 + 0 \times y_1 + 0 \times y_2 \dots + 0 \times y_n \\ &= y_0\end{aligned}$$

Put $x = x_1$, so

$$\begin{aligned}\mathrm{p}_n(x_1) &= \mathrm{L}_0(x_1)y_0 + \mathrm{L}_1(x_1)y_1 + \mathrm{L}_2(x_1)y_2 \dots + \mathrm{L}_n(x_1)y_n \\ &= 0 \times y_0 + 1 \times y_1 + 0 \times y_2 \dots + 0 \times y_n \\ &= y_1\end{aligned}$$

Similarly, put $x = x_i$ so

$$\begin{aligned}\mathrm{p}_n(x_i) &= \mathrm{L}_0(x_i)y_0 + \mathrm{L}_1(x_i)y_1 \dots + \mathrm{L}_i(x_i)y_i \dots + \mathrm{L}_n(x_i)y_n \\ &= 0 \times y_0 + 0 \times y_1 \dots + 1 \times y_i \dots + 0 \times y_n \\ &= y_i\end{aligned}$$

Hence, we have that

$$\mathrm{p}_n(x) = \mathrm{L}_0(x)y_0 + \mathrm{L}_1(x)y_1 \dots + \mathrm{L}_n(x)y_n$$

passes through each of the $n + 1$ points $(x_0, y_0), (x_1, y_1), \dots (x_n, y_n)$ and we have almost achieved our objective. What remains is to show that $\mathrm{p}_n(x)$ is a polynomial and to say something about the degree of this polynomial.

Step 3

We saw earlier that each of $\mathrm{L}_0(x), \mathrm{L}_1(x), \dots, \mathrm{L}_n(x)$ is a polynomial of degree n. Hence each of $\mathrm{L}_0(x)y_0, \mathrm{L}_1(x)y_1 \dots, \mathrm{L}_n(x)y_n$ is also a polynomial of degree n. Hence $\mathrm{p}_n(x)$ is the sum of $n + 1$ polynomials each of degree n. It is possible, however, that the x^n term (and possibly other terms) will disappear in the simplification of $\mathrm{L}_0(x)y_0 + \mathrm{L}_1(x)y_1 \dots + \mathrm{L}_n(x)y_n$. Hence we write: $\mathrm{p}_n(x)$ is a polynomial of degree not more than n, and the proof is complete.

2.6 Errors

It is clear from the examples that there can be a considerable error involved in using the Lagrange interpolating polynomial to approximate a function $\mathrm{f}(x)$. The next step is to derive an expression for the error. The mathematics involved in this process is necessarily pitched at a higher level and some readers may prefer to leave this section until a second reading.

In this book, all the proofs concerning errors will be based either on the Taylor polynomial (see Chapter 5) or on Rolle's theorem (see the Introduction). The proof that follows will be based on Rolle's theorem and the reader is advised to review that theorem before beginning the proof.

Given a function $f(x)$, select $n + 1$ distinct numbers $x_0, x_1, x_2, \dots, x_n$ all of which lie within an interval $[a,b]$. Let $p_n(x)$ denote the polynomial of degree at most n that interpolates $f(x)$ at the $n + 1$ data points $(x_0, f(x_0)), (x_1, f(x_1)), \dots, (x_n, f(x_n))$.

Write $\quad f(x) = p_n(x) + e(x) \qquad x \in [a,b]$.

Then $e(x)$ represents the error in using $p_n(x)$ to approximate $f(x)$ at the value x in the interval $[a,b]$.

Theorem 2.2

Under the above conditions, there exists a number $\eta \in [a,b]$ such that

$$f(x) = p_n(x) + \frac{f^{(n+1)}(\eta)}{(n+1)!}(x - x_0)(x - x_1)(x - x_2) \dots (x - x_n)$$

where $f^{(n+1)}(\eta)$ denotes the $(n + 1)$th derivative of $f(x)$ evaluated at $x = \eta$ and where, clearly,

$$e(x) = \frac{f^{(n+1)}(\eta)}{(n+1)!}(x - x_0)(x - x_1) \dots (x - x_n) \qquad \text{represents the error.}$$

The proof may appear at first sight to be very complicated. The essential argument is not difficult, however, and the proof will repay careful study.

Proof

The theorem will be proved for the case $n = 1$. The proof for general n proceeds along exactly the same lines and is the subject of Exercise 37 at the end of this chapter.

The idea is to set up the function $Q(t)$ written below. This function looks awful but is really not as bad as it looks. (For this reason it is called the Quasimodo function.)

Let $\quad Q(t) = f(t) - p_1(t) - [f(x) - p_1(x)]\dfrac{(t - x_0)(t - x_1)}{(x - x_0)(x - x_1)}$

We shall regard $Q(t)$ as a function of t. So x will be regarded as a constant and so, consequently will $f(x) - p_1(x)$ and $(x - x_0)(x - x_1)$.

The first step in the proof is to show that $Q(x_0) = 0$, that $Q(x_1) = 0$ and that $Q(x) = 0$.

When $t = x_0$

$$Q(x_0) = f(x_0) - p_1(x_0) - [f(x) - p_1(x)]\frac{(x_0 - x_0)(x_0 - x_1)}{(x - x_0)(x - x_1)}$$

Since $p_1(x)$ passes through the point $(x_0, f(x_0))$, we must have

$$f(x_0) = p_1(x_0)$$

that is $\quad f(x_0) - p_1(x_0) = 0$

The final term on the right-hand side contains the factor $(x_0 - x_0)$; hence this term is also zero. Consequently, $Q(x_0) = 0$.

Similarly, it may be shown that $Q(x_1) = 0$.

If $t = x$, then

$$Q(x) = f(x) - p_1(x) - [f(x) - p_1(x)]\frac{(x - x_0)(x - x_1)}{(x - x_0)(x - x_1)}$$

$$= f(x) - p_1(x) - [f(x) - p_1(x)] \times 1$$
$$= 0$$

Hence, $Q(x_0) = 0, \quad Q(x_1) = 0 \quad$ and $\quad Q(x) = 0$

The second step in the proof is to use Rolle's theorem. Since $Q(x_0)$ and $Q(x_1) = 0$ Rolle's theorem tells us that $Q'(r) = 0$ for some $r \in (x_0, x_1)$. Also, since $Q(x_1) = 0$ and $Q(x) = 0$, we must have $Q'(s) = 0$ for some $s \in (x_1, x)$. But now, since $Q'(r) = 0$ and $Q'(s) = 0$, Rolle's theorem tells us that $Q''(\eta) = 0$ for some $\eta \in [a,b]$.

Differentiating $Q(t)$ twice with respect to t gives:

$$Q''(t) = f''(t) - p_1''(t) - \frac{[f(x) - p_1(x)]}{(x - x_0)(x - x_1)} \frac{d^2}{dt^2}(t - x_0)(t - x_1)$$

since $f(x) - p_1(x)$ and $(x - x_0)(x - x_1)$ may be regarded as constants.

But note that $p_1(t)$ is a polynomial in t of degree at most one,

so $\quad p_1''(t) = 0$

and $\quad (t - x_0)(t - x_1) = t^2 - (x_0 + x_1)\,t + x_0x_1$

so $\quad \frac{d^2}{dt^2}(t - x_0)(t - x_1) = 2$

Hence $\quad Q''(t) = f''(t) - 0 - \frac{f(x) - p_1(x)}{(x - x_0)(x - x_1)} \times 2$

Letting $t = \eta$ gives

$$Q''(\eta) = f''(\eta) - \frac{f(x) - p_1(x)}{(x - x_0)(x - x_1)} \times 2$$

Since $Q''(\eta) = 0$ we have, on rearrangement:

$$[f(x) - p_1(x)]2 = f''(\eta)(x - x_0)(x - x_1)$$

or $\quad f(x) = p_1(x) + \frac{f''(\eta)}{2}(x - x_0)(x - x_1)$

as required. The difficulty with this result is that in practice the expression $f^{(n+1)}(\eta)$ cannot be evaluated. The theorem tells us only that η exists: it gives no indication of how the value of η could be found. Hence, in general, the value of η is unknown. Furthermore, on many occasions, we have no idea of the form of the function $f(x)$. This would be the case if, for example, we were fitting a polynomial to a set of data points. In such a situation, we could not differentiate $f(x)$ hence the form of $f^{(n+1)}(x)$

would also be unknown. There is, then, little hope of using the expression

$$\frac{f^{(n+1)}(\eta)}{(n+1)!}(x-x_0)(x-x_1)\ldots(x-x_n)$$

to calculate a precise value of the error. This result is, however, still of great value. Its usefulness lies principally in two areas.

(1) If the function $f(x)$ is known then the error term may allow us to calculate maximum bounds on the error. This will be illustrated in the example that follows.

(2) The error term allows us to investigate the errors involved in other types of approximation. We shall show how this may be achieved in Chapter 7.

Example 2.9

Find the polynomial of degree two that interpolates the function $f(x) = e^{-x}$ at the points given by $x_0 = 1$, $x_1 = 1.5$, $x_2 = 2$. Calculate the greatest value that could be taken by the error in the interval [1,2].

We have

x	1	1.5	2
y	0.36788	0.22313	0.13534

The interpolating polynomial $p_2(x)$ is given by:

$$\begin{aligned} p_2(x) &= \frac{(x-1.5)(x-2)}{(1-1.5)(1-2)} \times 0.367\,88 + \frac{(x-1)(x-2)}{(1.5-1)(1.5-2)} \times 0.223\,13 \\ &\quad + \frac{(x-1)(x-1.5)}{(2-1)(2-1.5)} \times 0.135\,34 \\ &= 0.735\,76(x^2-3.5x+3) - 0.892\,52(x^2-3x+2) \\ &\quad + 0.270\,68(x^2-2.5x+1.5) \\ &= 0.113\,92x^2 - 0.574\,30x + 0.828\,26 \end{aligned}$$

The error term is:

$$\frac{f'''(\eta)}{3!}(x-1)(x-1.5)(x-2)$$

We are interested in the absolute error, so we consider

$$\left|\frac{f'''(\eta)}{3!}(x-1)(x-1.5)(x-2)\right|$$

and calculate separately the maximum values that can be taken by

$$|f'''(\eta)| \qquad \text{and by} \qquad |(x-1)(x-1.5)(x-2)|$$

Step 1 $\quad f'''(\eta)$

Since
$$f(x) = e^{-x}$$
$$f'(x) = -e^{-x}$$
$$f''(x) = e^{-x}$$
$$f'''(x) = -e^{-x}$$

In the interval [1,2], we see from the graph (Figure 2.8) that

$$\left| f'''(x) \right| = \left| -e^{-x} \right| \leqslant \left| -e^{-1} \right| = 0.367\,88$$

So $\quad \left| f'''(\eta) \right| \leqslant 0.367\,88$

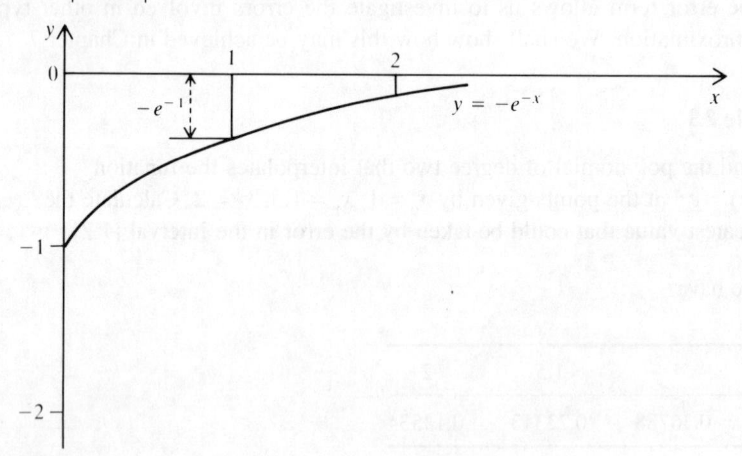

Figure 2.8

Step 2 $\quad (x - 1)(x - 1.5)(x - 2)$

The graph of $y = (x - 1)(x - 1.5)(x - 2)$ is shown in Figure 2.9.

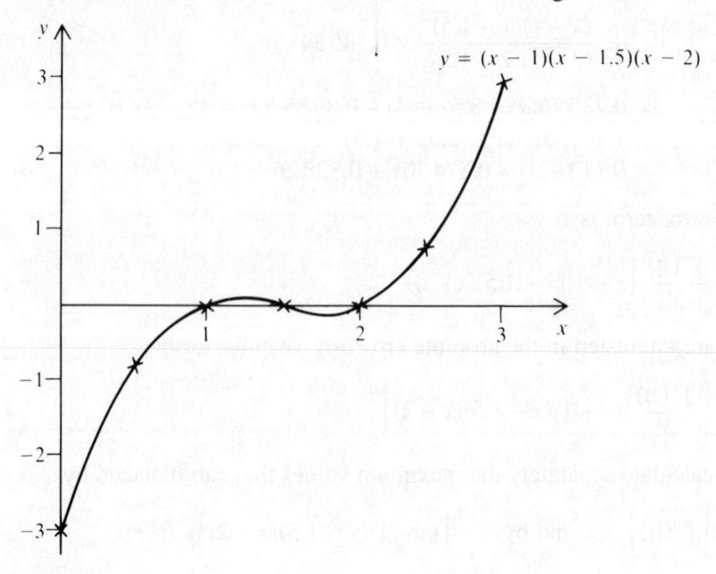

Figure 2.9

To calculate the maximum value of $(x-1)(x-1.5)(x-2)$, we use the max/min technique of the calculus.

$$y = x^3 - 4.5x^2 + 6.5x - 3$$

$$\frac{dy}{dx} = 3x^2 - 9x + 6.5$$

Putting $\frac{dy}{dx} = 0$ gives $3x^2 - 9x + 6.5 = 0$, so

$$x = \frac{9 \pm \sqrt{81-78}}{6}$$

$$= \begin{cases} 1.7887 \\ 1.2113 \end{cases}$$

When $x = 1.7887, \quad y = -0.048\,113$
$x = 1.2113, \quad y = 0.048\,113$

Hence $\left|(x-1)(x-1.5)(x-2)\right| \leqslant 0.048\,113$

Finally, we see that

$$\left|\frac{f'''(\eta)}{3!}(x-1)(x-1.5)(x-2)\right| \leqslant \frac{0.36788}{6} \times 0.048\,113$$
$$= 0.002\,950\,0$$

and the error in the interval [1,2] is less than 0.002 950 0.

Exercise 2

Introduction

1 Simplify the following expressions.
(a) $3(x-1) + 2(x+4)$
(b) $4(x^2 - 3x + 1) + 5(x+2)$
(c) $5(x^2 - 4x + 1) - 2(x^2 - x + 3)$
(d) $8x^2 + 2x - 1 - 8(x^2 - 1)$
(e) $2x^3 - 4x^2 + 3x + 1 - 2(x^2 + 4x - 1)$
(f) $\frac{5}{4}x^2 + \frac{3}{2}x - 1 + 4(\frac{3}{8}x^2 - \frac{1}{8}x + 1)$
(g) $3(2x^3 - 6x^2 + 2x - 1) + 5(2x^3 - 4x^2 - x + 3)$
(h) $5x^3 + x - 10 - 3(2x^3 - x^2 + 10)$

2 Multiply out the brackets and simplify the resulting expressions.
(a) $(x+1)(x+3)$
(b) $(x^2 - 2x + 1)(x - 3)$
(c) $(x-2)(x^2 - 4x + 1)$
(d) $(x-1)(x+3)(x-2)$
(e) $(x+3)(x-1)^2$
(f) $(x^2 - 2x + 1)(x^2 + 1)$
(g) $(2x-1)(3x^2 + 4x - 1)$
(h) $(2x+1)(2x-3)(x-4)$
(i) $(3x^2 + 2x - 1)(4x^2 - 2x + 3)$
(j) $(4x-1)^2(2x+1)^2$
(k) $(\frac{4}{3}x - 1)(x^2 + 2x - 1)$
(l) $(\frac{1}{5}x^2 + 2x - 1)(x^2 + \frac{1}{2})$

Section 2.1

3 Construct a polynomial of degree one to pass through the points A(1,3) and B(4,–1). Plot the points A and B on a graph and on the same axes, plot the graph of the polynomial.

4 Two points A(–2,–1) and B(3,2) are given. Plot the points A and B and draw the straight line AB on your graph. Write down the equation of the line AB. Now construct the polynomial of degree one that interpolates the points A and B. Check that the two equations are identical. If they are not, find your mistake.

5 Six points from a set of data are given as

x	1	1.5	2	2.5	3	3.5
y	4	8	7	5	1	6

Plot the points on a graph and join the points by a sequence of five straight lines. Calculate the Lagrange interpolating polynomial of each of the five straight lines.

Use your equations to calculate approximately the value of $\int_1^{3.5} y\,dx$.

6 $A(x_0, y_0)$ and $B(x_1, y_1)$ are two points in the plane. Let $p_1(x) = a_1x + a_0$ be the polynomial of degree one whose graph (a straight line) joins A and B.
Show that $y_0 = a_1x_0 + a_0$ and derive a similar equation in x_1 and y_1.
Eliminate a_0 from your equations and hence obtain an expression for a_1. Using this expression, obtain also an expression for a_0. Using these expressions for a_0 and a_1 show that $p_1(x) = a_1x + a_0$ may be written in the form

$$p_1(x) = \frac{x - x_1}{x_0 - x_1} y_0 + \frac{x - x_0}{x_1 - x_0} y_1$$

7 The function f(x) is known at two values of x: when $x = 2$, f(x) = 4 and when $x = 5$, f(x) = –3. The approximate value of x at which f(x) = 0 is to be calculated. Calculate the Lagrange polynomial of degree one which interpolates f(x) at the above points and by using this polynomial, calculate approximately the solution of the equation f(x) = 0.

8 In a table of the normal distribution probability function, it is given that

when $z = 1.94$ $p = 0.9738$
and when $z = 1.95$ $p = 0.9744$

Calculate an approximation to the value of p when

(a) $z = 1.945$ (b) $z = 1.942$ (c) $z = 1.946$

(This technique, of approximating the value of a function at an intermediate point by considering a straight line approximation to the function, is called linear interpolation.)

9 In the same table it shows that

when	$z = 1.28$	$p = 0.8997$
and when	$z = 1.29$	$p = 0.9015$

Use this information to calculate an estimate of the value of z that would give p the value 0.9000.

10 Again, in the normal distribution table,

when	$z = 1.03$	$p = 0.8485$
and when	$z = 1.04$	$p = 0.8508$

What value of z would cause p to take the value 0.8500?

11 The function f(x) is known at three points, given in the table.

x	2	3	4
f(x)	6.5	9.1	8.7

Find the polynomial of degree two that interpolates f(x) at these points. Use this polynomial to calculate approximately the value of f(x) when:

(a) $x = 2.8$ (b) $x = 3.8$ (c) $x = 4.8$

Which of your answers is likely to be the least accurate?

12 $\dfrac{dy}{dx}$ is to be calculated for the function $y = f(x)$. However, f(x) is known only at three values of x:

x	−1	0	1.5
f(x)	90	25	34

Construct the polynomial of degree two that interpolates f(x) at these three points. Use this polynomial to calculate approximately the value of $\dfrac{dy}{dx}$ when

(a) $x = 0$ (b) $x = -0.5$ (c) $x = 1.5$ (d) $x = 3$

Comment on the relative accuracy of these results.

13 Calculate the Lagrange interpolating polynomial of degree two that interpolates the data

x	3	4.5	6
f(x)	100	54	64

Use your polynomial to calculate approximately the value of $\int_3^6 f(x)\, dx$.

14 Projected profits from a certain company are given in the table:

year	1	3	5
profits in £100 000	2.5	5.1	4.7

Construct the Lagrange interpolating polynomial of degree two for this data. Using your polynomial, estimate
(a) when profits will equal £500 000,
(b) the greatest profit that can be expected over this period.

15 Estimate, as accurately as you can the solution(s) of the equation $f(x) = 0$ when values of $f(x)$ are given only at $x = 2$, $x = 3.5$, $x = 4$:

x	2	3.5	4
$f(x)$	2.17	–1.28	–1.43

Which (if any) of your solutions could be described as the most reliable?

Section 2.3

16 Construct the Lagrange interpolating polynomial of degree three for the data

x	1.5	2	3	4
y	6	9	2	4

Calculate as accurately as you can
(a) the value of y when $x = 2.5$,
(b) the value of $\frac{dy}{dx}$ when $x = 3.8$,
(c) the value of the integral $\int_2^4 f(x)dx$.

17 Construct a polynomial of degree three to interpolate the data

x	–3	–1	0	1
$f(x)$	–11	–2	4.5	3.1

Use your polynomial to calculate (a) $f(x)$ and (b) $f'(x)$ at

(i) $x = -2$ (ii) $x = -0.5$ (iii) $x = 0.5$

18 The following data records the value of the function $f(x) = \frac{1}{\sqrt{2\pi}} e^{-\frac{1}{2}x^2}$ when $x = 0.5$, $x = 1$, $x = 1.5$, $x = 2$.

x	0.5	1	1.5	2
f(x)	0.14045	0.096532	0.051670	0.021539

Calculate as accurately as possible the value of $\int_1^{1.5} \mathrm{f}(x)\mathrm{d}x$.

19 The values of the function f(x) at $x = 2.5$ and $x = 3.5$ are to be found when f(x) is known only from the data in the following table.

x	1	2	3	4
f(x)	0.7071	0.4472	0.3162	0.2425

Calculate approximately the values of f(2.5) and f(3.5) using interpolating polynomials of degrees one, two and three.

In fact, $\mathrm{f}(x) = \dfrac{1}{\sqrt{x^2+1}}$. Compare the accuracy of your approximations. What would you expect to happen if you used your interpolating polynomials to estimate (a) f(0) and (b) f(5)?

20 Construct a polynomial $\mathrm{p}_2(x)$ of degree two to interpolate the function $\mathrm{f}(x) = \ln x$ at the points $x = 1$, $x = 2$, $x = 2.5$. Draw, on the same axes, the graphs of $y = \ln x$ and $y = \mathrm{p}_2(x)$. Use your graphs to estimate the greatest error in $|\ln x - \mathrm{p}_2(x)|$ for x in the range $1 \leqslant x \leqslant 2.5$.

21 The function $\mathrm{f}(x) = \dfrac{1}{1+\sqrt{x}}$ is to be approximated in the interval [0,3] by a polynomial of degree three. Construct an interpolating polynomial $\mathrm{p}_3(x)$, choosing your own interpolating points, and draw on the same axes the graphs $y = \dfrac{1}{1+\sqrt{x}}$ and $y = \mathrm{p}_3(x)$. Comment on the error in $|\mathrm{f}(x) - \mathrm{p}_3(x)|$ for $0 \leqslant x \leqslant 3$ and use your graph to estimate the position and magnitude of the greatest error.

22 The function $\mathrm{f}(x) = \dfrac{4}{1+x^2}$ is to be approximated by a cubic polynomial in the interval [2,4]. Using four points of your choice, construct $\mathrm{p}_3(x)$ the Lagrange interpolating polynomial of degree three and draw, on the same axes, the graphs $y = \dfrac{4}{1+x^2}$ and $y = \mathrm{p}_3(x)$. Comment on the accuracy of the approximation.

Section 2.6

23 Construct $\mathrm{p}_2(x)$, a polynomial of degree two that interpolates the function $\mathrm{f}(x) = \dfrac{1}{1-x}$ at the points $x = 0$, $x = 1.5$, $x = 3$. Draw on the same axes, the graphs $y = \dfrac{1}{1-x}$ and $y = \mathrm{p}_2(x)$. What is happening to the error in the neighbourhood of $x = 1$? Consider the error term for $\mathrm{p}_2(x)$ and show how the error in the neighbourhood of $x = 1$ can occur.

24 Find $p_1(x)$, a polynomial of degree one to interpolate the function $f(x) = x\ln x$ at the points $x = 5$, $x = 7$.

(a) Write down $p_1(x)$ and find an expression for the error in using $p_1(x)$ as an approximation for $f(x)$ in the interval [5,7]. Determine the greatest value of the error $\left|f(x) - p_1(x)\right|$ in the interval [5,7].

(b) Draw, on the same axes, the graphs of $y = x\ln x$ and $y = p_1(x)$. Comment on your answer to part (a).

25 Find a polynomial $p(x)$ which will interpolate the function $f(x) = e^{-x}$ at the points given by $x = 1$ and $x = 2$. Write down an expression for the error term in the interpolation and find the greatest value taken by the error term in the interval [1,2]. Verify your results graphically.

26 Find a polynomial $p_2(x)$ which interpolates the function $f(x) = \sqrt{x}$ at the points given by $x = 1$, $x = 2$ and $x = 3$. Write down an expression for the error term and find the greatest value that can be taken by the error term in the interval [1,3].

27 Write down a polynomial of degree two that will interpolate the function $f(x) = \sin x$ at the points given by $x = 0.5$, $x = 1$ and $x = 2$ (x in radians). Write down also an expression for the error term in the interpolation and hence determine the greatest value that can be taken by the error term in the interval [0.5,2]. Draw the graphs of $y = \sin x$ and of your polynomial and comment on your answer.

28 (a) If $g(x) = (x - a)(x - b)$, show that the maximum value of $\left|g(x)\right|$ for $a \leqslant x \leqslant b$ occurs at $x = \frac{1}{2}(a + b)$.
Hence show that the maximum value of $\left|g(x)\right|$ for $a \leqslant x \leqslant b$ is $\frac{1}{4}(b - a)^2$.

(b) The function $f(x) = \dfrac{1}{x - 1}$ is interpolated by a polynomial of degree one at the points $x = 2$ and $x = 3$. Determine the greatest error that occurs in the interval [2,3] by using the polynomial to approximate $f(x)$.

29 (a) The points a, b, and c are equally spaced on the x axis with $a < b < c$. Consider the expression

$$g(x) = (x - a)(x - b)(x - c)$$

By writing $a = b - h$ and $c = b + h$, show that the maximum value of $\left|g(x)\right|$ with $x \in [a,b]$ is $\dfrac{2h^3}{3\sqrt{3}}$.

(b) Find $p_2(x)$, the polynomial of degree two that interpolates $f(x) = \ln(1 + x)$ at the points given by $x = 2$, $x = 3$ and $x = 4$. Write down the expression for the error term in the approximation and find the greatest value taken by the error term when $x \in [2,4]$.

30 The function $y = x^7 + 5x^2 + 3x - 2$ is to be interpolated by a polynomial of degree two at the points $x = 0$, $x = 0.5$ and $x = 1$. Estimate the greatest value of the error term in the interpolation.

Miscellaneous

31 (a) Using any means that seems to you to be sensible estimate f(4.72) from the data

x :	4	5	6	7	8
$f(x)$:	0.35	0.88	1.71	2.90	4.51

(b) Find the cubic polynomial in x which takes the values 1, –3, –1, 13 when $x = 1, 2, 3, 4$ respectively.

(MEI adapted)

32 Find the cubic polynomial $y(x)$ in x which takes the values –2, 1, 4 and 73 when $x = 1, 2, 3$ and 6 respectively. Show that there is only one real solution of the equation $y(x) = 0$ and find a first estimate for the solution. Find also, an estimate for $\frac{dy}{dx}$ at $x = 4.6$.

(MEI adapted)

33 Draw the graph of $y = x \sin\sqrt{x}$ for $0 \leqslant x \leqslant 2$ (where x is measured in radians). It is required to approximate $x \sin\sqrt{x}$ in the interval [0,2] by a polynomial. Choose suitable interpolating points and draw the graphs of your interpolating polynomials. Comment on your results.

34 Interpolate the function $f(x) = \sqrt{x+1}$ in the interval [2,3] by a polynomial of degree two. By taking the three interpolating points to have the form $t - h, t, t + h$ where $2 \leqslant t - h$ and $t + h \leqslant 3$, determine the value of h so that

$$|\text{error}| \leqslant 0.0005 \text{ for all } x \in [2,3]$$

35 A Lagrange interpolating polynomial of degree two is to be found for a function $f(x)$ in the interval [1,2]. Taking interpolating points $a_0 = 1$, $a_1 = 1.5$, $a_2 = 2$, write down an expression for the error in the interpolation at the point x.

If, for all $1 \leqslant x \leqslant 2$ the error is to be bounded by 10^{-3} show that $|f'''(x)| \leqslant 0.124\,71$ for all $1 \leqslant x \leqslant 2$.

Consider the three functions

(a) $f(x) = \sin x$ (b) $f(x) = e^{-(x+1)}$ (c) $f(x) = \frac{1}{x}$

Do any of these functions satisfy the above error bound?

36 A polynomial is required to approximate the function

$$f(x) = \frac{2}{1 + x^2} \qquad -1 \leqslant x \leqslant 1$$

(a) Find $p_1(x)$ the polynomial of degree one that interpolates $f(x)$ at $x = -1$ and $x = 1$.

(b) Find $p_2(x)$ the polynomial of degree two that interpolates $f(x)$ at the points $x = -1$, $x = 0$ and $x = 1$.

(c) Find $p_3(x)$ taking $x = -1$, $x = -\frac{1}{3}$, $x = \frac{1}{3}$, $x = 1$ as the interpolating points.

(d) Draw the graphs of $y = f(x)$, $y = p_1(x)$, $y = p_2(x)$ and $y = p_3(x)$.
Repeat (a), (b), (c) and (d) for (e), (f), (g) and redraw the graphs, taking interpolation points as follows.
(e) $x = \cos\frac{1}{4}\pi$, $x = \cos\frac{3}{4}\pi$
(f) $x = \cos\frac{1}{6}\pi$, $x = \cos\frac{3}{6}\pi$, $x = \cos\frac{5}{6}\pi$
(g) $x = \cos\frac{1}{8}\pi$, $x = \cos\frac{3}{8}\pi$, $x = \cos\frac{5}{8}\pi$, $x = \cos\frac{7}{8}\pi$
(These are known as the Chebechev interpolation points†.)
Comment on the differences in the interpolation.

37 Prove Theorem 2.2 for general n.

38 Consider the functions $L_0(x)$, $L_1(x)$, from Section 2.5.
Show that if $x_0, x_1, x_2, \ldots, x_n$ are interpolating points
(a) $L_1(x_3) = 0$ (b) $L_3(x_2) = 0$ (c) $L_2(x_2) = 1$

Show generally that $L_i(x_j) = \begin{cases} 0 & \text{if } i \neq j \\ 1 & \text{if } i = j \end{cases}$

Prove that $L_0(x) + L_1(x) + L_2(x) \ldots + L_n(x) = 1$ for all x.
(Hint: consider the constant function $f(x) \equiv 1$.)

39 The function $y = f(x)$ is known only at the values $x = 3$, $x = 3.5$, $x = 4$ and $x = 5$.

x	3	3.5	4	5
f(x)	12	2	−4	−1

Find the polynomial of degree three which interpolates $f(x)$ at these points. Use the techniques of Chapter 3 or Chapter 6 to estimate one of the solutions of the equation $f(x) = 0$.

40 The function $y = f(x)$ takes values as shown in the table

x	2	3	4	5
f(x)	1.8186	0.423 36	−3.0272	−4.7946

Construct interpolating polynomials of degree one, two and three and use these polynomials to solve the equation $f(x) = 0$.
The data is generated by the equation $f(x) = x \sin x$. Solve the equation $f(x) = 0$ correctly to five decimal digits (rounded) and comment on the accuracy of your earlier solutions.

41 Suppose that a projectile has an initial velocity of $20\,\text{m s}^{-1}$ and the gravitation constant is $g = 9.81\,\text{m s}^{-2}$. Then as a function of angle θ, and ignoring the effects of friction, the maximum height the projectile will reach is given by

$$h(\theta) = \frac{200 \sin^2(\theta)}{g} \qquad 0 < \theta < \frac{\pi}{2}$$

† More details are given in Burden and Faires *Numerical analysis* 3rd edition, Prindle, Weber and Schmidt.

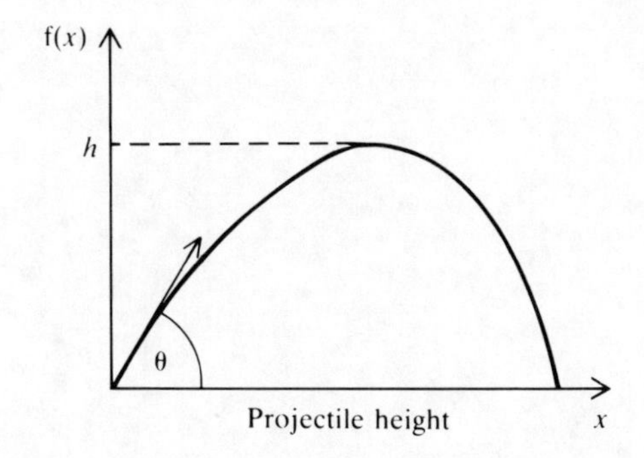

Approximate this function by an interpolation polynomial evaluated at the points $\frac{\pi j}{10}$, $j = 0, 1, 2, 3, 4, 5$. By computer evaluate the polynomial thus constructed at the angles $q_i = \frac{\pi i}{201}$, $i = 1, 2, \ldots, 100$, and print out the maximum absolute error

$$\max_{1 \leqslant i \leqslant 100} \left| h(\theta_i) - p(\theta_i) \right|$$

at these test points. (Y & S)

42 Evaluation of interpolation functions outside the interval determined by the data points is called 'extrapolation'. In the figure below, estimation of f(x) for $x < a$ or $x > b$ would constitute extrapolation.

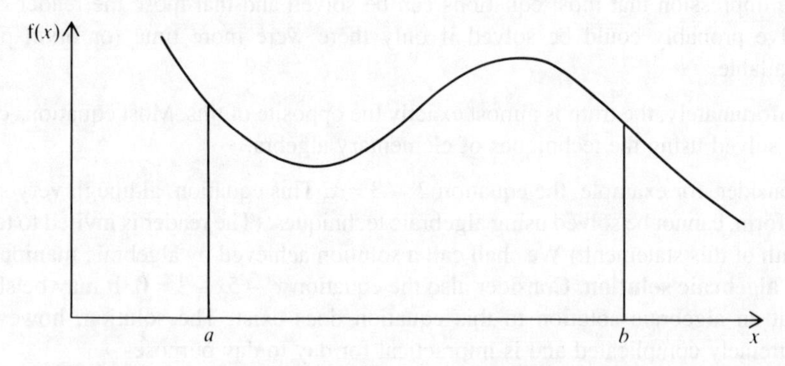

Explain the pitfalls of extrapolation. Your explanation should include examples of extrapolation, the errors in extrapolation and reference to the error term in the interpolating polynomial.

Chapter 3

Solution of equations 1

Introduction

One of the most important areas of mathematics is the solution of equations. Indeed, it is difficult to think of a topic in mathematics that does not involve equations and which does not require their solution.

If the reader is just beginning an A-level course, the idea that there is any difficulty in solving equations may come as a surprise. In the GCSE syllabus, and in much of the A-level syllabus, most of the equations that occur are capable of solution even though a great deal of effort may be required to achieve this solution. This may give the impression that most equations can be solved and that those the reader cannot solve probably could be solved if only there were more time (or brain power) available.

Unfortunately, the truth is almost exactly the opposite of this. Most equations cannot be solved using the techniques of elementary algebra.

Consider, for example, the equation $2^x - 3 = x$. This equation, although very simple in form, cannot be solved using algebraic techniques. (The reader is invited to test the truth of this statement.) We shall call a solution achieved by algebraic manipulation an **algebraic solution**. Consider also the equation $x^3 - 5x - 3 = 0$. It may be shown† that an algebraic solution to this equation does exist. The solution, however, is extremely complicated and is impractical for day to day purposes.

In this chapter we shall consider three methods of obtaining solutions to equations when either an algebraic solution does not exist or such a solution is extremely difficult to find. These solutions will be numerical and not algebraic expressions. These solutions are called **numerical solutions** and the methods that produce them are described as **numerical methods**. Before describing these methods, however, we should note some general remarks that will be of use throughout the chapter.

Remark 1

First, we need to be clear about what is meant by a solution. If the equation to be solved is written in the form $\mathrm{f}(x) = 0$ then λ is a solution if $\mathrm{f}(\lambda) = 0$. If, for example,

† See Hall and Knight 4th edn. *Higher algebra*, Chapter 35 pp 480–485.

the equation to be solved is $2^x - 3 = x$ then, when written in the form $f(x) = 0$, the equation becomes $2^x - x - 3 = 0$ and λ is a solution if $2^\lambda - \lambda - 3 = 0$.

Geometrically, a solution of the equation $f(x) = 0$ occurs when the graph of $y = f(x)$ crosses the x-axis. The reason for this is fairly obvious. Suppose that the graph of $y = f(x)$ crosses the x-axis at the point $(\lambda, 0)$, Figure 3.1. (Recall that every point on the x-axis has a y coordinate of zero). Since the point $(\lambda, 0)$ lies on the graph, the coordinates $x = \lambda$, $y = 0$ must satisfy the equation $y = f(x)$. Hence $0 = f(\lambda)$ and by the above definition, λ is a solution of the equation $f(x) = 0$.

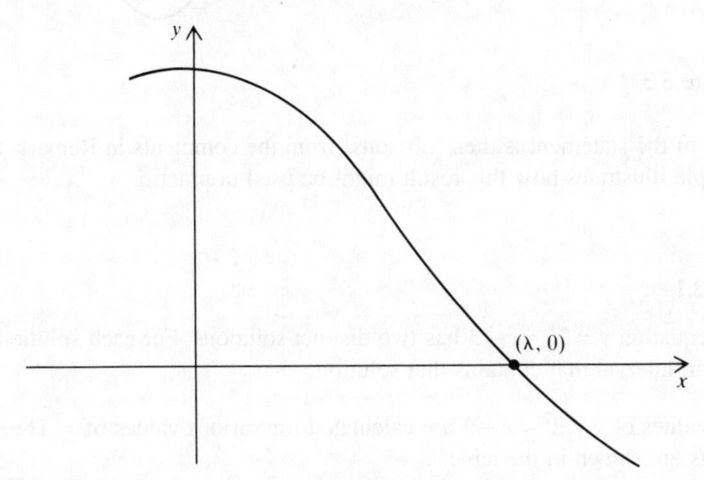

Figure 3.1

It is important to realise that this interpretation of a solution of an equation applies only when the equation is written in the form $f(x) = 0$.

Remark 2

Suppose that $x = a$ and $x = b$ are two points on the x-axis and that $f(a)$ and $f(b)$ are of opposite sign. It might be, for example, that $f(a)$ is positive while $f(b)$ is negative, as in Figure 3.2.

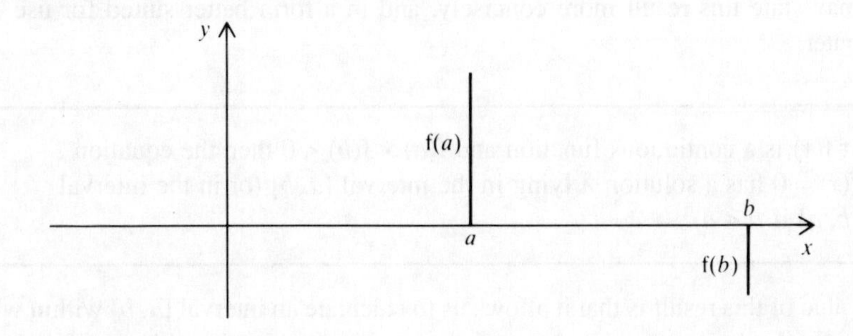

Figure 3.2

Then, if $f(x)$ is continuous, there is a solution to the equation $f(x) = 0$ lying between a and b. The graph of $y = f(x)$ might be like Figure 3.3.

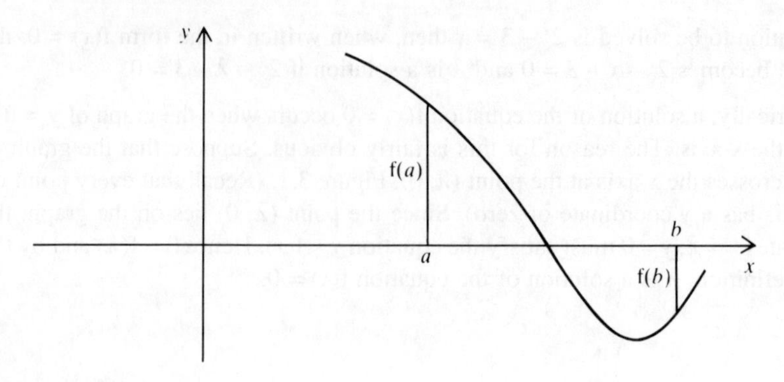

Figure 3.3

The truth of the statement is then 'obvious' from the comments in Remark 1. An example illustrates how this result might be used in practice.

Example 3.1

The equation $y = 2^x - x - 3$ has two distinct solutions. For each solution, find an interval that contains that solution.

The values of $y = 2^x - x - 3$ are calculated for various values of x. The results are shown in the table.

x	−4	−3	−2	−1	0	1	2	3	4
$y = 2^x - x - 3$	1.0625	0.125	−0.75	−1.5	−2	−2	−1	2	9

From the above comments, we know that a solution occurs every time there is a sign change in y. Hence solutions occur in the intervals [−3, −2] and [2, 3].

We may state this result more concisely, and in a form better suited for use on a computer.

> If $f(x)$ is a continuous function and $f(a) \times f(b) < 0$ then the equation $f(x) = 0$ has a solution λ lying in the interval $[a, b]$ (or in the interval $[b, a]$ if $b < a$).

The value of this result is that it allows us to calculate an interval $[a, b]$ within which the required solution must lie. As we shall see, this can often provide a useful starting point for our numerical methods.

Remark 3

In each of the numerical techniques described in this chapter, the first step will be to find an approximate value of the solution. This approximation will be called **the starting value** and will be written as x_0.

There are several ways in which a value for x_0 might be calculated.

(a) Draw the graph of the function and observe the point at which the graph crosses the x-axis.

(b) Find an interval that contains the solution and take x_0 to be the mid-point of this interval.

(c) Find an interval that contains the solution and take x_0 to be one of the end points of the interval.

Having found a starting value, each method will then seek to calculate a second (and better) approximation to the solution and then a third (and still better) approximation and so on until the required degree of accuracy has been achieved.

To illustrate the process, let x_0 denote the first approximation to the solution, x_1 the second approximation, x_2 the third approximation and so on. Then, if λ is the true value of the solution, the values x_0, x_1, x_2, ... might appear on the graph $y = f(x)$ as shown in Figure 3.4.

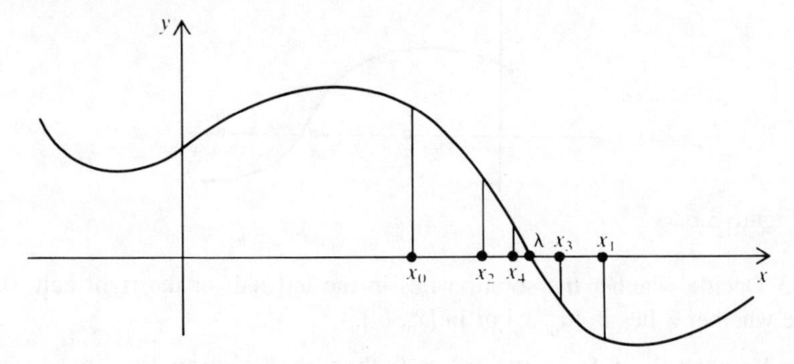

Figure 3.4

Three different methods for calculating the successive approximations x_1, x_2, x_3, ... will be described.

The common feature of all three methods is that, to calculate a particular approximation (x_3 say), all that is needed are the previous approximations (x_0, x_1, x_2) and the equation itself. Such methods are called **iterative methods.**

3.1 The bisection method

This is also known as the binary search method.

The idea is to locate an interval that contains the required solution. • Cut this interval into two equal parts (hence the name – bisection method) and decide which of the two parts contains the solution. Now take the part containing the solution to be the new interval of interest and repeat the entire procedure (go to •).

In more detail, the method consists of the following steps.

Step 1 Locate an interval that contains the required solution λ. Call this interval $[a_0, b_0]$.

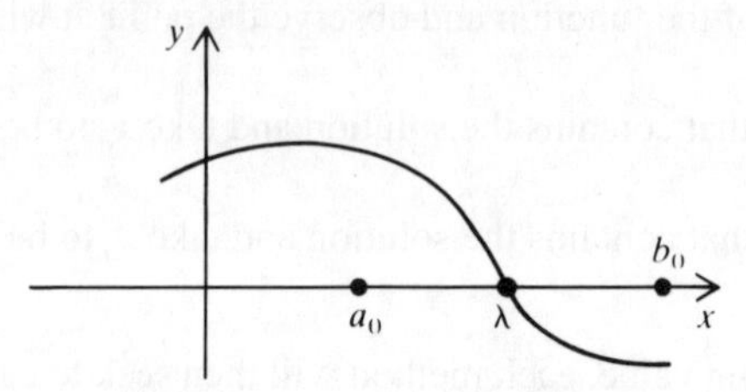

Figure 3.5

Step 2 Calculate the mid-point of the interval $[a_0, b_0]$. Call the mid-point x_0. Thus $x_0 = \frac{1}{2}(a_0 + b_0)$.

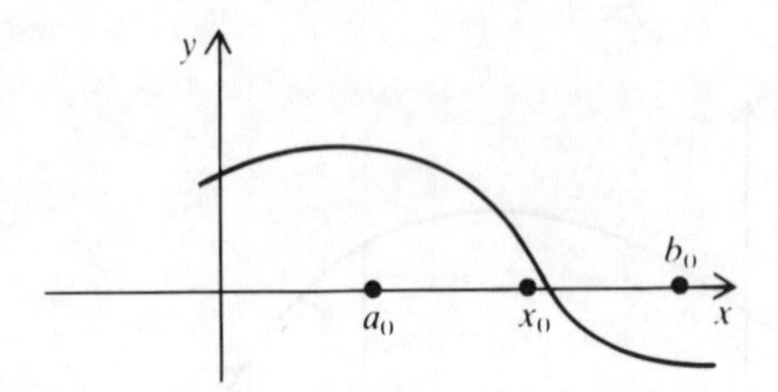

Figure 3.6

Step 3 Decide whether the solution lies in the left half or the right half, that is, decide whether λ lies in $[a_0, x_0]$ or in $[x_0, b_0]$.

Using Remark 2: if $f(a_0) \times f(x_0) < 0$ then the solution lies in $[a_0, x_0]$; if $f(x_0) \times f(b_0) < 0$, the solution lies in $[x_0, b_0]$; if $f(x_0) = 0$, then $\lambda = x_0$ and the procedure halts.

Step 4 Let the 'half' containing the solution be given the new name $[a_1, b_1]$. So, if the solution lies in $[a_0, x_0]$, take $a_1 = a_0$, $b_1 = x_0$; if the solution lies in $[x_0, b_0]$, take $a_1 = x_0$, $b_1 = b_0$.

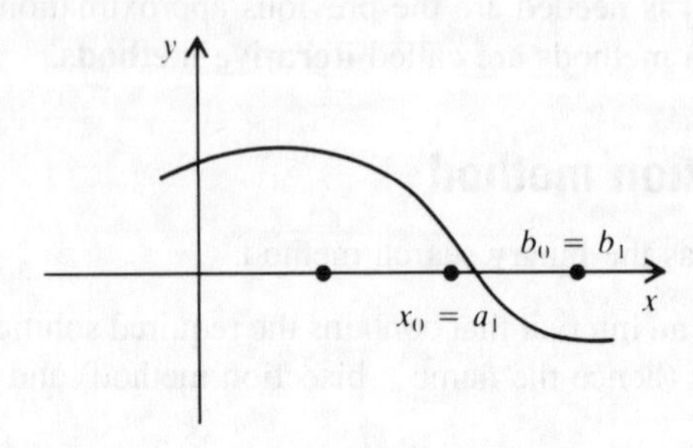

Figure 3.7

Step 5 Repeat Steps 2, 3 and 4, working successively with the intervals

$[a_1, b_1]$ with mid-point x_1
$[a_2, b_2]$ with mid-point x_2
...
$[a_i, b_i]$ with mid-point x_i

This produces the required sequence $x_1, x_2, x_3, \ldots$ Intuitively, it is clear that by repeatedly halving the length of an interval known to contain the solution, we can produce an arbitrarily small interval which must contain the solution. Hence the mid-points $x_0, x_1, x_2, \ldots$ of these intervals must converge to the solution to any required degree of accuracy. An example should help to make this clear.

Example 3.2

Use the bisection method to determine to three decimal digits (with rounding) the positive solution of the equation $2^x - x - 3 = 0$.

A part of the graph of $y = 2^x - x - 3$ is shown in Figure 3.8 and it is clear that the required solution lies in the interval [2, 3]. Hence $a_0 = 2$ and $b_0 = 3$.
The mid-point of this interval lies at $x = 2.5$; hence $x_0 = 2.5$.

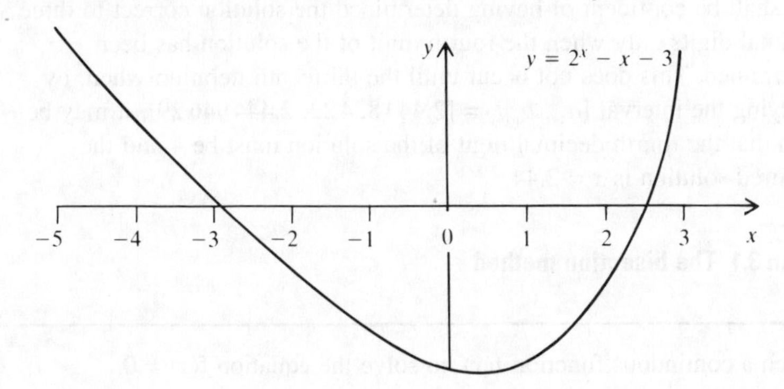

Figure 3.8

To decide whether the solution belongs to [2, 2.5] or [2.5, 3] calculate

$$f(2) \times f(2.5) = (-1) \times (0.1569)$$
$$= -0.1569 < 0$$

From Remark 2, we deduce that the solution lies in the interval [2, 2.5]; hence $a_1 = 2$ and $b_1 = 2.5$.
The mid-point of this interval lies at $x = 2.25$; hence $x_1 = 2.25$.

To decide whether the solution belongs to [2, 2.25] or to [2.25, 2.5] calculate

$$f(2) \times f(2.25) = (-1) \times (-0.4932)$$
$$= 0.4932 > 0$$

Since this product is greater than zero, we deduce that the solution belongs to the interval [2.25, 2.5]. Hence $a_2 = 2.25$ and $b_2 = 2.5$. The mid-point of this interval lies at $x = 2.375$; hence $x_2 = 2.375$.

It may be seen (using the algorithm that follows) that the successive intervals and their mid-points are given in the table.

$a_0 = 2$	$b_0 = 3$	$x_0 = 2.5$
$a_1 = 2$	$b_1 = 2.5$	$x_1 = 2.25$
$a_2 = 2.25$	$b_2 = 2.5$	$x_2 = 2.375$
$a_3 = 2.375$	$b_3 = 2.5$	$x_3 = 2.4375$
$a_4 = 2.4375$	$b_4 = 2.5$	$x_4 = 2.46875$
$a_5 = 2.4375$	$b_5 = 2.46875$	$x_5 = 2.453125$
$a_6 = 2.4375$	$b_6 = 2.453125$	$x_6 = 2.4453125$
$a_7 = 2.4375$	$b_7 = 2.4453125$	$x_7 = 2.44140625$
$a_8 = 2.44140625$	$b_8 = 2.4453125$	$x_8 = 2.44335938$
$a_9 = 2.44335938$	$b_9 = 2.4453125$	$x_9 = 2.44433594$
$a_{10} = 2.44433594$	$b_{10} = 2.4453125$	$x_{10} = 2.44482422$
$a_{11} = 2.44482422$	$b_{11} = 2.4453125$	$x_{11} = 2.44506836$
$a_{12} = 2.44482422$	$b_{12} = 2.44506836$	$x_{12} = 2.44494629$
$a_{13} = 2.44482422$	$b_{13} = 2.44494629$	$x_{13} = 2.44488525$

We shall be confident of having determined the solution correct to three decimal digits only when the fourth digit of the solution has been determined. This does not occur until the thirteenth iteration when, by studying the interval $[a_{13}, b_{13}] = [2.44482422, 2.44494629]$, it may be seen that the fourth decimal digit of the solution must be 4 and the required solution is $x = 2.44$.

Algorithm 3.1 The bisection method

Given a continuous function f(x), to solve the equation f(x) = 0.

Find an interval $[a_0, b_0]$ such that $f(a_0) \times f(b_0) < 0$.

```
input:  a_0, b_0
for n = 0,1, ..., until satisfied
    x_n := ½(a_n + b_n)
    if f(a_n) × f(x_n) < 0 then
        a_{n+1} := a_n; b_{n+1} := x_n
    otherwise
        a_{n+1} := x_n; b_{n+1} := b_n
    endif
endloop
output:  x_0, x_1, x_2, ...
a sequence that will converge to the required solution.
```

3.2 Stopping rules

A very important question that applies to all iterative methods is: 'When do we stop the procedure?' Earlier the correct, but generally unhelpful remark was made that the iteration continued until 'the required degree of accuracy had been achieved'. We now provide a more practical answer to this question and discuss stopping rules. There are three stopping rules in common use.

SR1

Stop when x_n is calculated for which the magnitude of $f(x_n)$ is less than some prescribed number. When this happens, take x_n to be the required solution. So given (for example) the prescribed tolerance 0.0005, the iteration would stop when the inequality $\left|f(x_n)\right| < 0.0005$ was obtained and x_n would be the required solution. Since the solution λ of the equation satisfies $f(\lambda) = 0$, there might seem to be considerable justification for this stopping rule, but see the comments that follow SR3.

SR2

Stop when two consecutive terms of the sequence $x_0, x_1, x_2, \ldots$ differ by less than the prescribed tolerance. So given the prescribed tolerance 0.0005, we would stop when a term x_{n+1} is calculated where the inequality $\left|x_{n+1} - x_n\right| < 0.0005$ was obtained. x_{n+1} would then be the required solution.

SR3

Stop when a term x_{n+1} is calculated for which the ratio $\left|\dfrac{x_{n+1} - x_n}{x_{n+1}}\right|$, $(x_{n+1} \neq 0)$ is less than the prescribed tolerance. Using again the tolerance 0.0005, the iteration would stop when the inequality $\left|\dfrac{x_{n+1} - x_n}{x_{n+1}}\right| < 0.0005$ was obtained and x_{n+1} would be the required solution.

Unfortunately, problems can arise using any of these stopping rules. It is possible to find examples where $\left|f(x_n)\right|$ becomes very small, but x_n is far from being an accurate solution of the equation. It is possible also for the terms $\left|x_{n+1} - x_n\right|$ to become very small, when again x_{n+1} is not an accurate approximation to the solution. In general, it is probably a good idea to use SR3. This rule provides a measure of the relative error of the solution, and is generally the most reliable of the three stopping rules.

Hence Algorithm 3.1 could be written as shown over the page:

Algorithm 3.1 The bisection method

To obtain a solution of the equation $f(x) = 0$.

Find an interval $[a_0, b_0]$ such that $f(a_0) \times f(b_0) < 0$.

```
input: a_0, b_0
x_0 := ½(a_0 + b_0)
For n = 0, 1, ... until satisfied
    if f(a_n) × f(x_n) < 0 then
        a_{n+1} := a_n; b_{n+1} := x_n
    otherwise
        a_{n+1} := x_n; b_{n+1} := b_n
    endif
    x_{n+1} := ½(a_{n+1} + b_{n+1})
    if |(x_{n+1} − x_n)/x_{n+1}| < ε  (where ε is the given tolerance) then stop
    endif
endloop
output: x_0, x_1, x_2, ... a sequence that is converging to the required
solution.
```

Take x_{n+1} to be the required approximation.

Example 3.3

Use the bisection method to obtain an approximation to the positive solution of the equation $x^3 - 5x - 3 = 0$. To obtain your answer, use each of the three stopping rules described above with tolerance $\varepsilon = 0.005$. Compare your answers.

The graph of $y = x^3 - 5x - 3$ is sketched in Figure 3.9.

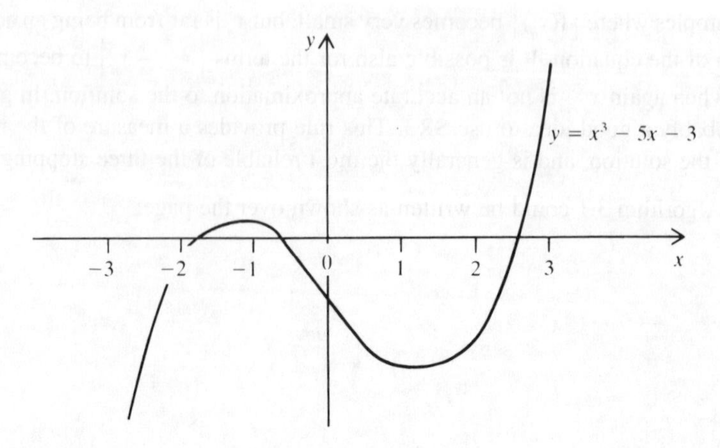

Figure 3.9

From the graph, it is clear that the required solution lies in the interval [2, 3]. Hence $a_0 = 2$ and $b_0 = 3$. The mid-point of this interval is 2.5. Hence $x_0 = 2.5$.

To decide whether the solution lies in [2, 2.5] or [2.5, 3] calculate $f(2) \times f(2.5) = (-5) \times (0.125) = -0.625$. Since the product is negative, we conclude that the solution lies in the interval [2, 2.5].

Hence $\quad a_1 = 2,\ b_1 = 2.5$ and $x_1 = \dfrac{2 + 2.5}{2} = 2.25.$

Now calculate $f(2) \times f(2.25) = (-5) \times (-2.859) = 14.295$. Since this product is positive, we conclude that the solution lies in the interval [2.25, 2.5].

Hence $\quad a_2 = 2.25,\ b_2 = 2.5$ and $x_2 = \dfrac{2.25 + 2.5}{2} = 2.375.$

The first fifteen terms of the sequence with their associated intervals are given in the table below.

$a_0 = 2$	$b_0 = 3$	$x_0 = 2.5$
$a_1 = 2$	$b_1 = 2.5$	$x_1 = 2.25$
$a_2 = 2.25$	$b_2 = 2.5$	$x_2 = 2.375$
$a_3 = 2.375$	$b_3 = 2.5$	$x_3 = 2.4375$
$a_4 = 2.4375$	$b_4 = 2.5$	$x_4 = 2.468\,75$
$a_5 = 2.468\,75$	$b_5 = 2.5$	$x_5 = 2.484\,375$
$a_6 = 2.484\,375$	$b_6 = 2.5$	$x_6 = 2.492\,187\,5$
$a_7 = 2.484\,375$	$b_7 = 2.492\,187\,5$	$x_7 = 2.488\,281\,25$
$a_8 = 2.488\,281\,25$	$b_8 = 2.492\,187\,5$	$x_8 = 2.490\,234\,38$
$a_9 = 2.490\,234\,38$	$b_9 = 2.492\,187\,5$	$x_9 = 2.491\,210\,94$
$a_{10} = 2.490\,234\,38$	$b_{10} = 2.491\,210\,94$	$x_{10} = 2.490\,722\,66$
$a_{11} = 2.490\,722\,66$	$b_{11} = 2.491\,210\,94$	$x_{11} = 2.490\,966\,8$
$a_{12} = 2.490\,722\,66$	$b_{12} = 2.490\,966\,8$	$x_{12} = 2.490\,844\,73$
$a_{13} = 2.490\,844\,73$	$b_{13} = 2.490\,966\,8$	$x_{13} = 2.490\,905\,76$
$a_{14} = 2.490\,844\,73$	$b_{14} = 2.490\,905\,76$	$x_{14} = 2.490\,875\,24$
$a_{15} = 2.490\,844\,73$	$b_{15} = 2.490\,875\,24$	$x_{15} = 2.490\,859\,99$

For the first stopping rule, the calculation stops when a term x_n is calculated for which $|f(x_n)| < 0.005$. It may be seen that

$$|f(x_8)| = |2.490\,234\,38^3 - 5 \times 2.490\,234\,38 - 3| \approx 8.56 \times 10^{-3}$$

while

$$|f(x_9)| = |2.491\,210\,94^3 - 5 \times 2.491\,210\,94 - 3| \approx 4.73 \times 10^{-3}$$

Hence when SR1 is used x_9 is the required solution and, in this case, the solution is $x_9 = 2.491\,210\,94$.

For the second stopping rule, the calculation stops when, for two consecutive terms x_{n+1} and x_n, $|x_{n+1} - x_n| < 0.005$. A straightforward calculation shows that $|x_6 - x_5| \approx 7.81 \times 10^{-3}$ while $|x_7 - x_6| \approx 3.91 \times 10^{-3}$. The solution, in this case, is $x_7 = 2.488\,281\,25$.

For the third stopping rule, the calculation stops when consecutive terms x_n, x_{n+1} are such that

$$\left| \frac{x_{n+1} - x_n}{x_{n+1}} \right| < 0.005$$

It may be shown that $\left| \dfrac{x_5 - x_4}{x_5} \right| \approx 6.29 \times 10^{-3}$

while $\left| \dfrac{x_6 - x_5}{x_6} \right| \approx 3.13 \times 10^{-3}$.

The solution, in this case, is $x_6 = 2.492\,187\,5$.

By using a tolerance of 0.005, we could hope, at best, for accuracy in the two places after the decimal point. This has been achieved in the first and third solutions and, if rounding is used, this accuracy has been achieved in all three solutions. If it is required to achieve accuracy in the first three digits of the solution, we have to wait until the interval

$$[a_9, b_9] = [2.490\,234\,38, 2.492\,187\,5]$$

Note that, although we stated earlier that it was necessary for the function f(x) to be continuous, in fact it is necessary only for f(x) to be continuous in the initial interval $[a_0, b_0]$. A discontinuity that occurs outside this interval will not affect the convergence of the sequence $x_0, x_1, x_2, \ldots$

3.3 Convergence

The beauty of the bisection method is that if the initial interval does contain the required solution, then $x_0, x_1, x_2, \ldots$ will converge to that solution. (We are assuming that there is only one solution in the initial interval.) The darker side to the method is that the convergence may be very slow. It was seen in Example 3.3 for instance that nine iterations were required to guarantee three decimal digit accuracy.

One approach to speeding up the convergence is to observe that the bisection method is extremely wasteful of information. In any iteration, if $[a, b]$ represents the interval containing the solution, then the values of f(a) and f(b) are known but are not used: only the signs are of interest. One way in which the values of f(a) and f(b) may be used is to argue that if the magnitude of (say) f(a) is much less than that of f(b) then the solution is likely to be closer to a than to b. Figure 3.10 will help to make this clear.

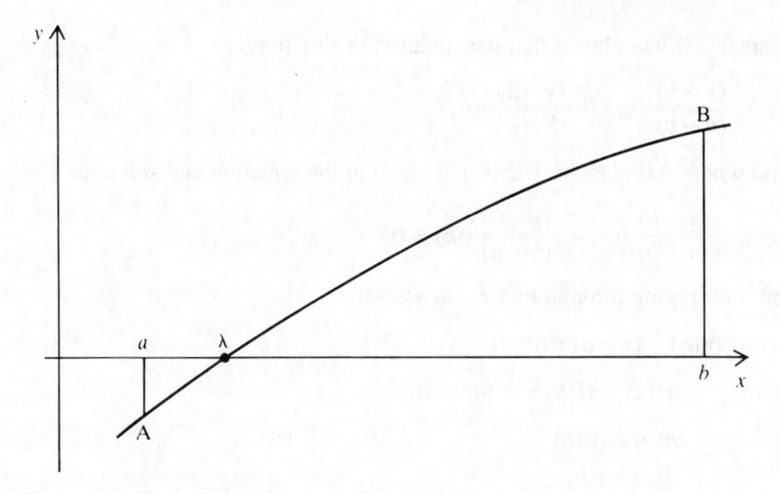

Figure 3.10

Here, $\left| f(a) \right|$ is considerably smaller than $\left| f(b) \right|$ and, as expected, the solution is closer to a than to b. To exploit this idea, draw the straight line joining A to B. The point C(c, 0) at which this line cuts the x-axis becomes the next approximation. From Figure 3.11, it is clear that there are grounds for believing that c will provide a better approximation to λ than would the mid-point of $[a, b]$.

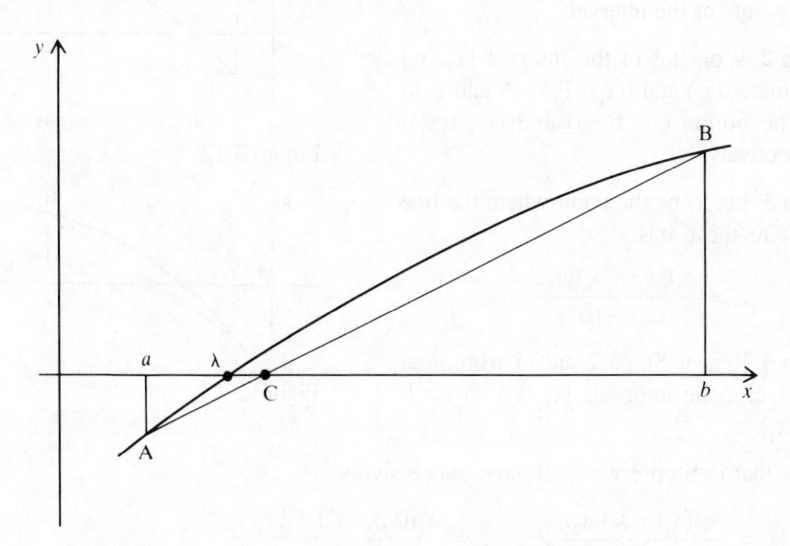

Figure 3.11

Hence, we can be optimistic that a method using this technique will converge faster than the bisection method. This technique is called **linear interpolation**, or **the secant method**.

We still have to calculate the value of c. To do this, observe that AB represents the polynomial of degree one which passes through the points A(a, f(a)) and B(b, f(b)).

In Chapter 2 it was shown that the equation of this line is

$$y = \frac{(x - b)}{(a - b)} \mathrm{f}(a) + \frac{(x - a)}{(b - a)} \mathrm{f}(b)$$

To find where AB cuts the x-axis, put $y = 0$ in the equation and solve for x.

Hence $$\frac{(x - b)}{(a - b)} \mathrm{f}(a) + \frac{(x - a)}{(b - a)} \mathrm{f}(b) = 0$$

and on multiplying through by $(a - b)$ we get

$$(x - b)\mathrm{f}(a) - (x - a)\mathrm{f}(b) = 0$$

Therefore $$x\mathrm{f}(a) - x\mathrm{f}(b) = b\mathrm{f}(a) - a\mathrm{f}(b)$$

and $$x = \frac{b\mathrm{f}(a) - a\mathrm{f}(b)}{\mathrm{f}(a) - \mathrm{f}(b)}$$

This discussion leads to the secant method (or the method of linear interpolation) for solving the equation $\mathrm{f}(x) = 0$.

3.4 The secant method

Step 1 Locate an interval that contains the required solution. Take x_0 and x_1 to be the end points of the interval.

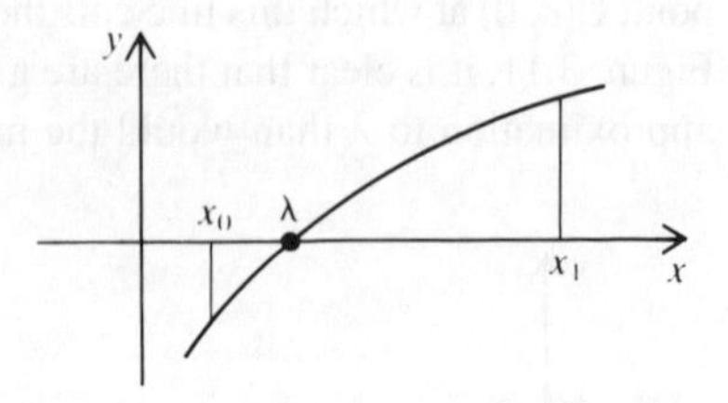

Figure 3.12

Step 2 Working in the interval $[x_0, x_1]$ calculate $\mathrm{f}(x_0)$ and $\mathrm{f}(x_1)$. Take A and B to be the points $(x_0, \mathrm{f}(x_0))$ and $(x_1, \mathrm{f}(x_1))$ respectively.

Step 3 Let x_2 be the point where the line AB cuts the x-axis.

So $$x_2 = \frac{x_1\mathrm{f}(x_0) - x_0\mathrm{f}(x_1)}{\mathrm{f}(x_0) - \mathrm{f}(x_1)}$$

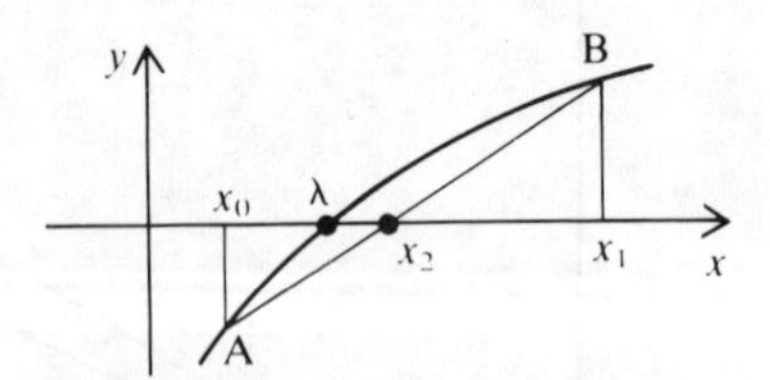

Figure 3.13

Step 4 Repeat **Steps 2** and **3** with, successively, the intervals $[x_1, x_2]$, $[x_2, x_3]$, $[x_3, x_4]$, ...

Note that in **Step 3** we shall have, successively

$$x_3 = \frac{x_2\mathrm{f}(x_1) - x_1\mathrm{f}(x_2)}{\mathrm{f}(x_1) - \mathrm{f}(x_2)}, \qquad x_4 = \frac{x_3\mathrm{f}(x_2) - x_2\mathrm{f}(x_3)}{\mathrm{f}(x_2) - \mathrm{f}(x_3)}$$

and generally $$x_{n+1} = \frac{x_n\mathrm{f}(x_{n-1}) - x_{n-1}\mathrm{f}(x_n)}{\mathrm{f}(x_{n-1}) - \mathrm{f}(x_n)}$$

Then the sequence $x_0, x_1, x_2, \ldots$ will (usually!) converge to the required solution.

Example 3.4 illustrates the secant method applied to the equation of Example 3.3.

Example 3.4

Use the secant method to obtain, correct to three decimal digits, the positive solution of the equation $x^3 - 5x - 3 = 0$.

From Example 3.3, we know that the required solution lies in the interval [2, 3]. Hence $x_0 = 2$ and $x_1 = 3$.

$$f(x_0) = f(2) = -5 \qquad f(x_1) = f(3) = 9$$

and from **Step 2** above:

$$x_2 = \frac{x_1 f(x_0) - x_0 f(x_1)}{f(x_0) - f(x_1)} = \frac{3 \times (-5) - 2 \times 9}{-5 - 9} = 2.357\,142\,86$$

Now, $x_1 = 3$, $x_2 = 2.357\,142\,86$, $f(x_1) = 9$ and $f(x_2) = -1.689\,139\,9$

and $$x_3 = \frac{x_2 f(x_1) - x_1 f(x_2)}{f(x_1) - f(x_2)} = \frac{2.357\,142\,86 \times 9 - 3 \times (-1.689\,139\,9)}{9 - (-1.689\,139\,9)}$$

$$= 2.458\,729\,67$$

Now, $x_2 = 2.357\,142\,86$, $x_3 = 2.458\,729\,67$

and $$x_4 = \frac{x_3 f(x_2) - x_2 f(x_3)}{f(x_2) - f(x_3)} = 2.493\,396\,22$$

The first six terms of the sequence are printed below and it may be seen that the required solution, correct to three decimal digits, is $x = 2.49$,

$x_0 = 2$
$x_1 = 3$
$x_2 = 2.357\,142\,86$
$x_3 = 2.458\,729\,67$
$x_4 = 2.493\,396\,22$
$x_5 = 2.490\,818\,39$
$x_6 = 2.490\,863\,55$

We mention also that the line AB represents the polynomial of degree one (a linear polynomial) that interpolates $f(x)$ at the points A and B, hence one of the names of the method. The line AB is called a secant to the curve, hence the other more popular name.

We conclude this section with a second example.

Example 3.5

Use the secant method to obtain an approximation to the positive solution of the equation $2^x - x - 3 = 0$. Obtain your approximations by using, in turn, each of the three stopping rules with tolerance $\varepsilon = 0.005$. Comment on your answers.

As we have seen, the required solution lies between $x = 2$ and $x = 3$. Hence, we take $x_0 = 2$ and $x_1 = 3$

and $$x_2 = \frac{x_1 f(x_0) - x_0 f(x_1)}{f(x_0) - f(x_1)} = \frac{3 \times f(2) - 2 \times f(3)}{f(2) - f(3)}$$

$$= \frac{3 \times (-1) - 2 \times (2)}{(-1) - (2)} = 2.333\,333\,3$$

Now, $x_1 = 3$ and $x_2 = 2.333\,333\,3$

and $$x_3 = \frac{x_2 f(x_1) - x_1 f(x_2)}{f(x_1) - f(x_2)} = \frac{2.333\,333\,3 \times f(3) - 3 \times f(2.333\,333\,3)}{f(3) - f(2.333\,333\,3)}$$

$$= 2.418\,684\,7$$

Continuing in the same way produces this table.

$x_0 = 2$
$x_1 = 3$
$x_2 = 2.333\,333\,3$
$x_3 = 2.418\,684\,7$
$x_4 = 2.446\,334\,28$
$x_5 = 2.444\,889\,81$
$x_6 = 2.444\,907\,54$

It may be seen that $|f(x_3)| \approx 0.071\,85$

while $|f(x_4)| \approx 3.96 \times 10^{-3}$

Hence the solution, if the first stopping rule is used, is

$x_4 = 2.446\,334\,28$

Also $|x_4 - x_3| \approx 0.027\,65$ while $|x_5 - x_4| \approx 1.44 \times 10^{-3}$

Hence the solution, if the second stopping rule is used, is

$x_5 = 2.444\,889\,81$

Finally $$\left| \frac{x_4 - x_3}{x_4} \right| \approx 0.0113$$

while $$\left| \frac{x_5 - x_4}{x_5} \right| \approx 5.908 \times 10^{-4}$$

Hence, if the third stopping rule is used, the solution is

$x_5 = 2.444\,889\,81$

The solution, correct to three decimal digits (rounding or chopping) is $x = 2.44$ and two of the stopping rules produce this solution. The first stopping rule, which is perhaps the least reliable of the three, produces the solution $x = 2.45$ (rounding to three decimal digits). We observe that in this example also, the secant method converges much more quickly than the bisection method.

3.5 Convergence

It is clear that the convergence is very much faster for the secant method than for the bisection method. We shall say more about this in Sections 3.9–3.11 when the errors arising from these methods will be discussed.

The secant method is not without its drawbacks, however, and problems can arise from the use of the equation

$$x_{n+1} = \frac{x_n \mathrm{f}(x_{n-1}) - x_{n-1}\mathrm{f}(x_n)}{\mathrm{f}(x_{n-1}) - \mathrm{f}(x_n)}$$

in **Step 3**. It is quite possible, for example, that $x_n\mathrm{f}(x_{n-1})$ and $x_{n-1}\mathrm{f}(x_n)$ are close in value. In this case, subtraction in the numerator could lead to loss in accuracy. Also, the denominator $\mathrm{f}(x_{n-1}) - \mathrm{f}(x_n)$ may become very small as the iteration continues. As we mentioned in Chapter 1, division by a very small number can also lead to a serious loss in accuracy. Hence, it is not unknown for problems to occur when the secant method is used (see Exercise 11 at the end of this chapter).

Algorithm 3.2 The secant method

To obtain a solution to the equation $\mathrm{f}(x) = 0$.

Find an interval $[x_0, x_1]$ that contains the required solution.

input: x_0, x_1
for $n = 1, 2, \ldots$, until satisfied

$$x_{n+1} := \frac{x_n \mathrm{f}(x_{n-1}) - x_{n-1}\mathrm{f}(x_n)}{\mathrm{f}(x_{n-1}) - \mathrm{f}(x_n)}$$

endloop
output: the sequence of approximations $x_0, x_1, x_2, \ldots$

3.6 The fixed point iterative method

The third iterative method is slightly different in character. The first step, in this method is to write the equation to be solved in the specific form $x = \mathrm{g}(x)$ for some function $\mathrm{g}(x)$.

So if we were solving the equation:

$$2^x - x - 3 = 0$$

we might write

$$x = 2^x - 3 \qquad \text{where} \qquad g(x) = 2^x - 3$$

Or, if the equation to be solved was

$$x^3 - 5x - 3 = 0$$

we might write

$$5x = x^3 - 3$$

giving $\quad x = \frac{1}{5}(x^3 - 3) \qquad$ where $\qquad g(x) = \frac{1}{5}(x^3 - 3)$

Having written the equation in the form $x = g(x)$, we then use one of the methods described above to obtain a first approximation x_0 to the solution.

The second approximation, x_1 is calculated from the equation

$$x_1 = g(x_0)$$

The third approximation x_2 is calculated from

$$x_2 = g(x_1)$$

x_3 is calculated from

$$x_3 = g(x_2)$$

and in general, x_n is calculated from

$$x_n = g(x_{n-1}) \qquad n \geqslant 1$$

The calculations continue until the terms of the sequence $x_0, x_1, x_2, \ldots$ become sufficiently close to the solution. This method is called the **fixed point iterative method**. An example will show how very simple it is to operate the fixed point iterative method.

Example 3.6

Use the fixed point iterative method to find correct to three decimal digits (using rounding) the negative solution of the equation $2^x - x - 3 = 0$.

To apply the fixed point iterative method, the equation $2^x - x - 3 = 0$ must be rearranged to be in the form $x = g(x)$. As shown above, the easiest way of doing this is to write;

$$x = 2^x - 3 \qquad \text{giving} \qquad g(x) = 2^x - 3$$

From Figure 3.14, the graph of the equation $y = 2^x - x - 3$,

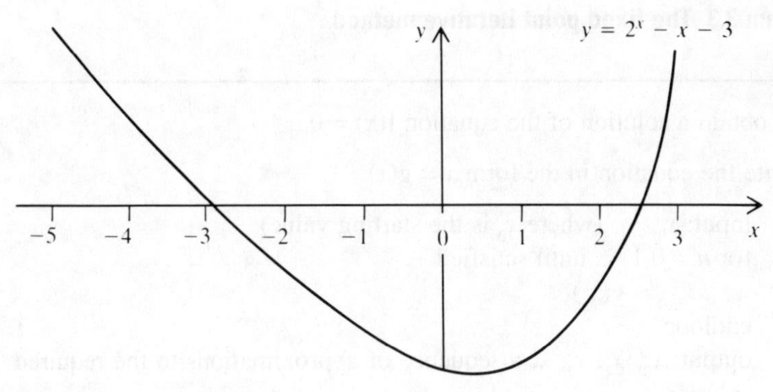

Figure 3.14

we see that the negative solution lies in the interval [–3, –2] and that this solution is closer to –3 than to –2. Hence, a suitable first approximation is $x_0 = -3$.

The successive approximations $x_1, x_2, x_3, \ldots$ are calculated from

$$x_1 = g(x_0) = g(-3) = 2^{-3} - 3 = -2.875$$
$$x_2 = g(x_1) = g(-2.875) = 2^{-2.875} - 3 = -2.863\,686\,53$$
$$x_3 = g(x_2) = g(-2.863\,686\,53) = -2.862\,613\,37$$

The following table gives the first ten terms of the sequence and from the table, it is clear that the value of the required solution is $x = -2.86$ (rounding to three decimal digits).

$x_0 = -3$
$x_1 = -2.875$
$x_2 = -2.863\,686\,53$
$x_3 = -2.862\,613\,38$
$x_4 = -2.862\,511\,14$
$x_5 = -2.862\,501\,4$
$x_6 = -2.862\,500\,47$
$x_7 = -2.862\,500\,38$
$x_8 = -2.862\,500\,37$
$x_9 = -2.862\,500\,37$
$x_{10} = -2.862\,500\,37$

One of the great advantages of the fixed point iterative method is that it is easy to perform the calculations on a calculator. If the current term (x_i say,) is stored in the memory, it is usually an easy matter to calculate $g(x_i)$, and hence x_{i+1}. When x_{i+1} is stored in the memory, it is easy to calculate $g(x_{i+1}) = x_{i+2}$ and so on. For convenience, we provide an algorithm for the fixed point iterative method.

Algorithm 3.3 The fixed point iterative method

To obtain a solution of the equation $f(x) = 0$.

Write the equation in the form $x = g(x)$.

input: x_0 (where x_0 is the starting value)
for $n = 0,1, \ldots$ until satisfied
$x_{n+1} = g(x_n)$
endloop
output: $x_0, x_1, x_2, \ldots$ a sequence of approximations to the required solution.

We shall call the function $g(x)$ the **iterative function**. A second and more interesting example is:

Example 3.7

Use the fixed point iterative method to find, correct to three decimal digits (using rounding), all the solutions of the equation $x^3 - 5x - 3 = 0$.

A portion of the graph of $y = x^3 - 5x - 3$ is shown in Figure 3.15. It is clear from the graph that the three solutions lie in the intervals [–2, –1] [–1, 0], [2, 3]

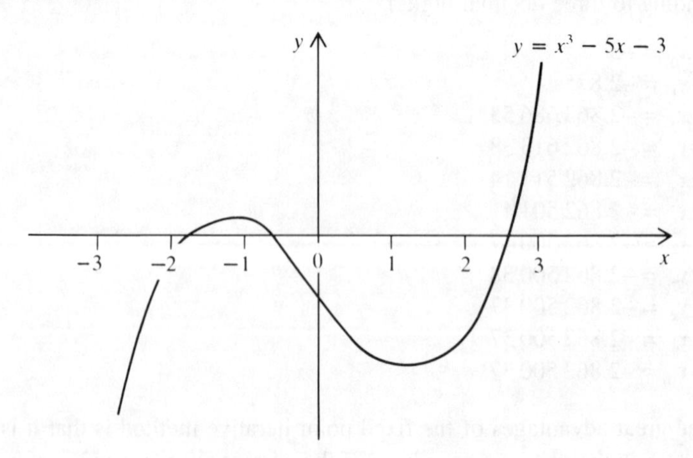

Figure 3.15

We begin with the solution lying in the interval [–1, 0]. Since the solution appears to be closer to –1 than to 0, we take $x_0 = -1$ as the starting value. To write $x^3 - 5x - 3 = 0$ in the form $x = g(x)$, write

$$5x = x^3 - 3$$

$$x = \tfrac{1}{5}(x^3 - 3) \qquad \text{so} \qquad g(x) = \tfrac{1}{5}(x^3 - 3)$$

Now,

$$x_1 = g(x_0) = g(-1) = -0.8$$
$$x_2 = g(x_1) = g(-0.8) = -0.7024$$
$$x_3 = g(x_2) = g(-0.7024) = -0.669\,308\,022$$

and so on. A table giving the values of x_0, x_1, ... x_{10} appears below.

$$x_0 = -1$$
$$x_1 = -0.8$$
$$x_2 = -0.7024$$
$$x_3 = -0.669\,308\,022$$
$$x_4 = -0.659\,966\,415$$
$$x_5 = -0.657\,490\,423$$
$$x_6 = -0.656\,845\,788$$
$$x_7 = -0.656\,678\,749$$
$$x_8 = -0.656\,635\,519$$
$$x_9 = -0.656\,624\,334$$
$$x_{10} = -0.656\,621\,441$$

It is clear that the required solution is $x = -0.657$.

Consider now the solution lying in the interval [–2, –1]. This solution is close to $x = -2$ so we take $x_0 = -2$ as the starting value. Proceeding as before

$$x_1 = g(x_0) = g(-2) = -2.2$$
$$x_2 = g(x_1) = g(-2.2) = -2.7296$$
$$x_3 = g(x_2) = g(-2.7296) = -4.667\,494\,97$$

and it may be shown that

$$x_5 = -1836.114\,68$$
$$x_6 \approx -1.24 \times 10^9$$

and it is clear that this sequence is not going to converge.

If, instead, we looked carefully at the graph, and chose $x_0 = -1.8$ to be the starting value, we would calculate

$$x_1 = g(-1.8) = -1.7664$$
$$x_2 = g(-1.7664) = -1.702\,293\,29$$
$$x_3 = g(-1.702\,293\,29) = -1.586\,581\,93$$

The values x_0, x_1, ... , x_{10} are printed over the page and it is clear that the sequence is converging to the solution that has already been calculated!

$x_0 = -1.8$
$x_1 = -1.7664$
$x_2 = -1.702\,293\,29$
$x_3 = -1.586\,581\,93$
$x_4 = -1.398\,762\,2$
$x_5 = -1.147\,345\,64$
$x_6 = -0.902\,073\,623$
$x_7 = -0.746\,810\,104$
$x_8 = -0.683\,302\,983$
$x_9 = -0.663\,807\,238$
$x_{10} = -0.658\,500\,011$

Hence, with the starting value $x_0 = -2$, the sequence diverges, or goes off to infinity, while with the second starting value $x_0 = -1.8$, the sequence converges, but to the wrong solution.

One explanation for this rather eccentric behaviour is that when the equation $x^3 - 5x - 3 = 0$ was written in the form $x = g(x)$, the form of the iterative function $g(x)$ was not chosen sensibly. Our description of the method requires only that the equation be written in the form $x = g(x)$; no guidance has been given on how the function $g(x)$ is to be chosen. We are, then, free to choose the form of $g(x)$ in any way that we please.

Suppose, instead, that we had written

$$x^3 = 5x + 3$$

and $$x = \frac{(5x + 3)}{x^2}$$

Now, $g(x) = \dfrac{(5x + 3)}{x^2}$ and with $x_0 = -2$ we have

$$x_1 = g(-2) = \frac{-10 + 3}{(-2)^2} = -1.75$$

$$x_2 = g(-1.75) = \frac{5(-1.75) + 3}{(-1.75)^2} = -1.877\,551\,02$$

and so on. The first ten terms of the sequence are shown in the table and it is clear that, with this form of $g(x)$, the sequence is converging to the solution $x = -1.83$ (rounding to three decimal digits).

$x_0 = -2$
$x_1 = -1.75$
$x_2 = -1.877\,551\,02$
$x_3 = -1.812\,027\,41$
$x_4 = -1.845\,665\,19$
$x_5 = -1.828\,375\,91$
$x_6 = -1.837\,258\,71$
$x_7 = -1.832\,693\,73$
$x_8 = -1.835\,039\,44$
$x_9 = -1.833\,834\,02$
$x_{10} = -1.834\,453\,44$

It remains to calculate the value of the solution lying in the interval [2,3]. The following tables show some of the terms obtained by using the two iterative functions described above with different starting values $x_0 = 2$ and $x_0 = 3$.

$$x = \frac{x^3 - 3}{5}$$

$x_0 = 2$	$x_0 = 3$
$x_1 = 1$	$x_1 = 4.8$
$x_2 = -0.4$	$x_2 = 21.5184$
$x_3 = -0.6128$	$x_3 = 1992.182\,61$
$x_4 = -0.646\,024\,202$	$x_4 = 1.581\,31 \times 10^9$
$x_5 = -0.653\,923\,287$	$x_5 = 7.908\,28 \times 10^{26}$
$x_6 = -0.655\,925\,569$	$x_6 = 3.402\,82 \times 10^{37}$
$x_7 = -0.656\,440\,867$	$x_7 = 3.402\,82 \times 10^{37}$
$x_8 = -0.656\,573\,993$	$x_8 = 3.402\,82 \times 10^{37}$
$x_9 = -0.656\,608\,419$	
$x_{10} = -0.656\,617\,324$	
$x_{11} = -0.656\,619\,627$	

$$x = \frac{5x + 3}{x^2}$$

$x_0 = 2$	$x_0 = 3$
$x_1 = 3.25$	$x_1 = 2$
$x_2 = 1.822\,485\,21$	$x_2 = 3.25$
$x_3 = 3.646\,725\,84$	$x_3 = 1.822\,485\,21$
$x_4 = 1.596\,680\,42$	$x_4 = 3.646\,725\,84$
$x_5 = 4.308\,249\,86$	$x_5 = 1.596\,680\,42$
$x_6 = 1.322\,193\,15$	$x_6 = 4.308\,249\,86$
$x_7 = 5.497\,651\,75$	$x_7 = 1.322\,193\,15$
$x_8 = 1.008\,737\,51$	$x_8 = 5.497\,651\,75$
$x_9 = 7.904\,944\,99$	$x_9 = 1.008\,737\,51$
$x_{10} = 0.680\,524\,57$	$x_{10} = 7.904\,944\,99$

Clearly, none of these sequences is going to converge to the required solution. To find a way out of this difficulty, consider the following useful remark.

Useful remark

Since $x^3 - 5x - 3 = 0$ then $k(x^3 - 5x - 3) = 0$

where k is any number.

Hence
$$x + k(x^3 - 5x - 3) = x + 0$$
$$= x$$

Reversing this equation:

$$x = x + k(x^3 - 5x - 3)$$

and now $\quad g(x) = x + k(x^3 - 5x - 3)$

For reasons to be explained in Section 3.13, let $k = -0.072\,73$.

Then $\quad g(x) = x - 0.072\,73(x^3 - 5x - 3)$

Proceeding with this form of the iterative equation:

$x_0 = 2$	$x_0 = 3$
$x_1 = 2.363\,65$	$x_1 = 2.345\,43$
$x_2 = 2.480\,957\,13$	$x_2 = 2.478\,150\,71$
$x_3 = 2.490\,712\,16$	$x_3 = 2.490\,649\,92$
$x_4 = 2.490\,862\,1$	$x_4 = 2.490\,861\,47$
$x_5 = 2.490\,863\,6$	$x_5 = 2.490\,863\,59$
$x_6 = 2.490\,863\,62$	$x_6 = 2.490\,863\,62$
$x_7 = 2.490\,863\,62$	$x_7 = 2.490\,863\,62$
$x_8 = 2.490\,863\,62$	$x_8 = 2.490\,863\,62$
$x_9 = 2.490\,863\,62$	$x_9 = 2.490\,863\,62$
$x_{10} = 2.490\,863\,62$	$x_{10} = 2.490\,863\,62$

Readers should discover for themselves what happens if the starting values $x_0 = 0$, $x_0 = -1$, $x_0 = -2$ are taken with this form of g(x).

It is clear that choosing the form of the iterative function g(x) is extremely important. There are many candidates for function g(x), some of which are printed next:

$$x = \tfrac{1}{5}(x^3 - 3)$$

$$x = \frac{5x + 3}{x^2}$$

$$x = x + k(x^3 - 5x - 3)$$

$$x = \sqrt{\frac{5x + 3}{x}}$$

$$x = \sqrt[3]{5x + 3}$$

$$x = \frac{3}{x^2 - 5}$$

A full discussion of how we can choose sensibly the form of g(x) is given in Section 3.12 when the errors in the fixed point iterative method are discussed. The next stage is to illustrate geometrically how the method works (or fails to work).

3.7 Geometrical interpretation of the fixed point iterative method

It was noted earlier in this chapter that a solution of the equation f(x) = 0 occurs when the graph y = f(x) cuts the x-axis. We may regard such a solution as the point of intersection of two lines: one line having the equation $y = 0$ and the other having the equation $y = f(x)$. In the fixed point iterative method, we may still regard the solution as being the point of intersection of two lines, but now the equations of these lines are:

$$y = x \qquad \text{and} \qquad y = g(x)$$

If the graphs of $y = x$ and $y = g(x)$ are as shown in Figure 3.16, then at the point of intersection P, the equations $y = x$ and $y = g(x)$ are satisfied simultaneously. At this point, we have $x = g(x)$, hence the x coordinate of P is the required solution. If the solution is $x = \lambda$ then we have this diagram:

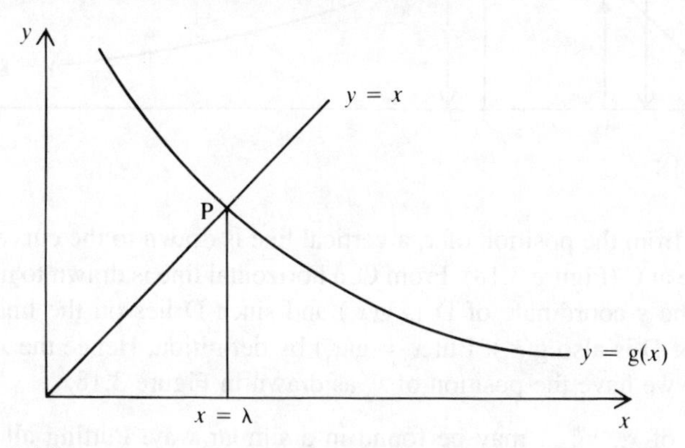

Figure 3.16

Note that λ satisfies the equation $\lambda = g(\lambda)$. This equation gives the method its name. Because $g(x)$ takes the value λ when $x = \lambda$ we may think of $x = \lambda$ as being a 'fixed point'. The iteration $x_{n+1} = g(x_n)$ is meant to converge to the fixed point $x = \lambda$: hence the name of the method.

To observe the behaviour of the sequence x_0, x_1, x_2, ... consider first the position of x_1. A vertical line drawn from x_0 to the curve $y = g(x)$ meets the curve at A (Figure 3.17). The length of this vertical line represents $g(x_0)$. Now draw a horizontal line from A to meet the line $y = x$ in the point B. Clearly, the y coordinate of B is $g(x_0)$. Since at every point on the line $y = x$ the y coordinate is equal to the x coordinate, the x coordinate of B is also $g(x_0)$. But by definition, $x_1 = g(x_0)$ so x_1 appears on the diagram as the x coordinate of B.

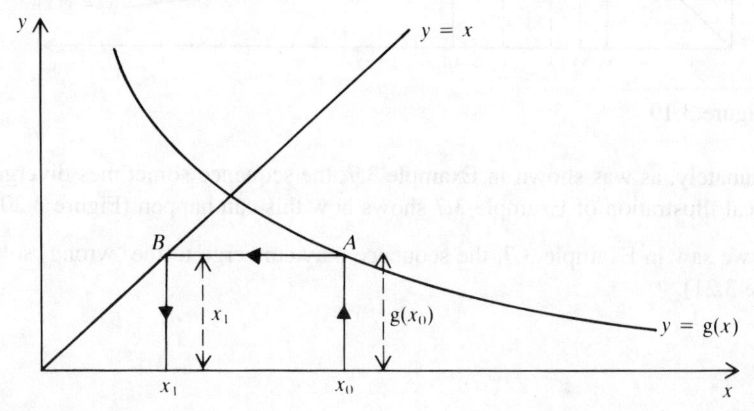

Figure 3.17

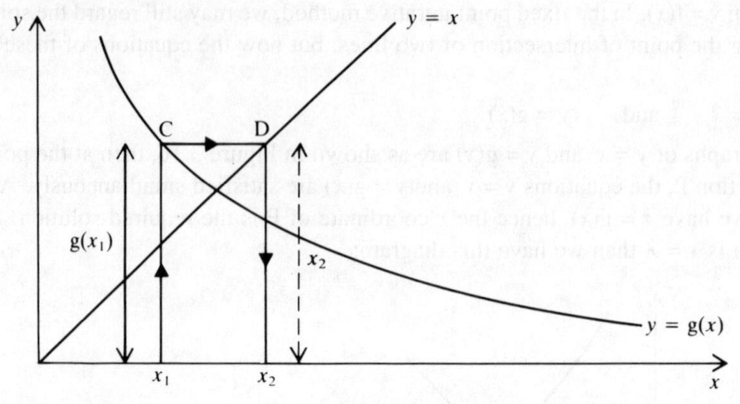

Figure 3.18

Now, starting from the position of x_1 a vertical line is drawn to the curve $y = g(x)$ to meet the curve at C (Figure 3.18). From C, a horizontal line is drawn to meet the line $y = x$ at D. The y coordinate of D is $g(x_1)$ and since D lies on the line $y = x$, the x coordinate of D is also $g(x_1)$. But $x_2 = g(x_1)$ by definition. Hence the x coordinate of D is x_2 and we have the position of x_2 as drawn in Figure 3.18.

The positions of x_3, x_4, ... may be found in a similar way. Putting all these steps together in one diagram gives Figure 3.19 and we can see the sequence converging to λ.

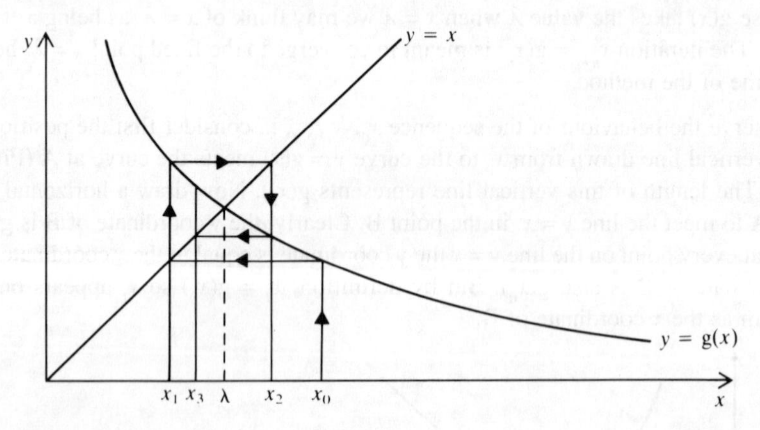

Figure 3.19

Unfortunately, as was shown in Example 3.7, the sequence sometimes diverges. A graphical illustration of Example 3.7 shows how this can happen (Figure 3.20).

Or, as we saw in Example 3.7, the sequence may converge to the 'wrong' solution (Figure 3.21).

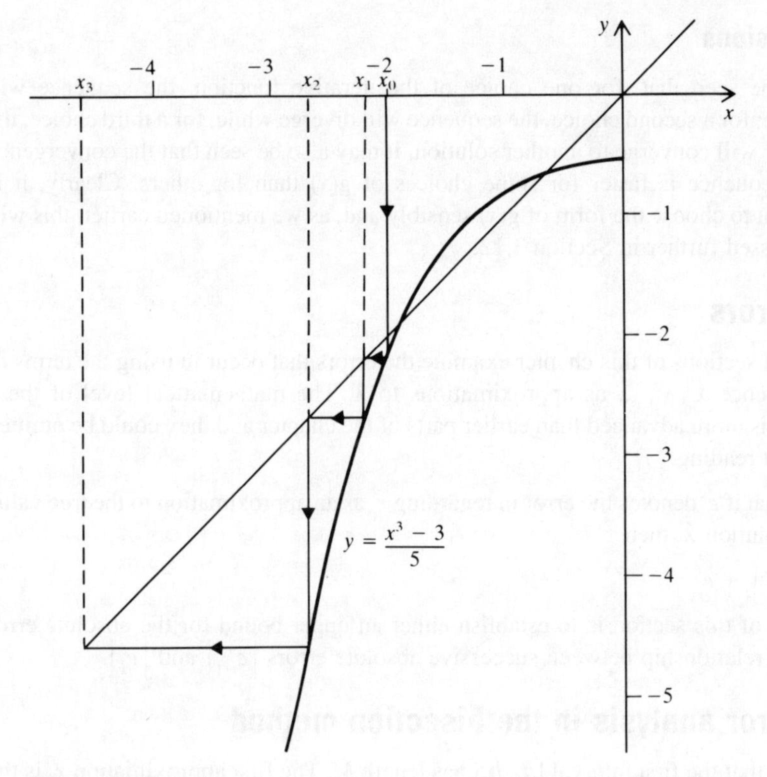

Figure 3.20

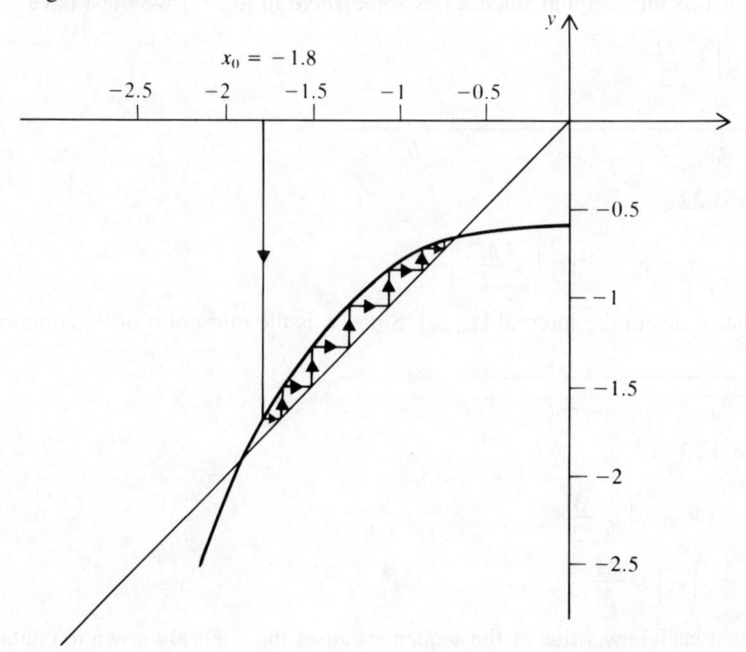

Figure 3.21

Conclusions

It may be seen that, for one choice of the iterative function, the sequence will converge; for a second choice, the sequence will diverge while, for a third choice, the sequence will converge to another solution. It may also be seen that the convergence of the sequence is faster for some choices of $g(x)$ than for others. Clearly, it is important to choose the form of $g(x)$ sensibly and, as we mentioned earlier, this will be discussed further in Section 3.12.

3.8 Errors

The final sections of this chapter examine the errors that occur in using the terms of the sequence $x_0, x_1, \ldots$ as approximations to λ. The mathematical level of these sections is more advanced than earlier parts of the chapter and they could be omitted on a first reading.

Recall that if e_i denotes the error in regarding x_i as an approximation to the true value of the solution λ, then

$$\lambda = x_i + e_i$$

The aim of this section is to establish either an upper bound for the absolute error $|e_i|$ or a relationship between successive absolute errors $|e_{i+1}|$ and $|e_i|$.

3.9 Error analysis in the bisection method

Suppose that the first interval $[a_0, b_0]$ has length M. The first approximation x_0 is the mid-point of this interval and since λ lies somewhere in $[a_0, b_0]$ we must have

$$|\lambda - x_0| \leqslant \frac{M}{2}$$

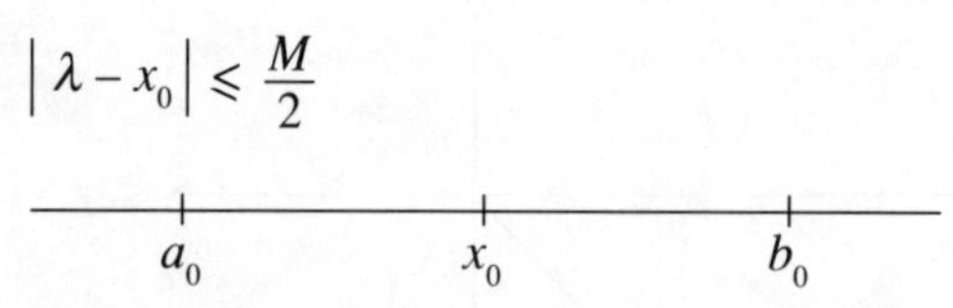

Figure 3.22

Since $\quad \lambda - x_0 = e_0 \qquad |e_0| \leqslant \dfrac{M}{2}$

Suppose that λ lies in the interval $[x_0, b_0]$. Since x_1 is the mid-point of this interval,

$a_0 \qquad x_0 \qquad x_1 \qquad b_0$

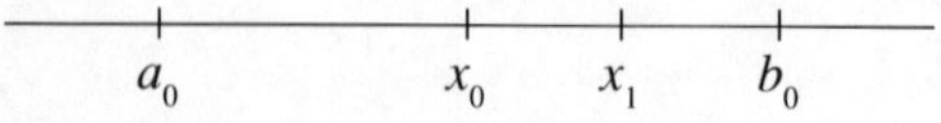

Figure 3.23

we have $\quad |\lambda - x_1| \leqslant \dfrac{M}{2^2}$

Therefore $\quad |e_1| \leqslant \dfrac{M}{2^2}$

It is clear that each new value of the sequence causes the interval known to contain λ to be halved. Hence we obtain the inequality

$$|e_i| \leqslant \frac{M}{2^{i+1}}$$

and this provides the required bound for the error. It is clear that for large values of i the error bound will be small and x_i will be close in value to λ.

An interesting question is, 'How large does i have to be in order that x_i should provide a satisfactory approximation to λ?' The answer is reasonably straightforward when the stopping rule SR2 is used and x_n is regarded as a satisfactory approximation to λ if $|x_n - x_{n-1}| < \varepsilon$ where ε is the given tolerance. From Figure 3.23, it is clear that

$$|x_1 - x_0| = \frac{M}{2^2} \qquad |x_2 - x_1| = \frac{M}{2^3}$$

and generally that $\quad |x_i - x_{i-1}| = \dfrac{M}{2^{i+1}}$

Hence, if $\quad |x_i - x_{i-1}| < \varepsilon$

then $\quad \dfrac{M}{2^{i+1}} < \varepsilon \quad$ and $\quad 2^{i+1} > \dfrac{M}{\varepsilon}$

Taking the natural logarithm of both sides

$$(i+1)\ln 2 > \ln\left(\frac{M}{\varepsilon}\right) \qquad \text{and} \qquad i + 1 > \frac{\ln\left(\dfrac{M}{\varepsilon}\right)}{\ln 2}$$

Since ε is likely to be small relative to M, write

$$\ln\left(\frac{M}{\varepsilon}\right) = -\ln\left(\frac{\varepsilon}{M}\right)$$

then $\quad i + 1 > -\dfrac{\ln\left(\dfrac{\varepsilon}{M}\right)}{\ln 2} \quad$ and $\quad i > -1 - \dfrac{\ln\left(\dfrac{\varepsilon}{M}\right)}{\ln 2} \qquad \ldots (1)$

Example 3.8

Use the bisection method to find the solution of $2^x - x - 3 = 0$ lying in the interval $[-3, -2]$. Provide an upper bound for the error after ten iterations, and show that seven iterations are necessary before $|x_n - x_{n-1}| < 0.005$.

The error term e_i satisfies $|e_i| \leqslant \dfrac{M}{2^{i+1}}$

where M is the length of the interval $[a_0, b_0]$. In this case, $M = 1$ and $|e_i| \leqslant \dfrac{1}{2^{i+1}}$. Hence, the error term e_{10} satisfies

$$|e_{10}| \leqslant \frac{1}{2^{11}} \approx 4.88 \times 10^{-4}$$

If $\varepsilon = 0.005$ then Inequality (1) gives

$$n > -1 - \frac{\ln\left(\dfrac{0.005}{1}\right)}{\ln 2} = 6.64 \qquad \text{to three decimal digits}$$

and seven iterations are necessary before $|x_n - x_{n-1}| < 0.005$.

3.10 Convergence

In the other two methods discussed in this chapter (and also the Newton-Raphson method, to be described in Chapter 6) it is not so straightforward to produce a bound on each error $|e_n|$. Instead, the aim of the error analysis will be to compare successive errors, that is, to compare $|e_n|$ and $|e_{n-1}|$. If $|e_n|$ is appreciably smaller than $|e_{n-1}|$ for all n, then we have good reason to believe that the sequence $x_0, x_1, \ldots$ is converging reasonably quickly. If, however, $|e_n|$ is close in value to $|e_{n-1}|$ then the convergence is likely to be slow. If $|e_n|$ is greater than $|e_{n-1}|$ then the sequence will diverge.

To achieve such a comparison, we shall produce, for each method, a statement of the form:

> When n is large, there exists a constant C such that
>
> $$|e_n| \approx C|e_{n-1}|^p$$

We shall produce such a statement for both the secant method and the fixed point iterative method (and, in Chapter 6 for the Newton-Raphson method). The interest lies in the value taken by p. Since it may be assumed that $|e_{n-1}| < 1$ (otherwise the errors are very large indeed and the sequence will diverge), the larger the value of p, the smaller the value of $|e_n|$ relative to $|e_{n-1}|$ and the more rapid the convergence of the sequence. Thus, by observing the values taken by p in the different iterative methods, the above statement may be used to compare the rates of convergence of the methods.

3.11 Error analysis in the secant method

> For the secant method, as n becomes large, there exists a constant C such that
>
> $$|e_n| \approx C|e_{n-1}|^{1.618}$$

A proof of this result lies beyond the scope of this book†.

3.12 Error analysis in the fixed point iterative method

In this section, it will be convenient to express the error in x_n in a slightly different form. We shall write:

$$\lambda = x_n - e_n \qquad \text{and} \qquad \lambda = x_{n+1} - e_{n+1}$$

So the change involves only the signs of the errors e_n and e_{n+1}. Since the result that

† A full proof is given in Conte and de Boor 1981, *Elementary numerical analysis* 3rd ed. McGraw-Hill.

we are seeking involves the absolute values $|e_n|$ and $|e_{n+1}|$, this slight change in notation will not cause any great difficulty.

Now we write:

$$x_n = \lambda + e_n$$
$$x_{n+1} = \lambda + e_{n+1}$$

Substituting for x_{n+1} and x_n in the iterative equation $x_{n+1} = g(x_n)$

gives $\lambda + e_{n+1} = g(\lambda + e_n)$

But from elementary calculus $\dfrac{g(\lambda + e_n) - g(\lambda)}{e_n} \approx g'(\lambda)$

giving $g(\lambda + e_n) \approx g(\lambda) + e_n g'(\lambda)$

(Or use the Taylor series, see Chapter 5.)

Hence $\lambda + e_{n+1} \approx g(\lambda) + e_n g'(\lambda)$

Since λ is the exact value of the solution, $\lambda = g(\lambda)$, then

$$e_{n+1} \approx e_n g'(\lambda)$$

We have thus arrived at the expression

$$|e_{n+1}| \approx |g'(\lambda)|\,|e_n|$$

This is the form of the statement in Section 3.10 for the fixed point iterative method and we see that in this case, $C = |g'(\lambda)|$ and $p = 1$. Since the value of p for the fixed point iterative method is less than the value of p for the secant method, we deduce that the convergence properties of the fixed point iterative method are not as good as those of the secant method.

This is important. By this piece of analysis, we have established that the secant method will generally converge more rapidly than the fixed point iterative method. We can, however, deduce more.

The above analysis applies also for the terms x_n and x_{n-1}.

Hence $|e_n| \approx |g'(\lambda)|\,|e_{n-1}|$

In the same way,

$$|e_{n-1}| \approx |g'(\lambda)|\,|e_{n-2}|$$
$$|e_{n-2}| \approx |g'(\lambda)|\,|e_{n-3}|$$

and so on.

Using these approximations, we obtain

$$\begin{aligned} |e_{n+1}| &\approx |g'(\lambda)|\,|e_n| \\ &\approx |g'(\lambda)|\,|g'(\lambda)|\,|e_{n-1}| = |g'(\lambda)|^2\,|e_{n-1}| \\ &\approx |g'(\lambda)|^2\,|g'(\lambda)|\,|e_{n-2}| = |g'(\lambda)|^3\,|e_{n-2}| \end{aligned}$$

and so on until

$$|e_{n+1}| \approx |g'(\lambda)|^{n+1}\,|e_0|$$

So, if $|g'(\lambda)| < 1$ it may be expected that for large values of n the errors $|e_n|$ will become small and the sequence $x_0, x_1, \ldots$ will converge to λ. Observe also that the smaller the value of $|g'(\lambda)|$ the more rapid will be the convergence of $x_0, x_1, x_2, \ldots$ to λ.

These considerations help us to decide whether or not a sensible form for the iterative function, g(x), has been chosen and we state formally the criteria for convergence of the fixed point iterative method.

Convergence and the fixed point iterative method

> Let the equation $x = g(x)$ have solution $x = \lambda$.
>
> If $|g'(\lambda)| < 1$ and x_0 is chosen to be sufficiently close to λ, then the sequence $x_0, x_1, x_2, \ldots$ generated by the fixed point iterative method will converge to the solution λ.

There is, however, one major problem with this criterion. To decide whether or not $|g'(\lambda)| < 1$ we need to know the value of λ. But it is to determine the value of λ that we are trying to decide whether or not $|g'(\lambda)| < 1$.

One way out of this circular argument is as follows:

if $\quad |g'(\lambda)| < 1$

then $\quad |g'(x)| < 1 \quad$ for all x sufficiently close to λ

So to decide whether or not $|g'(\lambda)| < 1$, a reasonable 'working rule' might be:

(a) find a value of x ($x = \eta$, say) close to λ,

(b) if $|g'(\eta)| < 1$ then we have grounds for believing that $|g'(\lambda)| < 1$.

If $|g'(\eta)| \geqslant 1$, then either try to find a value of x closer to λ or find a different iterative function.

This working rule is not, of course, infallible. It might happen that η is close in value to λ and $|g'(\eta)| < 1$ but still $|g'(\lambda)| > 1$. However, the rule works well in practice as we will now see, applying it to the examples considered earlier[†].

3.13 Using the convergence property of the fixed point iterative method

Example 3.9

Consider the equation $2^x - x - 3 = 0$ which we wrote in the form

$$x = 2^x - 3 \qquad \text{giving} \qquad g(x) = 2^x - 3$$

[†] The reader is referred to Conte and de Boor 1981 *Elementary numerical analysis* 3rd edn. McGraw-Hill for a rigorous presentation of convergence and the fixed point iterative method.

To find g'(x), write $u = 2^x$ and take natural logarithms of both sides:

$$\begin{aligned}\ln u &= \ln 2^x \\ &= x \ln 2\end{aligned}$$

Now differentiate both sides, remembering that u is a function of x.

$$\frac{1}{u}\frac{du}{dx} = \ln 2$$

Hence $\dfrac{du}{dx} = u \ln 2$

$$= 2^x \ln 2$$

and $\quad g'(x) = 2^x \ln 2$

We saw earlier that the solutions to the equation are $x = -2.86$ and $x = 2.44$. Now, $g'(-2.86) = 0.0955$, and $g'(2.44) = 3.76$.

Hence we would expect that, given a suitable starting value, the fixed point iterative method would converge to the negative solution. We would not, however, expect the method to be successful with the positive solution – and so it turns out to be.

$x_0 = -2$	$x_0 = 2$
$x_1 = -2.75$	$x_1 = 1$
$x_2 = -2.851\,349\,11$	$x_2 = -1$
$x_3 = -2.861\,433\,46$	$x_3 = -2.5$
$x_4 = -2.862\,398\,65$	$x_4 = -2.823\,223\,31$
$x_5 = -2.862\,490\,68$	$x_5 = -2.858\,705\,55$
$x_6 = -2.862\,499\,45$	$x_6 = -2.862\,138\,22$
$x_7 = -2.862\,500\,28$	$x_7 = -2.862\,465\,85$
$x_8 = -2.862\,500\,36$	$x_8 = -2.862\,497\,08$
$x_9 = -2.862\,500\,37$	$x_9 = -2.862\,500\,06$
$x_{10} = -2.862\,500\,37$	$x_{10} = -2.862\,500\,34$
$x_{11} = -2.862\,500\,37$	$x_{11} = -2.862\,500\,37$

A graph of $y = g'(x)$ is shown in Figure 3.24.

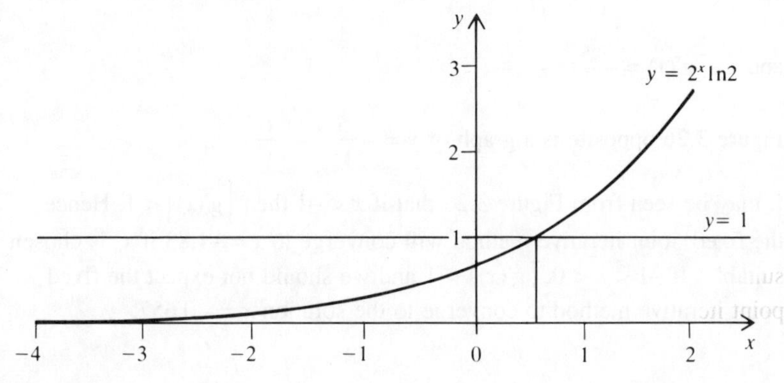

Figure 3.24

The line $y = 1$ is drawn on the graph and it may be seen clearly that

for $\quad x \in [-3, -2], \quad |g'(x)| < 1$

for $\quad x \in [2, 3], \quad |g'(x)| > 1$

Example 3.10

Consider the second equation $x^3 - 5x - 3 = 0$. The solutions of this equation are (to three decimal digits) $x = -1.83$, $x = -0.657$ and $x = 2.49$.

If the equation is written in the form $x = \frac{1}{5}(x^3 - 3)$

then $\quad g(x) = \frac{1}{5}(x^3 - 3) \quad$ and $\quad g'(x) = \frac{3}{5}x^2$

The graphs of $y = g'(x)$ and $y = 1$ are shown in Figure 3.25 and it is clear that $|g'(x)| < 1$ only for $-1.29 < x < 1.29$. Hence the fixed point iterative method, with this particular form of the iterative function will converge only for the solution $x = -0.657$ (with suitable choice of x_0).

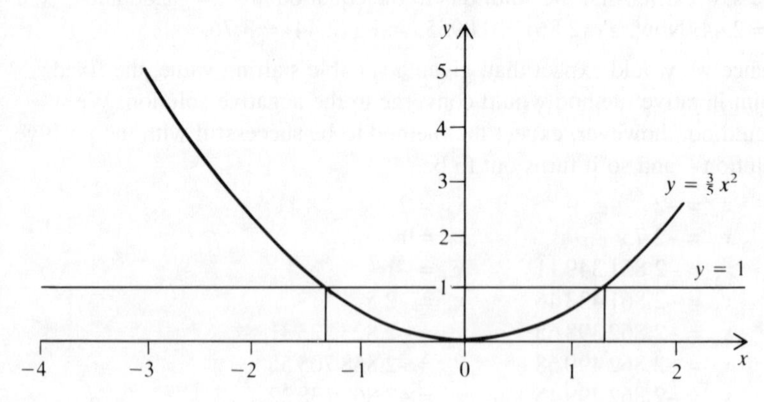

Figure 3.25

If, instead, we had written $x = \dfrac{5x+3}{x^2}$, then

$$g(x) = \frac{5x+3}{x^2}$$

$$= \frac{5}{x} + \frac{3}{x^2}$$

and $\quad g'(x) = -\dfrac{5}{x^2} - \dfrac{6}{x^3}$

Figure 3.26 opposite is a graph of $y = -\dfrac{5}{x^2} - \dfrac{6}{x^3}$.

It may be seen from Figure 3.26 that if $x < -1$ then $|g'(x)| < 1$. Hence the fixed point iterative method will converge to $x = -1.83$ if x_0 is chosen suitably. If $-1 < x < 0$, $|g'(x)| > 1$ and we should not expect the fixed point iterative method to converge to the solution $x = -0.657$.

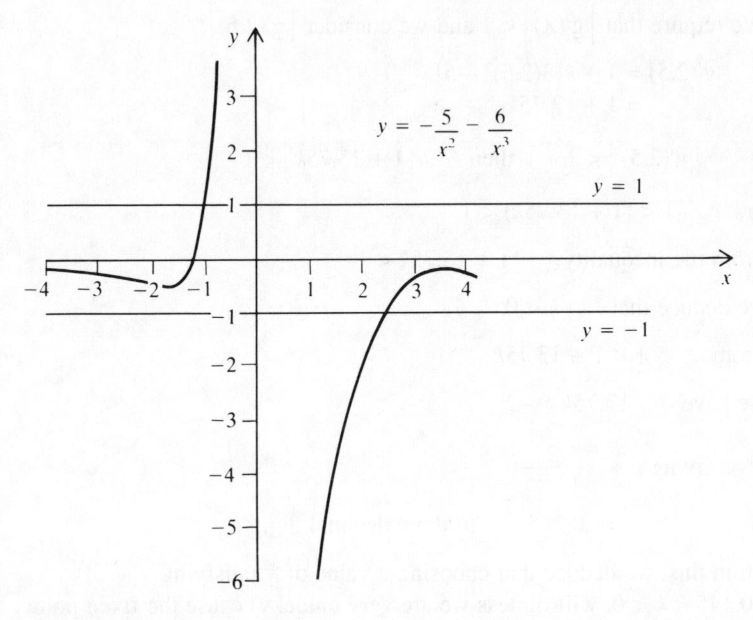

Figure 3.26

For the third solution we have $|g'(2.49)| = 1.195$ (to four decimal digit accuracy) and the fixed point iterative method should not be expected to converge to this solution.

Finally, if we write

$$x = x + k(x^3 - 5x - 3)$$

then $\quad g(x) = x + k(x^3 - 5x - 3)$

and $\quad g'(x) = 1 + k(3x^2 - 5)$

The next question is, 'What is the best value that we can choose for k?' The answer, in general terms, must be, 'Choose k so that $|g'(x)|$ is as small as possible in some region surrounding the required solution.' To see how this might be applied in practice, suppose that we do not know the value of the third solution. We observe from the graph of $y = x^3 - 5x - 3$ (Figure 3.15) that $x = 2.5$ provides a good approximation to this solution.

We require that $|g'(\lambda)| < 1$ and we consider $|g'(2.5)|$.

$$\begin{aligned} g'(2.5) &= 1 + k[3(2.5)^2 - 5] \\ &= 1 + 13.75k \end{aligned}$$

If $|g'(2.5)| < 1$ then $|1 + 13.75k| < 1$

or, $-1 < (1 + 13.75k) < 1$

From the inequality $1 + 13.75k < 1$

we deduce that $k < 0$

From $-1 < 1 + 13.75k$

we have $13.75k > -2$

$$\begin{aligned} \text{giving } k &> -\frac{2}{13.75} \\ &= -0.145 \quad \text{to three decimal digits} \end{aligned}$$

From this, we deduce that choosing a value of k satisfying $-0.145 < k < 0$, will (unless we are very unlucky) cause the fixed point iterative method to converge. Ideally, though, we would like $|g'(\lambda)|$ to be as small as possible.

So, consider $|1 + 13.75k| = 0$

This gives

$$\begin{aligned} k &= -\frac{1}{13.75} \\ &= -0.07273 \quad \text{to four decimal digits} \end{aligned}$$

Hence our choice of k in Example 3.7.

Finally, we will now see how the convergence of the fixed point iterative method may be accelerated.

3.14 Acceleration of convergence

We saw in Section 3.12 that

$$e_{n+1} \approx g'(\lambda)e_n$$

Write, for convenience $k = g'(\lambda)$

then $e_{n+1} \approx ke_n$

also $e_{n+2} \approx ke_{n+1}$

giving

$$\begin{aligned} e_{n+2} &\approx k(ke_n) \\ &= k^2 e_n \end{aligned}$$

In Section 3.12 we wrote the error terms in the form

$$\begin{aligned} x_n &= \lambda + e_n \\ x_{n+1} &= \lambda + e_{n+1} \\ x_{n+2} &= \lambda + e_{n+2} \end{aligned}$$

Using the above expressions for e_{n+1} and e_{n+2}, we have

$$\begin{aligned} x_n &= \lambda + e_n \\ x_{n+1} &\approx \lambda + ke_n \\ x_{n+2} &\approx \lambda + k^2e_n \end{aligned} \qquad \ldots (1)$$

Subtracting

$$\begin{aligned} x_{n+1} - x_n &\approx (k-1)e_n \\ x_{n+2} - x_{n+1} &\approx k(k-1)e_n \end{aligned} \qquad \ldots (2)$$

and $\quad (x_{n+2} - x_{n+1}) - (x_{n+1} - x_n) \approx k(k-1)e_n - (k-1)e_n$

After a little algebra, we have:

$$x_{n+2} - 2x_{n+1} + x_n \approx (k-1)^2e_n \qquad \ldots (3)$$

From (2) and (3):

$$\frac{(x_{n+2} - x_{n+1})^2}{x_{n+2} - 2x_{n+1} + x_n} \approx k^2e_n$$

Substituting this expression for k^2e_n in (1) gives

$$\lambda \approx x_{n+2} - \frac{(x_{n+2} - x_{n+1})^2}{x_{n+2} - 2x_{n+1} + x_n}$$

and we obtain the iterative equation

$$\boxed{x_{n+3} = x_{n+2} - \frac{(x_{n+2} - x_{n+1})^2}{x_{n+2} - 2x_{n+1} + x_n}}$$

To derive this equation, we substituted an expression for the error term k^2e_n and ignored error terms proportional to e_n^2. This suggests that the above iterative equation has an error term proportion to e_n^2, which, if true, will cause the sequence generated by the above iterative equation to converge much faster than the fixed point iterative method.

The procedure for using this equation is as follows:

To solve the equation $x = g(x)$

Step 1 Calculate a suitable starting value, x_0.

Step 2 Let $x_1 = g(x_0)$, let $x_2 = g(x_1)$

Step 3 $x_3 = x_2 - \dfrac{(x_2 - x_1)^2}{x_2 - 2x_1 + x_0}$

Step 4 Let $x_4 = g(x_3)$, let $x_5 = g(x_4)$

Step 5 $x_6 = x_5 - \dfrac{(x_5 - x_4)^2}{x_5 - 2x_4 + x_3}$

and so on until sufficient accuracy has been obtained.

Algorithm 3.4 Aitken's process

> To obtain a solution of the equation $x = g(x)$.
>
> input: a starting value x_0
> for $n = 0, 3, 6, \ldots$ until satisfied
> $x_{n+1} := g(x_n)$
> $x_{n+2} := g(x_{n+1})$
> $$x_{n+3} := x_{n+2} - \frac{(x_{n+2} - x_{n+1})^2}{x_{n+2} - 2x_{n+1} + x_n}$$
> endloop
> output: $x_0, x_1, x_2, \ldots$ a sequence of approximations to the required solution.

This method is called **Aitken's process**[†]. It may be used to speed up the convergence of any process where the error at any stage is given by $e_{n+1} \approx ke_n$.

To illustrate the effectiveness of Aitken's method, consider the equation $x^2e^{3x} = 2$. There is a solution lying in the interval [0, 1]. Using the fixed point iterative method, with $x = \sqrt{2e^{-3x}}$, we have the two sequences (corresponding to the starting values $x_0 = 0$ and $x_0 = 1$).

$x_0 = 0$	$x_0 = 1$
$x_1 = 1.414\,213\,56$	$x_1 = 0.315\,553\,699$
$x_2 = 0.169\,526\,376$	$x_2 = 0.880\,947\,754$
$x_3 = 1.096\,676\,26$	$x_3 = 0.377\,249\,676$
$x_4 = 0.272\,957\,055$	$x_4 = 0.803\,080\,016$
$x_5 = 0.939\,073\,064$	$x_5 = 0.423\,989\,565$
$x_6 = 0.345\,751\,142$	$x_6 = 0.748\,704\,606$
$x_7 = 0.841\,934\,447$	$x_7 = 0.460\,020\,917$
$x_8 = 0.399\,985\,061$	$x_8 = 0.709\,313\,424$
$x_9 = 0.776\,154\,251$	$x_9 = 0.488\,021\,144$
$x_{10} = 0.441\,464\,449$	$x_{10} = 0.680\,138\,976$
$x_{11} = 0.729\,334\,297$	$x_{11} = 0.509\,851\,959$
$x_{12} = 0.473\,583\,112$	$x_{12} = 0.658\,227\,708$
$x_{13} = 0.695\,029\,438$	$x_{13} = 0.526\,887\,634$
$x_{14} = 0.498\,590\,298$	$x_{14} = 0.641\,620\,763$
$x_{15} = 0.669\,441\,259$	$x_{15} = 0.540\,177\,465$
$x_{16} = 0.518\,099\,33$	$x_{16} = 0.628\,956\,862$
$x_{17} = 0.650\,134\,897$	$x_{17} = 0.550\,536\,676$
$x_{18} = 0.533\,322\,616$	$x_{18} = 0.619\,259\,157$
$x_{19} = 0.635\,457\,331$	$x_{19} = 0.558\,603\,62$
$x_{20} = 0.545\,194\,643$	$x_{20} = 0.611\,811\,017$

[†] A full and rigorous description can be found in Conte and de Boor 1981, *Elementary numerical analysis* 3rd edn. McGraw-Hill.

It is reasonable to suppose that both sequences will converge, eventually, to the solution $x = 0.586\,63$ (to five decimal digits). The convergence is, however, very slow indeed.

When Aitken's accelerated convergence is used, we have:

$x_0 = 0$	$x_0 = 1$
$x_1 = 1.414\,213\,56$	$x_1 = 0.315\,553\,699$
$x_2 = 0.169\,526\,376$	$x_2 = 0.880\,947\,754$
$x_3 = 0.752\,190\,543$	$x_3 = 0.625\,178\,739$
$x_4 = 0.457\,621\,789$	$x_4 = 0.553\,665\,527$
$x_5 = 0.711\,870\,623$	$x_5 = 0.616\,359\,612$
$x_6 = 0.594\,085\,642$	$x_6 = 0.587\,072\,469$
$x_7 = 0.580\,099\,945$	$x_7 = 0.586\,234\,668$
$x_8 = 0.592\,398\,146$	$x_8 = 0.586\,971\,853$
$x_9 = 0.586\,643\,834$	$x_9 = 0.586\,626\,808$
$x_{10} = 0.586\,611\,71$	$x_{10} = 0.586\,626\,692$
$x_{11} = 0.586\,639\,977$	$x_{11} = 0.586\,626\,794$
$x_{12} = 0.586\,626\,746$	$x_{12} = 0.586\,626\,746$
$x_{13} = 0.586\,626\,746$	$x_{13} = 0.586\,626\,746$

It is clear that by using Aitken's method the convergence has been accelerated considerably.

This error analysis and also the analysis of the equation $x = x + k(x^3 - 5x - 3)$ shows how results can be improved enormously by applying often quite elementary mathematics. It could be argued that this is precisely what numerical analysis is all about.

Exercise 3

Section 3.1

1 Explain carefully what is meant by 'a solution of an equation'. Give an example of a situation where it would be useful to obtain the solution of an equation. Write down an equation that you cannot solve algebraically. By drawing a graph, obtain an approximate solution to your equation.

2 By using the fact that if $f(a) \times f(b) < 0$ then the equation $f(x) = 0$ has a solution in the interval $[a, b]$, show that the following equations have a solution in the intervals indicated.

(a) $3x - \ln x - 5 = 0$ $[1, 2]$
(b) $e^{2x} + 3x - 5 = 0$ $[0, 1]$
(c) $x^3 - 3x^2 + 3 = 0$ $[-1, 0], [1, 2], [2, 3]$
(d) $4x + 3\sin x = \cos x$ $[0, 1]$ in radians
(e) $\dfrac{x^2}{\sqrt{x-1}} = 5$ $[2, 3]$

Sketch the graphs of the functions
(a) $y = 3x - \ln x - 5$
(b) $y = x^3 - 3x^2 + 3$
(c) $y = 4x + 3\sin x - \cos x$
and verify the above results.

3 Solve the following equations using the bisection method. In each case, give the solutions correct to four decimal digits (rounded). The brackets adjacent to each equation provide an interval within which a solution lies.

(a) $3x - \ln x - 5 = 0$ [1, 2]
(b) $e^{2x} + 3x - 5 = 0$ [0, 1]
(c) $4x + 3\sin x = \cos x$ [0, 1] in radians
(d) $\dfrac{x^2}{\sqrt{x-1}} = 5$ [2, 3]

4 Use the bisection method to solve the following equations. Give answers rounded to five decimal digits. For each equation, there is a solution lying in each of the brackets given.

(a) $\dfrac{15}{2+x} - \sqrt{x} = 3$ [1, 2]
(b) $x^3 - 3x^2 + 3 = 0$ [–1, 0], [1, 2], [2, 3]
(c) $\dfrac{x}{\sqrt{x+1}} = x - 2$ [–0.95, 0], [3, 4]
(why could the first interval of (c) not have been given as [–1, 0]?)
(d) $x\tan x - 3x = 1$ [–3, –1], [–1, 0], [1, 2] in radians

5 It is required to calculate the solution of the equation $x\ln x + x - 3 = 0$ that lies in the interval [1, 2]. Calculate the value of this solution using the bisection method and with tolerance $\varepsilon = 0.0001$ and

(a) stopping rule SR1
(b) stopping rule SR2 and
(c) stopping rule SR3 (see page 101)

6 Use the bisection method and

(a) stopping rule SR1
(b) stopping rule SR2 and
(c) stopping rule SR3

each with tolerance $\varepsilon = 0.005$ to obtain the solution of the equation $x = 3\sin x$ that lies in the interval [2, 3] (x is measured in radians).

Section 3.4

7 Each of the following equations has a solution lying in the interval(s) written next to the equation. Use the secant method to calculate the value of this solution correct to five decimal digits (using rounding).

(a) $x^2 \ln x - 2 = 0$ [1, 2]
(b) $xe^x - 1 - x = 0$ [–2, –1], [0, 1]
(c) $x^4 + 5x^2 - 3 = 0$ [–1, 0], [0, 1]
(d) $x^3 - 4x + 1 = 0$ [–3, –2], [0, 1], [1, 2]

8 Each of the equations written below has a solution in the interval(s) written next to the equation. Use the secant method to find these solutions correct to four decimal digits (rounded).

(a) $4x^3 + 3x^2 - 8x - 1 = 0$ $[-2, -1], [-1, 0], [1, 2]$
(b) $4 - xe^{2x} = 0$ $[0, 1]$
(c) $\ln(x + 1) = 5$ $[140, 150]$
(d) $2e^{-x} - \cos x + 0.5 = 0$ $[5, 6], [7, 8]$ in radians

9 It is required to solve the equation

$$\frac{x}{5} - \ln(x^2) = 0$$

(a) Use the secant method and the stopping rule SRI (page 101) with a tolerance of 0.0005 to obtain your answer.
(b) Calculate the solutions that result from using stopping rules SR2 and SR3 with the same tolerance, and compare your answers.

10 Show that the iterative equation for the secant method may be written in the form

$$x_{n+1} = x_n - \frac{f(x_n)(x_{n-1} - x_n)}{f(x_{n-1}) - f(x_n)}$$

Comment on any advantages that might be gained from writing the iterative equation in this form.

11 The equation $(x^3 - 3x^2 - x + 9)e^{-2x} = 0$ has as solution lying between $x = -2$ and $x = -1$. Applying the secant method to this equation taking $x_0 = -1$ and $x_1 = -2$. Describe what happens. Draw the graph of the equation

$$y = (x^3 - 3x^2 - x + 9)e^{-2x}$$

and illustrate on your graph why the secant method diverges.

Section 3.6

12 Show that the equation $2x^7 + x^2 - 4 = 0$ can be written in the form $x = \left(\frac{4 - x^2}{2}\right)^{\frac{1}{7}}$.

Hence use the fixed point iterative method to find the solution of the equation $2x^7 + x^2 - 4 = 0$ that lies in the interval $[1, 2]$. Give your answer to four decimal digits using rounding.

13 Show that the equation $3x - 2e^{-x} = 0$ can be written in the form $x = \frac{2e^{-x}}{3}$. Hence use the fixed point iterative method to find, correct to three decimal digits (rounded) the solution of the equation $3x - 2e^{-x} = 0$ that lies in the interval $[0, 1]$.

14 Write each of the equations below in the form $x = g(x)$ for some iterative function $g(x)$. Use the fixed point iterative method to obtain the solution which lies in the interval shown. Give all solutions correct to four decimal digits (rounded).

(a) $x - \cos x = 0$ [0, 1] in radians
(b) $x + 5 = 3^x$ [–5, –4]
(c) $(x + 2)^{\frac{1}{2}} = 3x$ [0, 1]
(d) $x \sin \sqrt{x} = 1$ [1, 2] in radians

15 Use the fixed point iterative method to obtain solutions (rounded to two decimal digits) to the following equations.

(a) $5(x + 1) = (x + 2)^2$ [–1, 0]
(b) $3x \sin x + 5x^2 - 10 = 0$ [1, 2] in radians
(c) $\dfrac{\cos x}{x} - 4 = 0$ [0, 1] in radians

16 Use the fixed point iterative method to obtain the solution of the equation $\dfrac{x + 1}{\ln x} - 5 = 0$ which lies in the interval [10, 11]. Using a tolerance of 0.0005, calculate the solution that arises from the use of

(a) stopping rule SR1
(b) stopping rule SR2
(c) stopping rule SR3

Compare your solutions and comment on your answers.

17 Obtain both solutions of the equation $2x^2 + 2x - 3 = 0$ (rounded to three decimal digits) using the fixed point iterative method. (You may have to use two forms of the iterative function to calculate both solutions.)

18 By choosing suitable forms for the iterative function $g(x)$, use the fixed point iterative method to find, correct to four decimal digits (rounded), the three solutions of the equation $x^3 - 7x - 2 = 0$.
The solutions lie in the intervals [–3, –2], [–1, 0], [2, 3].

19 Sketch the curve $y = x^3 - 5x^2 + 4x + 1$. Hence obtain rough approximate values for the three solutions of the equation $x^3 - 5x^2 + 4x + 1 = 0$. By writing $x^3 - 5x^2 + 4x + 1 = 0$ in the form $x = g(x)$ in three different ways, calculate the values of the solutions correct to four decimal digits (rounded).

Section 3.7

20 It is required to obtain the solution of the equation $x - 3\sin x = 0$ which lies in the interval [2, 3].

(a) Verify that a solution does lie in this interval.
(b) Use the fixed point iterative method and attempt to calculate this solution.
(c) Describe graphically what has gone wrong.

21 The equation $x^3 - 5x^2 + 2 = 0$ has a solution lying in the interval [–1, 0] (verify this statement). Write the equation in the form $x = g(x)$ and attempt to calculate this solution. Draw a graph to show how your method is succeeding or failing. Repeat the question for the solution lying in the interval [4, 5].

22 Obtain, correct to four decimal digits (rounded) all solutions of the following equations that lie in the given intervals. Use (i) the bisection method (ii) the secant method and (iii) the fixed point iterative method. Compare the speeds of convergence of the methods.

(a) $\dfrac{\cos x}{x} - 4 = 0$ [0, 1] in radians

(b) $x \ln x + x - 3 = 0$ [1, 2]

(c) $\sqrt{x+1} - \sqrt{x} = 0.2$ [4, 6]

Section 3.9

23 The equation $x \sin \sqrt{x} = 1$ has a solution lying in the interval [1, 2]. If the bisection method is used to calculate a sequence of approximations to the solution x_0, x_1, x_2, ... , what will be the largest error that can occur in

(a) x_0 (b) x_1 (c) x_2 (d) x_5 (e) x_{10}?

24 The equation $x^4 + 5x^2 - 3 = 0$ has solutions lying in the intervals [–1, 0] and [0, 1]. For each solution, use the bisection method to generate a sequence $x_0, x_1, x_2, \ldots$ of approximations to that solution. Write down the greatest error that can occur in

(a) x_3 (b) x_6 (c) x_{12} (d) x_n

25 It is required to solve the equation $\dfrac{x+2}{\ln x} = 7$ using the bisection method. If an absolute error of 0.0005 is permissible in the solution, how many iterations are necessary before the required accuracy is achieved?

26 The bisection method is to be used to calculate approximations to the solutions of the equation

$$xe^x - 3 - x = 0$$

which lie in the intervals [–4, –3] and [1, 2]. If a maximum absolute error of 0.0001 is acceptable in each solution, how many iterations will be necessary to calculate each solution to this degree of accuracy?

Section 3.12

27 Each of the following equations is written in the form $x = \mathrm{g}(x)$. Determine an interval within which the solution may be expected to lie if the fixed point iterative method with iterative function $\mathrm{g}(x)$ is to converge to that solution. If the equation has a solution lying in that interval, use the fixed point iterative method to calculate the value of that solution rounded to four decimal digits.

(a) $x = 4\cos x$ (work in radians)

(b) $x = 3 - \dfrac{4x-1}{x}$

(c) $x = \sqrt{\dfrac{e^x}{3}} - 1$

(d) $x = \dfrac{7}{x+10} + 3x$

28 We wish to calculate, using the fixed point iterative method, the solution of the equation

$$3x^2 - 7x + 3 = 0$$

which lies in the interval [1, 2]. Three forms of the iterative equation are proposed:

(i) $x_{n+1} = \dfrac{3x_n^2 + 3}{7}$

(ii) $x_{n+1} = \dfrac{7x_n - 3}{3x_n}$

(iii) $x_{n+1} = \sqrt{\dfrac{7x_n - 3}{3}}$

(a) Verify that each of these iterative equations is valid for the equation $3x^2 - 7x + 3 = 0$.
(b) Decide which of the three iterative equations will produce a sequence of approximations which will converge to the required solution.
(c) Which of the three iterative equations will give the fastest convergence? (Give reasons.)

29 The equation $4x^3 - 5x - 1 = 0$ has a solution in the interval [–1, 0]. The following iterative functions are proposed

(i) $g(x) = \dfrac{4x^3 - 1}{5}$

(ii) $g(x) = \dfrac{5x + 1}{4x^2}$

(iii) $g(x) = \sqrt[3]{\dfrac{5x + 1}{4}}$

(a) Verify that each iterative function is valid for the equation $4x^3 - 5x - 1 = 0$.
(b) Decide (and give reasons for your answer) which of the three iterative functions should be chosen for the iterative equation.

30 (a) Obtain approximately the solutions of the equation

$$x^2 - 5x + 2 = 0$$

(b) To write this equation in the form $x = g(x)$, we write $x = x + k(x^2 - 5x + 2)$. Calculate the range of values of k for which the associated iterative equation may be expected to converge.
(c) For what value of k will the associated sequence $x_0, x_1, x_2, \ldots$ achieve most rapid convergence?

31 (a) Describe carefully why the equation

$$x = x + k(\ln x - 6x^2 + 3)$$

could be used to form an iterative equation for the equation

$$\ln x - 6x^2 + 3 = 0$$

(b) The equation $\ln x - 6x^2 + 3 = 0$ has a solution in [0, 1]. By finding an approximate value for this solution:

(i) Calculate a value for k so that the associated sequence $x_0, x_1, x_2, \ldots$ converges as rapidly as possible.

(ii) Use the iterative equation suggested above with your value of k to calculate the solution to five decimal digits (using rounding).

32 (a) The equation $f(x) = 0$ has a solution at $x = a$. Let the equation be written in the form $x = g(x)$ and let $x_0, x_1, x_2, \ldots$ be the sequence of approximations generated by the iterative equation $x_{n+1} = g(x_n)$ If $a = x_n + e_n$ (where e_n represents the error in x_n) show that

$$|e_{n+1}| \approx |g'(a)|\ |e_n|$$

(b) Draw the graph $y = x^3 - 3x + 1$ and hence obtain a rough approximation to each of the three solutions of the equation $x^3 - 3x + 1 = 0$. Write $x^3 - 3x + 1 = 0$ in the form $x = g(x)$ in five different ways. For each of the three solutions explore the convergence properties of each of the five associated iterative equations.

Section 3.14

33 For each of the following equations

(a) $5x - 3e^{-x} = 1$ [0, 1]

(b) $3x^4 - 5x - 2 = 0$ [–1, 0]

(c) $x^2e^{3x} = 2$ [0, 1]

(d) $\sqrt{x} = \sqrt{x+1} - 0.2$ [5, 6]

(e) $2x^3 - 9x + 4 = 0$ [–3, –2], [0, 1], [1, 2]

use (i) the fixed point iterative method (ii) Aitken's method, both with stopping rule SR3 and a tolerance of 0.0005 to obtain approximations to the solutions indicated in the brackets.

34 Write down an equation of your own choice. Solve the equation using

(a) the fixed point iterative method

(b) the secant method

(c) accelerated convergence (Aitken's method)

Comment on the speed of convergence of the three methods.

35 The equation $4x^3 + 3x^2 - 8x - 1 = 0$ has solutions in the intervals [–2, –1], [–1, 0], [1, 2]. For each solution, find the form of the iterative function $g(x)$ that will achieve fastest convergence when the iterative equation $x_{n+1} = g(x_n)$ is used to calculate that solution. Now obtain solutions to the same degree of accuracy using Aitken's method. Compare the convergence properties of the two methods.

Miscellaneous

36 Show that the equation $x^3 - 5x + 3 = 0$ has a root between $x = 0$ and $x = 1$. Using the iteration formula $x_{n+1} = \dfrac{3 + x_n^3}{5}$, find the value of the root to two decimal places. (MEI)

37 The equation $e^x = \tan x$ has solutions in the intervals $[-4, -3]$ and $[1, 2]$. Use
(a) the bisection method
(b) the secant method
to obtain these solutions correct to four decimal digits.
Compare your two methods of solution, giving advantages and disadvantages of each method.

38 The equation $x + \ln x = 0$ has a solution in the interval $[0, 1]$. Prove that one and only one of the following iterative schemes will converge to that solution.

$$x_{n+1} = -\ln x_n$$
$$x_{n+1} = e^{-x_n}$$

Determine that solution correct to four digit accuracy.

39 Show that the equation $4x^3 - 3x^2 + 2 = 0$ has exactly one real solution and that this solution lies in the interval $[-1, 0]$. Use the secant method, with starting value x_0, to generate successive approximations to the solution $x_1, x_2, x_3, \ldots$ Calculate the value of the solution to five digit accuracy.
Draw the graph of $y = 4x^3 - 3x^2 + 2$ and, on your graph, illustrate carefully the positions of x_0, x_1, x_2. Describe one advantage of using the secant method and one disadvantage.

40 Show that the cubic equation $x^3 + 2x - 11 = 0$ has only one real root and further that the root lies between $x = 1$ and $x = 2$.
Two possible iterative schemes for finding the root are

(a) $x_{n+1} = \dfrac{11 - x_n^3}{2}$ and

(b) $x_{n+1} = (11 - 2x_n)^{\frac{1}{3}}$

Show that only one of these schemes converges from an initial estimate of $x = 2$ and hence find the root correct to three decimal places, justifying the accuracy of your answer. (MEI)

41 Show that the equation $2^{-x} + e^x - 5 = 0$ has solutions in the intervals $[-3, -2]$, $[1, 2]$.
Use the secant method to obtain these solutions correct to four decimal digits.
Why is it preferable to use the secant method to solve this equation rather than the bisection method or the fixed point iterative method?

42 An equation can be written in the form $x = \mathrm{F}(x)$ and it is known that it has a root in the neighbourhood of $x = x_1$. Explain with the aid of a diagram how the iterative formula

$$x_{r+1} = \mathrm{F}(x_r)$$

may give the root to whatever degree of accuracy is required and state a condition for convergence.
The cubic equation $x^3 - 2x - 5 = 0$ has one real root which is near $x = 2$. The equation can be written in any one of the forms

(a) $x = \frac{1}{2}(x^3 - 5)$
(b) $x = 5(x^2 - 2)^{-1}$
(c) $x = (2x + 5)^{\frac{1}{3}}$.
Choose the form for which the previous iterative formula will converge to the real root and use the formula to find this root as accurately as possible.
(MEI)

43 Show that the equation $x^3 - 5 \tan x + 1 = 0$ has a solution in the interval [0, 1]. Taking $[a_0, b_0] = [0,1]$ as the first interval, use the bisection method to generate successively smaller intervals $[a_1, b_1]$, $[a_2, b_2]$, ... each of which contains the required solution.

Write $\quad x_0 = \dfrac{a_0 + b_0}{2} \qquad x_1 = \dfrac{a_1 + b_1}{2} \qquad x_2 = \dfrac{a_2 + b_2}{2} \quad \ldots$

What is the greatest value of the error when the solution is approximated by
(a) x_0 (b) x_1 (c) x_2 (d) x_n?
If, in this scheme, the error must be less than 0.001, how many iterations must be made, and what will then be the approximate solution?

44 It is required to obtain a solution for each of the following equations
(a) $x^3 - x - 1 = 0$
(b) $2x - \tan x = 0$ (consider the solution lying in the interval [–2, –1])
(c) $e^{-x} - \sin x = 0$ (consider the solution lying in the interval [3, 4])
For each equation, write down an iterative equation and a starting value x_0 and use your iterative equation to generate a sequence x_1, x_2, x_3, ... that converges to the required solution. Can you suggest a method by which the convergence could be accelerated?

45 The volume of a sphere is $\frac{4}{3}\pi r^3$ where r is the radius of the sphere. A sphere of radius 4 cm is partially submerged so that 25% of its volume is under water. The greatest depth of the sphere below the water line is h (as in the figure).

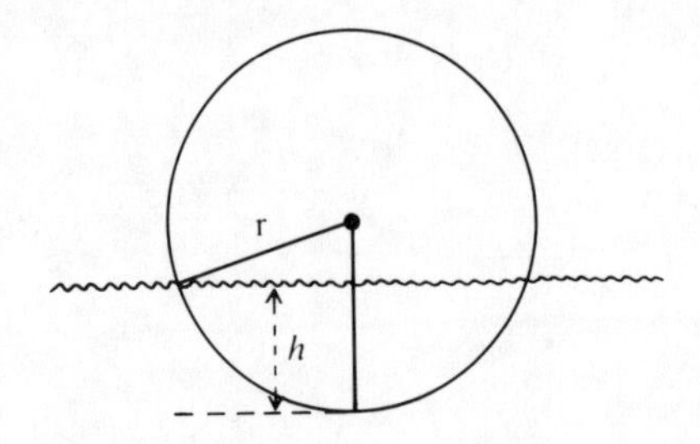

Find the height of the top of the sphere above the surface of the water.

46 Show by a sketch that for a given value of k (where $k > 1$) there is just one root of $\tan x - kx = 0$ in the interval $0 < x < \frac{1}{2}\pi$.
It is required to tabulate values of this root for

$$k = 1.4, 1.5, ..., 2.0.$$

Draw a graph of $\frac{\tan x}{x}$ for $0.82 \leqslant x \leqslant 1.22$ and so find the approximate values of the root. Calculate the value of the root to three significant figures when $k = 1.6$.

(MEI)

47 The equation $x^3 + 3x = 2$ has a root X between 0 and 1 which is to be found using the iterative process

$$x_{n+1} = x_n - \frac{x_n^3 + 3x_n - 2}{3(x_n^2 + 1)}, \qquad x_0 = 0$$

Show that if x_n differs from X by a small quantity δ, x_{n+1} differs from X by approximately

$$\frac{X\delta^2}{X^2 + 1}$$

Find X correct to four decimal places using the above process. Give all intermediate results and justify the accuracy of your answer.
Comment on the difference, from the point of view of determining the numerical value of X to any given accuracy, between the above process and the algebraically equivalent one defined by

$$x_{n+1} = \frac{2(x_n^3 + 1)}{3(x_n^2 + 1)}, \qquad x_0 = 0$$

(MEI)

Chapter 4

Solution of systems of linear equations

Introduction

In many areas of mathematics, one frequently encounters systems of equations such as those shown below

(a)
$$\begin{aligned} x_1 + 2x_2 - x_3 + x_4 &= 10 \\ 3x_1 + 8x_2 + 2x_3 + x_4 &= 41 \\ 2x_1 - 4x_2 + x_3 - 2x_4 &= -11 \\ 4x_1 + 2x_2 - 3x_3 + 5x_4 &= 6 \end{aligned}$$

(b)
$$\begin{aligned} 3x_1 - 4x_2 - x_3 &= 9 \\ x_1 + x_2 + 3x_3 &= 14 \\ 2x_1 - 2x_2 - x_3 &= 4 \end{aligned}$$

Such systems of equations are natural extentions to simultaneous equations like

$$\begin{aligned} 3x + 2y &= 7 \\ 2x - 3y &= -4 \end{aligned}$$

The aim is to solve these systems of equations, that is, (in (a) for example) to find values for x_1, x_2, x_3 and x_4 that satisfy simultaneously the four equations. Observe that in each of these systems, there are the same number of equations as there are unknown quantities to be found. This will be true of all the systems that we study in this chapter.

We remark that systems of equations such as these occur frequently in physics, engineering, probability theory, statistics, economics and in numerical analysis. Since it is certainly not unknown for systems of 100000 equations in 100000 unknowns to occur, it is essential to have an efficient method for solving such systems and it is the major purpose of this chapter to describe in detail one such method.

But what makes a method 'efficient'? This is not an easy question to answer, but an efficient method should be

(i) easy to program on a computer,
(ii) reasonably accurate with regard to rounding errors.

We shall see how the method described behaves with respect to these requirements and we shall illustrate some of the pitfalls that await the unwary.

We begin with two definitions and a method of solving a special system of equations.

Definition

> Two systems of equations are said to be equivalent if they have the same solution.

The two systems

$$\begin{aligned} 3x_1 + 4x_2 &= 9 \\ 7x_1 + x_2 &= -4 \end{aligned} \qquad \begin{aligned} 2x_1 - x_2 &= -5 \\ 4x_1 + 3x_2 &= 5 \end{aligned}$$

each have the solution $x_1 = -1$ and $x_2 = 3$. Hence the two systems are equivalent. This definition may appear to be obvious. We shall see later, however, that equivalence is a very useful concept.

Definition

> A system of equations such as
>
> $$\begin{aligned} 3x_1 - 2x_2 - 4x_3 &= 1 \\ 5x_2 - 8x_3 &= -1 \\ 4x_3 &= 8 \end{aligned}$$
>
> is called a triangular system. In a triangular system, the first equation may contain all the unknowns, the second equation may contain all the unknowns except x_1, the third equation may contain all the unknowns except x_1 and x_2 and so on until the final equation (the nth equation) contains only x_n.

It is clear that a triangular system of equations is extremely easy to solve. In the triangular system written above, use the last equation to find the value of x_3.

$$\begin{aligned} 4x_3 &= 8 \\ x_3 &= 2 \end{aligned}$$

Now use the next to the last equation (and the known value of x_3) to find the value of x_2.

$$\begin{aligned} 5x_2 - 8 \times 2 &= -1 \\ 5x_2 &= -1 + 16 = 15 \\ x_2 &= 3 \end{aligned}$$

Now use the first equation (and the known values of x_3 and x_2) to find the value of x_1.

$$\begin{aligned}3x_1 - 2 \times 3 - 4 \times 2 &= 1\\ 3x_1 &= 1 + 6 + 8 = 15\\ x_1 &= 5\end{aligned}$$

Hence the solution is $x_1 = 5$, $x_2 = 3$ and $x_3 = 2$.

This method of solving a triangular system of equations is called, for obvious reasons, **backward substitution.**

Example 4.1

Use the method of backward substitution to solve the triangular system of equations

$$\begin{aligned}-x_1 + 3x_2 - x_3 + x_4 &= 0\\ 5x_2 + 3x_3 - 2x_4 &= 1\\ 2x_3 - x_4 &= 2.5\\ 6x_4 &= 9\end{aligned}$$

From the last equation,

$$6x_4 = 9 \quad \text{so} \quad x_4 = 1.5$$

From the third equation

$$\begin{aligned}2x_3 - 1.5 &= 2.5\\ 2x_3 &= 4\\ x_3 &= 2\end{aligned}$$

From the second equation

$$\begin{aligned}5x_2 + 3 \times 2 - 2 \times 1.5 &= 1\\ 5x_2 &= 1 - 6 + 3 = -2\\ x_2 &= -0.4\end{aligned}$$

From the first equation

$$\begin{aligned}-x_1 + 3 \times (-0.4) - 2 + 1.5 &= 0\\ -x_1 &= 0 + 1.2 + 2 - 1.5 = 1.7\\ x_1 &= -1.7\end{aligned}$$

Hence $\quad x_1 = -1.7$, $x_2 = -0.4$, $x_3 = 2$, $x_4 = 1.5$

We have established that it is very easy indeed to solve triangular systems of equations. This fact is central to our method of solving a general system of equations. The method that we shall describe is called **Gaussian elimination**.

4.1 The method of Gaussian elimination for solving a system of equations

This consists of two steps:

Step 1 Reduce the given system of equations to an equivalent triangular system.

Step 2 Use backward substitution to solve the resulting system of triangular equations.

It remains, then, to show how a general system of equations may be reduced to an equivalent triangular system. To achieve this reduction we use the following three operations, called **elementary operations**.

A Interchange two equations

In the equation system (b) given at the beginning of the chapter, interchanging equation 2 and equation 3 gives the system

$$\begin{aligned} 3x_1 - 4x_2 - x_3 &= 9 \\ 2x_1 - 2x_2 - x_3 &= 4 \\ x_1 + x_2 + 3x_3 &= 14 \end{aligned}$$

We write this operation as $E_2 \leftrightarrow E_3$.

B Multiply one of the equations by a constant

Multiplying the first equation of system (b) by 4 would produce the system

$$\begin{aligned} 12x_1 - 16x_2 - 4x_3 &= 36 \\ 2x_1 - 2x_2 - x_3 &= 4 \\ x_1 + x_2 + 3x_3 &= 14 \end{aligned}$$

We write this type of operation as $4E_1$.

C Replace an equation by the sum of that equation and a multiple of a second equation

We may replace equation 2 of system (b) by
equation 2 + (–3) × equation 1

Now, (–3) × equation 1 is

$$-9x_1 + 12x_2 + 3x_3 = -27$$

So equation 2 + (–3) × equation 1 is

$$\begin{aligned} 2x_1 - 2x_2 - x_3 &= 4 \\ -9x_1 + 12x_2 + 3x_3 &= -27 \quad + \\ \hline -7x_1 + 10x_2 + 2x_3 &= -23 \end{aligned}$$

Replacing equation 2 by equation 2 + (–3) × equation 1 gives the system

$$\begin{aligned} 3x_1 - 4x_2 - x_3 &= 9 \\ -7x_1 + 10x_2 + 2x_3 &= -23 \\ x_1 + x_2 + 3x_3 &= 14 \end{aligned}$$

We write this operation as $E_2 + (-3)E_1$

Readers should convince themselves that the solution of a system of equations will not be altered by the performance of any of these operations. Hence, performing one or more elementary operations will produce an equivalent system of equations. Examples will now be used to show how the repeated use of elementary operations will reduce a system of equations to a triangular system (in fact, to an equivalent triangular system).

The method of Gaussian elimination is illustrated on the system (b) given at the beginning of the chapter.

Step 1

Using elementary operations of type C, eliminate all occurrences of x_1 from the second and third equations. To eliminate x_1 from the second equation, replace the second equation by $E_2 + (-\frac{1}{3})E_1$ (observe that the multiplying factor $(-\frac{1}{3})$ was chosen so that in the resulting simplification, the x_1 terms will disappear).

$$\begin{array}{ll} E_2 & x_1 + x_2 + 3x_3 = 14 \\ (-\frac{1}{3})E_1 & -x_1 + \frac{4}{3}x_2 + \frac{1}{3}x_3 = -3 \quad + \\ \hline & \frac{7}{3}x_2 + \frac{10}{3}x_3 = 11 \end{array}$$

When the second equation is replaced by $E_2 + (-\frac{1}{3})E_1$ the system becomes

$$\begin{aligned} 3x_1 - 4x_2 - x_3 &= 9 \\ \tfrac{7}{3}x_2 + \tfrac{10}{3}x_3 &= 11 \\ 2x_1 - 2x_2 - x_3 &= 4 \end{aligned}$$

To eliminate x_1 from the third equation, replace the third equation by $E_3 + (-\frac{2}{3})E_1$

$$\begin{array}{ll} E_3 & 2x_1 - 2x_2 - x_3 = 4 \\ (-\frac{2}{3})E_1 & -2x_1 + \frac{8}{3}x_2 + \frac{2}{3}x_3 = -6 \quad + \\ \hline & \frac{2}{3}x_2 - \frac{1}{3}x_3 = -2 \end{array}$$

When the third equation is replaced by $E_3 + (-\frac{2}{3})E_1$ the system becomes

$$\begin{aligned} 3x_1 - 4x_2 - x_3 &= 9 \\ \tfrac{7}{3}x_2 + \tfrac{10}{3}x_3 &= 11 \\ \tfrac{2}{3}x_2 - \tfrac{1}{3}x_3 &= -2 \end{aligned}$$

and Step 1 is complete.

The equation that is used to eliminate some x_i from subsequent equations is called the **pivotal equation**. In Step 1 the pivotal equation was equation 1.

Step 2

Using elementary operations of type C, eliminate all occurrences of x_2 from the third equation. To achieve this, replace the third equation by $E_3 + (-\frac{2}{7})E_2$†.

$$\begin{array}{ll} E_3 & \frac{2}{3}x_2 - \frac{1}{3}x_3 = -2 \\ (-\frac{2}{7})E_2 & -\frac{2}{3}x_2 - \frac{20}{21}x_3 = -\frac{22}{7} \quad + \\ \hline & -\frac{27}{21}x_3 = -\frac{36}{7} \end{array}$$

† The multiplier $(-\frac{2}{7})$ is obtained by asking, 'What number multiplies $\frac{7}{3}$ to give $-\frac{2}{3}$?' The answer is $-\frac{2}{7}$.

When the third equation is replaced by $E_3 + (-\frac{2}{7})E_2$ the system becomes

$$\begin{aligned} 3x_1 - 4x_2 - \quad x_3 &= 9 \\ \tfrac{7}{3}x_2 + \tfrac{10}{3}x_3 &= 11 \\ -\tfrac{27}{21}x_3 &= -\tfrac{36}{7} \end{aligned}$$

and it is clear that the original system of equations has been reduced to an equivalent triangular system. Observe, before we move on to backward substitution, that the pivotal equation used in Step 2 was the second equation.

We now obtain the solution to the above system of equations by using backward substitution.

From the last equation

$$x_3 = \tfrac{36}{7} \times \tfrac{21}{27} = 4$$

From the second equation

$$\begin{aligned} \tfrac{7}{3}x_2 + \tfrac{10}{3} \times 4 &= 11 \\ \tfrac{7}{3}x_2 &= 11 - \tfrac{40}{3} = -\tfrac{7}{3} \\ x_2 &= -1 \end{aligned}$$

From the first equation

$$\begin{aligned} 3x_1 - 4 \times (-1) - 4 &= 9 \\ 3x_1 &= 9 - 4 + 4 = 9 \\ x_1 &= 3 \end{aligned}$$

Hence the solution is $\quad x_1 = 3, x_2 = -1, x_4 = 4.$

This, then, is Gaussian elimination. The explanation has been lengthy, but the ideas, once seen, are straightforward. As a further exercise, we solve the system of equations (a) given at the beginning of the chapter.

Example 4.2

Use the method of Gaussian elimination to solve the system of equations

$$\begin{aligned} x_1 + 2x_2 - \ x_3 + \ x_4 &= 10 \\ 3x_1 + 8x_2 + 2x_3 + \ x_4 &= 41 \\ 2x_1 - 4x_2 + \ x_3 - 2x_4 &= -11 \\ 4x_1 + 2x_2 - 3x_3 + 5x_4 &= 6 \end{aligned}$$

Use the first equation as the pivotal equation to eliminate x_1 from the second, third and fourth equations. Performing successively the operations

$$E_2 + (-3)E_1 \qquad E_3 + (-2)E_1 \qquad E_4 + (-4)E_1$$

gives

$$\begin{aligned} x_1 + 2x_2 - x_3 + x_4 &= 10 \\ 2x_2 + 5x_3 - 2x_4 &= 11 \\ -8x_2 + 3x_3 - 4x_4 &= -31 \\ -6x_2 + x_3 + x_4 &= -34 \end{aligned}$$

Now use the second equation of this system as the pivotal equation to eliminate x_2 from the third and fourth equations, performing successively the operations

$$E_3 + 4E_2 \qquad E_4 + 3E_2$$

$$\begin{aligned} x_1 + 2x_2 - x_3 + x_4 &= 10 \\ 2x_2 + 5x_3 - 2x_4 &= 11 \\ 23x_3 - 12x_4 &= 13 \\ 16x_3 - 5x_4 &= -1 \end{aligned}$$

Finally, use the third equation as the pivotal equation to eliminate x_3 from the fourth equation, performing the operation

$$E_4 + (-\tfrac{16}{23})E_3$$

$$\begin{aligned} x_1 + 2x_2 - x_3 + x_4 &= 10 \\ 2x_2 + 5x_3 - 2x_4 &= 11 \\ 23x_3 - 12x_4 &= 13 \\ \tfrac{77}{23}x_4 &= -\tfrac{231}{23} \end{aligned}$$

The system has now been reduced to an equivalent triangular system and backward substitution may be used to obtain the solution.

From the last equation

$$x_4 = -\tfrac{231}{77} = -3$$

From the third equation

$$\begin{aligned} 23x_3 - 12(-3) &= 13 \\ 23x_3 &= 13 - 36 = -23 \\ x_3 &= -1 \end{aligned}$$

From the second equation

$$\begin{aligned} 2x_2 + 5(-1) - 2(-3) &= 11 \\ 2x_2 &= 11 + 5 - 6 = 10 \\ x_2 &= 5 \end{aligned}$$

From the first equation

$$\begin{aligned} x_1 + 2(5) - (-1) + (-3) &= 10 \\ x_1 &= 10 - 10 - 1 + 3 = 2 \\ x_1 &= 2 \end{aligned}$$

Hence the solution is $\quad x_1 = 2, x_2 = 5, x_3 = -1, x_4 = -3.$

We have not, so far, used elementary operations A and B. The next example illustrates when elementary operations of these types should be used.

Example 4.3

Use Gaussian elimination to solve the equations

$$\begin{aligned} 2x_2 - x_3 &= -3 \\ 4x_1 - 3x_2 + 2x_3 &= 13 \\ -x_1 + 4x_2 &= -6 \end{aligned}$$

With the equations in their present form, the first equation cannot be used as a pivotal equation to eliminate x_1. Performing the elementary operation of type A, $E_1 \leftrightarrow E_2$ gives the equivalent system

$$\begin{aligned} 4x_1 - 3x_2 + 2x_3 &= 13 \\ 2x_2 - x_3 &= -3 \\ -x_1 + 4x_2 &= -6 \end{aligned}$$

Using the 'new' first equation as the pivotal equation, the elementary operation $E_3 + (\frac{1}{4})E_1$ gives

$$\begin{aligned} 4x_1 - 3x_2 + 2x_3 &= 13 \\ 2x_2 - x_3 &= -3 \\ \tfrac{13}{4}x_2 + \tfrac{1}{2}x_3 &= -\tfrac{11}{4} \end{aligned}$$

To produce integer coefficients in the third equation, we perform the elementary operation of type B, $4 \times E_3$. This gives

$$\begin{aligned} 4x_1 - 3x_2 + 2x_3 &= 13 \\ 2x_2 - x_3 &= -3 \\ 13x_2 + 2x_3 &= -11 \end{aligned}$$

Finally, to eliminate x_2 from the third equation, we use $E_3 + (-\frac{13}{2})E_2$. This gives

$$\begin{aligned} 4x_1 - 3x_2 + 2x_3 &= 13 \\ 2x_2 - x_3 &= -3 \\ \tfrac{17}{2}x_3 &= \tfrac{17}{2} \end{aligned}$$

Backward substitution then yields

from the third equation

$$x_3 = 1$$

from the second equation

$$\begin{aligned} 2x_2 - 1 &= -3 \\ 2x_2 &= -3 + 1 = -2 \\ x_2 &= -1 \end{aligned}$$

from the first equation

$$\begin{aligned} 4x_1 - 3(-1) + 2(1) &= 13 \\ 4x_1 &= 13 - 3 - 2 = 8 \\ x_1 &= 2 \end{aligned}$$

Hence the solution is $x_1 = 2, x_2 = -1, x_3 = 1$.

4.2 Gaussian elimination applied to an augmented matrix

In many cases, and certainly when using a computer, it is more convenient to use Gaussian elimination in a matrix setting.

For example, we may rewrite the system of equations

$$\begin{aligned} -x_1 + 2x_2 - x_3 &= 1 \\ 4x_1 + x_3 &= -5 \\ 4x_2 - 5x_3 &= 2 \end{aligned}$$

in the form

$$\begin{pmatrix} -1 & 2 & -1 \\ 4 & 0 & 1 \\ 0 & 4 & -5 \end{pmatrix} \begin{pmatrix} x_1 \\ x_2 \\ x_3 \end{pmatrix} = \begin{pmatrix} 1 \\ -5 \\ 2 \end{pmatrix}$$

The matrix

$$\begin{pmatrix} -1 & 2 & -1 \\ 4 & 0 & 1 \\ 0 & 4 & -5 \end{pmatrix}$$

is called the matrix of coefficients of the system (since the numbers appearing in the matrix are called the coefficients of the system). Since Gaussian elimination operates on the coefficients of a system of equations we may, with great economy, consider only the coefficients of the system and operate directly on the matrix:

$$\begin{pmatrix} -1 & 2 & -1 \\ 4 & 0 & 1 \\ 0 & 4 & -5 \end{pmatrix}$$

But the right-hand sides of the equations, stored in the matrix $\begin{pmatrix} 1 \\ -5 \\ 2 \end{pmatrix}$ are also involved in the calculations. An efficient way of writing, or storing, all the coefficients of the system is to produce the augmented matrix:

$$\begin{pmatrix} -1 & 2 & -1 & 1 \\ 4 & 0 & 1 & -5 \\ 0 & 4 & -5 & 2 \end{pmatrix}$$

where the right-hand sides of the equations are written as an extra column of the matrix. This warrants a definition.

Definition

The augmented matrix of a system of equations is the matrix of coefficients augmented by an extra column which consists of the numbers occurring on the right-hand sides of the equations.

Example 4.4

Write down the augmented matrix for the equations

$$\begin{aligned} 4x_1 - 3x_2 &= -2 \\ 3x_1 - 4x_3 &= 0 \\ 7x_2 &= -3 \end{aligned}$$

The matrix of coefficients for this system is

$$\begin{pmatrix} 4 & -3 & 0 \\ 3 & 0 & -4 \\ 0 & 7 & 0 \end{pmatrix}$$

When the numbers on the right-hand sides of the equations are added as an extra column, we obtain the augmented matrix

$$\begin{pmatrix} 4 & -3 & 0 & -2 \\ 3 & 0 & -4 & 0 \\ 0 & 7 & 0 & -3 \end{pmatrix}$$

To apply the method of Gaussian elimination to an augmented matrix, proceed exactly as before.

There are some minor changes of language, however, and for completeness, we restate the method of Gaussian elimination in a matrix setting. First, a definition.

Definition

An augmented matrix **M** is said to be in (upper) triangular form if **M** is the augmented matrix associated with a triangular system of equations.

For example, the equations

$$\begin{aligned} -3x_1 + 2x_3 &= -1 \\ 4x_2 - 5x_3 &= 2 \\ 6x_3 &= 12 \end{aligned}$$

form a triangular system of equations. Hence the associated augmented matrix

$$\begin{pmatrix} -3 & 0 & 2 & -1 \\ 0 & 4 & -5 & 2 \\ 0 & 0 & 6 & 12 \end{pmatrix}$$

is in (upper) triangular form.

Thus, in an upper triangular matrix, there is a triangle of zeros lying below the top-left to bottom-right diagonal of the matrix.

The aim of Gaussian elimination is to reduce the augmented matrix of a system to upper triangular form. Then, backward substitution is used on the associated triangular system of equations.

To achieve the reduction to upper triangular form, the three elementary operations A, B and C may be applied to the augmented matrix. These operations are now called **elementary row operations** but we shall continue to use the notation $E_3 + (-1)E_1$ to denote the operation of replacing the third row by the sum of the third row and $(-1) \times$ (first row). Also, instead of the pivotal equation, we shall now speak of the pivotal row. An example should make all this clear.

Example 4.5

Use Gaussian elimination applied to the augmented matrix to solve the system of equations

$$\begin{aligned} 2x_1 + 4x_2 - 3x_3 &= 20 \\ x_1 - 2x_2 + x_3 &= -11 \\ 4x_1 + 3x_2 - 2x_3 &= 12 \end{aligned}$$

The augmented matrix of the system is

$$\begin{pmatrix} 2 & 4 & -3 & 20 \\ 1 & -2 & 1 & -11 \\ 4 & 3 & -2 & 12 \end{pmatrix}$$

To sweep in the zeros in the first column, use row 1 as the pivotal row. Performing successively the elementary row operations

$$E_2 + (-\tfrac{1}{2})E_1;\ E_3 + (-2)E_1$$

produces $\begin{pmatrix} 2 & 4 & -3 & 20 \\ 0 & -4 & \frac{5}{2} & -21 \\ 0 & -5 & 4 & -28 \end{pmatrix}$

To produce a zero in the third row of the second column, use row 2 as the pivotal row. Performing the elementary row operation

$$E_3 + (-\tfrac{5}{4})E_2$$

produces $\begin{pmatrix} 2 & 4 & -3 & 20 \\ 0 & -4 & \frac{5}{2} & -21 \\ 0 & 0 & \frac{7}{8} & -\frac{7}{4} \end{pmatrix}$

The augmented matrix has thus been reduced to upper triangular form. The associated triangular system of equations is

$$\begin{aligned} 2x_1 + 4x_2 - 3x_3 &= 20 \\ -4x_2 + \tfrac{5}{2}x_3 &= -21 \\ \tfrac{7}{8}x_3 &= -\tfrac{7}{4} \end{aligned}$$

and backward substitution produces the required solution.

From the third equation

$$x_3 = -\tfrac{7}{4} \times \tfrac{8}{7} = -2$$

From the second equation

$$\begin{aligned} -4x_2 + \tfrac{5}{2} \times (-2) &= -21 \\ -4x_2 &= -21 + 5 = -16 \\ x_2 &= 4 \end{aligned}$$

From the first equation

$$\begin{aligned} 2x_1 + 4(4) - 3(-2) &= 20 \\ 2x_1 &= 20 - 16 - 6 = -2 \\ x_1 &= -1 \end{aligned}$$

Hence the solution is $\quad x_1 = -1,\ x_2 = 4,\ x_3 = -2.$

4.3 An algorithm for Gaussian elimination

It is anticipated that systems of equations will normally be solved with the aid of a computer. Hence we provide an algorithm for such a purpose.

Algorithm 4.1 Gaussian elimination

To solve the system of n equations in n unknowns

$$\begin{aligned} a_{11}x_1 + a_{12}x_2 + \ldots + a_{1n}x_n &= b_1 \\ a_{21}x_1 + a_{22}x_2 + \ldots + a_{2n}x_n &= b_2 \\ \ldots \quad \ldots \quad \ldots \quad \ldots \quad &\ \ldots \\ a_{n1}x_1 + a_{n2}x_2 + \ldots + a_{nn}x_n &= b_n \end{aligned}$$

Let **A** be the augmented matrix with n rows and $n + 1$ columns. Store the coefficients a_{ij} in $\mathbf{A}(i, j)$ $(1 \leqslant i, j \leqslant n)$ and store b_i in $\mathbf{A}(i, n+1)$ $(1 \leqslant i \leqslant n)$.

input: n and the augmented matrix of coefficients **A**
for $k = 1, 2, \ldots, n - 1$
  for $i = k + 1, k + 2, \ldots, n$

$$m := \frac{a_{ik}}{a_{kk}}$$

```
            for j = k + 1, k + 2, ..., n + 1
                a_ij := a_ij - m a_kj
            endloop
        endloop
    endloop
```

At this point the array **A** stores the required upper triangular matrix. Backward substitution will now yield the solution.

$$x_n := \frac{a_{n\,n+1}}{a_{nn}}$$

for $k = n - 1, n - 2, \ldots, 1$

$s := 0$

for $j = k + 1, k + 2, \ldots, n$

$s := a_{kj} x_j + s$

endloop

$$x_k := \frac{a_{k\,n+1} - s}{a_{kk}}$$

endloop

output: the solution $x_1, x_2, x_3, \ldots, x_n$

4.4 Sum check in Gaussian elimination

If a computer is not to be used to solve the system of equations, then it is probably a good idea to perform a sum check. This is a procedure which at each stage in a calculation performs a check on whether or not an arithmetic mistake has been made. The method consists of adding an extra column to the augmented matrix. In this column are stored the sums of the numbers in the corresponding row of the augmented matrix. When elementary row operations are performed, the operations are also performed on the extra column. If the operations have been performed accurately, the numbers in the extra column will continue to be the sums of the corresponding rows. If, however, a mistake has been made, then the numbers in the extra column will not, in general, be equal to the sum of the numbers in the rows. By observing such a discrepancy, it is possible to deduce not only that a mistake has been made but also where the error occurred. We illustrate with an example, in which an arithmetic error will be made.

Example 4.6

Using a sum check, solve the equations

$$\begin{aligned} 2x_1 + x_2 - x_3 &= 0 \\ x_1 \quad\quad - 3x_3 &= 1 \\ 4x_1 + 3x_2 \quad\quad &= 1 \end{aligned}$$

The augmented matrix for this system is

$$\begin{pmatrix} 2 & 1 & -1 & 0 \\ 1 & 0 & -3 & 1 \\ 4 & 3 & 0 & 1 \end{pmatrix}$$

Writing an extra column in which each entry gives the sum of the numbers in the corresponding row:

$$\left(\begin{array}{cccc|c} & & & & \text{Sum} \\ 2 & 1 & -1 & 0 & 2 \\ 1 & 0 & -3 & 1 & -1 \\ 4 & 3 & 0 & 1 & 8 \end{array}\right)$$

Now perform the operations of Gaussian elimination to reduce this matrix to upper triangular form. Using row 1 as the pivotal row

$$\begin{array}{l} \\ \\ E_2 + (-\tfrac{1}{2})E_1 \\ E_3 + (-2)E_1 \end{array} \left(\begin{array}{cccc|c} & & & & \text{Sum} \\ 2 & 1 & -1 & 0 & 2 \\ 0 & -\tfrac{1}{2} & -\tfrac{5}{2} & 1 & -2 \\ 0 & 1 & 2 & 1 & 4 \end{array}\right)$$

In row 1 the sum of the numbers in the augmented matrix is $2 + 1 + (-1) + 0 = 2$ and there is a 2 in the first row of the sum column. We see that in the other rows also the number in the sum column represents the sum of the numbers in the corresponding row. Hence, we are confident that no arithmetic errors have been made so far.

To produce integers in the second row, perform the elementary row operation $2E_2$

$$\left(\begin{array}{cccc|c} & & & & \text{Sum} \\ 2 & 1 & -1 & 0 & 2 \\ 0 & -1 & -5 & 2 & -4 \\ 0 & 1 & 2 & 1 & 4 \end{array}\right)$$

Using row 2 as the pivotal row

$$\begin{array}{l} \\ \\ \\ E_3 + (1)E_2 \end{array} \left(\begin{array}{cccc|c} & & & & \text{Sum} \\ 2 & 1 & -1 & 0 & 2 \\ 0 & -1 & -5 & 2 & -4 \\ 0 & 0 & 3 & 3 & 0 \end{array}\right)$$

We see that the numbers in the first two rows of the sum column do represent the sums of the numbers in the first two rows. But in the third row, this is not the case. The sum of the numbers in the third row of the augmented matrix is $0 + 0 + 3 + 3 = 6$ while the number in the sum column is 0. What has gone wrong?

In the third column of the third row, we have written +3; the correct entry should be –3. When this has been corrected, we have

$$\left(\begin{array}{cccc|c} & & & & \text{Sum} \\ 2 & 1 & -1 & 0 & 2 \\ 0 & -1 & -5 & 2 & -4 \\ 0 & 0 & -3 & 3 & 0 \end{array}\right)$$

and the entries in the sum column now represent the sums of the numbers in the corresponding rows.

It sometimes happens that the form of a system of equations leads to a simplified method of solution. We saw this with triangular systems at the beginning of the chapter. We now examine tridiagonal systems which arise naturally in several areas of numerical analysis.

4.5 Tridiagonal systems of equations

A tridiagonal system has the form

$$\begin{array}{lr} d_1x_1 + r_1x_2 & = b_1 \\ l_2x_1 + d_2x_2 + r_2x_3 & = b_2 \\ \quad l_3x_2 + d_3x_3 + r_3x_4 & = b_3 \\ \quad\quad \dots \quad \dots \quad \dots & \dots \\ \quad\quad\quad l_{n-1}x_{n-2} + d_{n-1}x_{n-1} + r_{n-1}x_n & = b_{n-1} \\ \quad\quad\quad\quad l_nx_{n-1} + d_nx_n & = b_n \end{array}$$

where in each equation, there is at most one term on either side of the diagonal. The diagonal terms are denoted by d_ix_i, those to the left by l_ix_{i-1} and those to the right by r_ix_{i+1}. The system below is an example of a tridiagonal system.

$$\begin{array}{lr} 2x_1 - x_2 & = 1 \\ -x_1 + 2x_2 - x_3 & = 0 \\ \quad 3x_2 + x_3 - x_4 & = 2 \\ \quad\quad 3x_3 + 2x_4 & = 7 \end{array}$$

To find the solution to a tridiagonal system, we use Gaussian elimination, but the form of a tridiagonal system allows the method to be considerably simplified. In fact, we need only clear the terms involving l_i to reduce the system to triangular form. We illustrate the method on the system of equations written above. The augmented matrix is

$$\begin{pmatrix} 2 & -1 & 0 & 0 & 1 \\ -1 & 2 & -1 & 0 & 0 \\ 0 & 3 & 1 & -1 & 2 \\ 0 & 0 & 3 & 2 & 7 \end{pmatrix}$$

Performing successively the elementary row operations

$$E_2 + (\tfrac{1}{2})E_1 \qquad \begin{pmatrix} 2 & -1 & 0 & 0 & 1 \\ 0 & \frac{3}{2} & -1 & 0 & \frac{1}{2} \\ 0 & 3 & 1 & -1 & 2 \\ 0 & 0 & 3 & 2 & 7 \end{pmatrix}$$

$$E_3 + (-2)E_2 \quad \begin{pmatrix} 2 & -1 & 0 & 0 & 1 \\ 0 & \frac{3}{2} & -1 & 0 & \frac{1}{2} \\ 0 & 0 & 3 & -1 & 1 \\ 0 & 0 & 3 & 2 & 7 \end{pmatrix}$$

$$E_4 + (-1)E_3 \quad \begin{pmatrix} 2 & -1 & 0 & 0 & 1 \\ 0 & \frac{3}{2} & -1 & 0 & \frac{1}{2} \\ 0 & 0 & 3 & -1 & 1 \\ 0 & 0 & 0 & 3 & 6 \end{pmatrix}$$

reduces the tridiagonal system to an equivalent triangular system. Backward substitution gives,

from the fourth equation

$$\begin{aligned} 3x_4 &= 6 & x_4 &= 2 \\ 3x_3 - 2 &= 1 & x_3 &= 1 \\ \tfrac{3}{2}x_2 - 1 &= \tfrac{1}{2} & x_2 &= 1 \\ 2x_1 - 1 &= 1 & x_1 &= 1 \end{aligned}$$

Hence the solution is $x_1 = 1; x_2 = 1; x_3 = 1; x_4 = 2.$

Algorithm 4.2 Solving a tridiagonal system

To solve the system of equations

$$\begin{aligned} d_1x_1 + r_1x_2 \qquad\qquad\qquad\qquad &= b_1 \\ l_2x_1 + d_2x_2 + r_2x_3 \qquad\qquad\quad &= b_2 \\ l_3x_2 + d_3x_3 + r_3x_4 \qquad\quad &= b_3 \\ l_4x_3 + d_4x_4 + r_4x_5 \quad &= b_4 \\ \dots \qquad \dots \qquad &\dots \\ l_nx_{n-1} + d_nx_n &= b_n \end{aligned}$$

input: $n; l_2, \dots, l_n; d_1, \dots, d_n; r_1, \dots, r_{n-1}; b_1, \dots, b_n$

Step 1

for $k = 2, 3, \dots, n$

if $d_{k-1} = 0$ then

print 'No solution': stop

endif

$m := \dfrac{l_k}{d_{k-1}}$

$d_k := d_k - m \times r_{k-1}$

$b_k := b_k - m \times b_{k-1}$

endloop

Step 2

if $d_n = 0$ then
print 'No solution': stop
endif

$$x_n := \frac{b_n}{d_n}$$

Step 3

for $k = n - 1, n - 2, ..., 1$

$$x_k := \frac{b_k - r_k \times x_{k+1}}{d_k}$$

endloop
output: $x_1, x_2, ..., x_n$

This completes the discussion of the method of Gaussian elimination. We have seen how Gaussian elimination can easily be programed for implementation on a computer and, in the following two sections, we shall examine some matters relating to the accuracy of the solution.

4.6 Errors using Gaussian elimination

Consider the system of equations

$$0.0003x_1 + 1.313x_2 = 1.316$$
$$0.451x_1 - 2.381x_2 = 2.129$$

We shall solve these equations working throughout to rounded four decimal digit accuracy. Proceeding as above, we take the first equation to be the pivotal equation and use the elementary operation $E_2 + \left(-\dfrac{0.451}{0.0003}\right)E_1$ to eliminate x_1 from the second equation.

Observing that $\dfrac{0.451}{0.0003} = 1503$, when written to four decimal digit accuracy, we arrive at the equations:

$$0.0003x_1 \quad + \quad 1.313x_2 = 1.316$$
$$[-2.381 + (-1503) \times 1.313]x_2 = 2.129 + (-1503) \times 1.316$$

To four digit accuracy,

$$1503 \times 1.313 = 1973 \quad \text{and} \quad 1503 \times 1.316 = 1978$$

Hence, the second equation becomes

$$[-2.381 - 1973]x_2 = 2.129 - 1978$$

or

$$[-1975]x_2 = -1976$$

when the coefficients are written to four digit accuracy.

Back substitution gives

$$x_2 = \frac{-1976}{-1975}$$
$$= 1.001$$

using four decimal digit accuracy.

And $\quad 0.0003x_1 + 1.313 \times 1.001 = 1.316$

Since $\quad 1.313 \times 1.001 = 1.314$

to four digit accuracy, we have

$$x_1 = \frac{1.316 - 1.314}{0.0003}$$
$$= \frac{0.002}{0.0003}$$
$$= 6.667 \quad \text{(to four digit accuracy)}$$

Hence the solution (with all calculations rounded to four decimal digits) is $x_1 = 6.667$, $x_2 = 1.001$.

Sadly, this is simply not true. The correct solution is $x_1 = 10$, $x_2 = 1$ as may be seen by direct substitution into the equations.

This is alarming – to say the least. A valid method of solution, when applied to one of the simplest systems imaginable, has produced a wildly inaccurate solution. How could this inaccuracy have been prevented?

The reader may believe (correctly) that the problem lies with the decimal 0.0003. Perhaps one way out of the difficulty is to remove this small decimal from the first equation by using an elementary operation of type B and multiplying the first equation by 10^4. This would give

$$\begin{array}{ll} 10^4\mathrm{E}_1 & 3x_1 + 13130x_2 = 13160 \\ & 0.451x_1 - 2.381x_2 = 2.129 \end{array}$$

Proceeding as before: perform the elementary operation $\mathrm{E}_2 + \left(-\frac{0.451}{3}\right)\mathrm{E}_1$. But to four decimal digits $\frac{0.451}{3} = 0.1503$ so working to this level of accuracy produces the equations

$$\begin{array}{rl} 3x_1 + & 13130x_2 = 13160 \\ & [-2.381 + (-0.1503) \times 13130]x_2 = 2.129 + (-0.1503) \times 13160 \end{array}$$

$$\text{Or,} \quad \begin{array}{rl} 3x_1 + & 13130x_2 = 13160 \\ & -1975x_2 = -1976 \end{array}$$

to four decimal digit accuracy.

Backward substitution gives

$$x_2 = \frac{-1976}{-1975} = 1.001 \quad \text{(to four decimal digit accuracy)}$$

and $3x_1 + 13\,130 \times 1.001 = 13\,160$

So, $x_1 = \dfrac{13\,160 - 13\,140}{3}$

giving $x_1 = \dfrac{20}{3} = 6.667$

to four decimal digit accuracy.
This is exactly the same (inaccurate) answer that we obtained previously and the strategy of multiplying the first equation by 10^4 has not helped in the least.

It is not the smallness of the decimal 0.0003 that is causing the inaccuracy in the value of x_1. Rather it is the fact that 0.0003 is small relative to the coefficient of x_2, 1.313. To see how the inaccuracy arises, consider the equation from which the value of x_1 is calculated

$$0.0003x_1 + 1.313x_2 = 1.316$$

Write $x_2 = 1 + e$

where e represents the error in the calculated value of x_2.

The above equation then becomes

$$0.0003x_1 + 1.313(1 + e) = 1.316$$
$$0.0003x_1 = 1.316 - 1.313 - 1.313e$$
$$0.0003x_1 = 0.003 - 1.313e$$
$$x_1 = \frac{0.003}{0.0003} - \frac{1.313e}{0.0003}$$
$$= 10 - \frac{1.313e}{0.0003}$$

and it may be seen that $-\dfrac{1.313e}{0.0003}$ represents the error in the calculated value of x_1.
But $e = 0.001$, therefore

$$\text{error in } x_1 = -\frac{1.313e}{0.0003} = -4376.666\,667 \times 0.001$$
$$= -4.376\,666\,667 \quad \text{to ten decimal digits}$$

Hence, division of the coefficient of x_2 by the relatively small number 0.0003 has caused the small error in x_2 to be exaggerated dramatically. In fact the error in x_2 (0.001) has been increased by a factor of more than 4000.

Note, however, that the error was not quite so large in the original calculation of x_1. There, the rounding to four decimal digits had the effect of reducing the magnitude of the error in x_1.

One way of preventing such a proliferation of error is to use the second equation as the pivotal equation. This involves an elementary operation of type A, $E_1 \leftrightarrow E_2$ and we write

$$0.451x_1 - 2.381x_2 = 2.129$$
$$0.0003x_1 + 1.313x_2 = 1.316$$

Solving these equations in the usual way, and working again to four decimal digit accuracy

$$0.451x_1 - 2.381x_2 = 2.129$$
$$1.315x_2 = 1.315$$

Using backward substitution

$$x_2 = \frac{1.315}{1.315} = 1$$
$$0.451x_1 - 2.381 \times 1 = 2.129$$
$$0.451x_1 = 2.129 + 2.381 = 4.51$$
$$x_1 = \frac{4.51}{0.451} = 10$$

Hence $x_1 = 10$, $x_2 = 1$ and the problem appears to be solved.

This example illustrates that Gaussian elimination as we have so far described it can fail to produce an accurate solution. The example illustrates also that the accuracy can be improved considerably by choosing the pivotal equation carefully. Techniques that indicate which equation is to be chosen as the pivotal equation are known as **pivoting strategies**. We now describe in detail one such pivoting strategy.

4.7 Pivoting strategy

The aim of Gaussian elimination, as we have seen, is to eliminate x_1 from the second and all subsequent equations, to eliminate x_2 from the third and all subsequent equations and so on. Consider the position when x_i is about to be eliminated from the $(i+1)$th and all subsequent equations.

For example: in the system shown below, we are about to eliminate x_3 from the fourth and all subsequent equations.

$$\begin{aligned}
3x_1 - 2x_2 + 3x_3 - x_4 + x_5 &= 1\\
2x_2 - 2x_3 + 5x_4 - x_5 &= -3\\
-x_3 - 3x_4 + 2x_5 &= 4\\
4x_3 + 2x_4 - x_5 &= 6\\
-7x_3 - 3x_4 + 3x_5 &= -1
\end{aligned}$$

One of the simplest pivotal strategies is to examine the ith and all subsequent equations and select as the pivotal equation that equation having the largest (absolute) coefficient of x_i.

In the example, examine the third and all subsequent equations and select the equation having the largest (absolute) coefficient of x_3. The coefficient of x_3 in the third, fourth and fifth equations are –1, 4 and –7. The largest of these, in absolute terms, is –7, so the fifth equation is chosen to be the pivotal equation.

Now perform an elementary operation of type A and interchange the ith equation and the pivotal equation. Elementary operations are now used, as described above, to eliminate x_i from the $(i + 1)$ and all subsequent equations and the entire process is then repeated with x_{i+1}.

In the above system, interchange the third and fifth equations to produce the following equivalent system

$$\begin{aligned}
3x_1 - 2x_2 + 3x_3 - \;\; x_4 + \;\; x_5 &= \;\; 1 \\
2x_2 - 2x_3 + 5x_4 - \;\; x_5 &= -3 \\
-7x_3 - 3x_4 + 3x_5 &= -1 \\
4x_3 + 2x_4 - \;\; x_5 &= \;\; 6 \\
-x_3 - 3x_4 + 2x_5 &= \;\; 4
\end{aligned}$$

Elementary operations are now used to eliminate x_3 from the fourth and fifth equations; the process is then repeated with x_4.

This particular pivoting strategy is called **partial pivoting**.

It will be seen that partial pivoting was used to achieve an accurate solution to the system of two equations with which we began the previous section. It may be necessary, however, to use a pivoting strategy in rather more ordinary circumstances as the following example shows.

Example 4.7

Solve the system of equations

$$\begin{aligned}
x + 2y + 3z &= \;\; 8 \\
x + 2y + 4z &= 11 \\
2x + 3y + \;\; z &= \;\; 2
\end{aligned}$$

(a) If pivoting is not used then the solution would be as follows. Using equation 1 as the pivotal equation:

$$\begin{array}{ll}
 & x + 2y + 3z = \;\; 8 \\
E_2 + (-1)E_1 & \qquad\qquad z = \;\; 3 \\
E_3 + (-2)E_1 & \quad -y - 5z = -14
\end{array}$$

But now it is clear that the second equation cannot be used as the pivotal equation to eliminate y from the third equation. If this system of equations were being solved using a computer (using Algorithm 4.1) then a 'division by zero' error would be produced (why?) and the program would halt. Since this is not a desirable state of affairs we consider how it might have been prevented.

(b) If pivoting is used, it is obvious that the third equation has the largest (absolute) coefficient of x. Making the third equation the pivotal equation involves the elementary operation $E_1 \leftrightarrow E_3$

$$\begin{array}{ll}
E_1 \leftrightarrow E_3 & 2x + 3y + \;\; z = \;\; 2 \\
 & \;\; x + 2y + 4z = 11 \\
 & \;\; x + 2y + 3z = \;\; 8
\end{array}$$

Performing successively $E_2 + (-\frac{1}{2})E_1$, $E_3 + (-\frac{1}{2})E_1$ gives

$$\begin{aligned}
2x + \;\; 3y + \qquad z &= \;\; 2 \\
0.5y + 3.5z &= 10 \\
0.5y + 2.5z &= \;\; 7
\end{aligned}$$

But now the coefficients of y and z in the second and third equations are equal. Hence we might use either equation as the pivotal equation. There seems little point in performing an elementary operation just for the sake of it and so we perform $E_3 + (-1)E_2$ to get

$$\begin{aligned} 2x + 3y + z &= 2 \\ 0.5y + 3.5z &= 10 \\ -z &= -3 \end{aligned}$$

Backward substitution gives the solution $z = 3$, $y = -1$, $x = 1$. This example illustrates the need to modify Algorithm 4.1 to include a pivoting strategy. The following amended version of Algorithm 4.1 achieves this.

Algorithm 4.3 Gaussian elimination with a pivoting strategy

```
input: n and the augmented matrix of coefficient A
for k = 1, 2, ..., n − 1
    w := 0
    for r = k, k + 1, ..., n
        if abs(a_rk) > w then
            w := abs(a_rk) and s := r
        endif
    endloop
    if s > k then
        exchange rows s and k
    endif
    for i = k + 1, k + 2, ..., n
        m := a_ik / a_kk
        for j = k + 1, k + 2, ..., n + 1
            a_ij = a_ij − m a_kj
        endloop
    endloop
endloop
x_n := a_{n n+1} / a_nn
for k = n − 1, n − 2, ..., 1
    s := 0
    for j = k + 1, k + 2, ..., n
        s := a_kj x_j + s
    endloop
    x_k := (a_{k n+1} − s) / a_kk
endloop
output: the solution x_1, x_2, x_3, ..., x_n
```

Although partial pivoting is satisfactory for most systems, it is not difficult to produce a system in which partial pivoting will fail to produce an accurate solution. Many other pivoting strategies have been proposed. Unfortunately, none has proved wholly satisfactory and, under certain conditions, all will fail to produce an accurate solution. The 'perfect' pivoting strategy has yet to be discovered†.

Before leaving this topic, we should mention that a pivoting strategy is unnecessary if all calculations are performed with perfect accuracy. It is the effect of roundoff error that causes the great inaccuracy and if all calculations are exact (as in the systems considered earlier in the chapter) then there is no roundoff error and usually no need for pivoting. It should be remembered though that if Gaussian elimination is to be performed on a computer then roundoff errors will occur and a pivoting strategy should be used.

4.8 Ill-conditioned systems

A second and potentially more dangerous situation is illustrated in the example below.

Example 4.8

Solve the equations:

$$\begin{aligned} 5x + y &= 3 \\ 2x + 0.39y &= -8 \end{aligned}$$

Gaussian elimination produces

$$\begin{aligned} & 5x + y = 3 \\ E_2 + (-\tfrac{2}{5})E_1 \qquad & -0.01y = -9.2 \end{aligned}$$

Backward substitution gives

$$y = \frac{-9.2}{-0.01} = 920$$

$$\begin{aligned} 5x + 920 &= 3 \\ 5x &= -917 \\ x &= \frac{-917}{5} = -183.4 \end{aligned}$$

Hence $x = -183.4$ and $y = 920$.

In this example, all the calculations were carried out exactly: there was no rounding and the solution $x = -183.4$, $y = 920$ is exact. In the absence of further information, however, we may wonder whether the decimal 0.39 is exact or whether it is the result of rounding to two decimal digits. If the latter, then the true value of the number represented by 0.39 must lie between 0.385 and 0.395. If the 'real' solution is the solution corresponding to the true value of this number, then it is reasonable to

† For a wider discussion of this topic, see Conte and de Boor 1981 *Elementary numerical analysis* 3rd edn. McGraw-Hill.

assume that we can calculate bounds on the real solution by considering the two systems of equations

$$\begin{aligned} 5x + \quad y &= 3 \\ 2x + 0.385y &= -8 \end{aligned} \qquad \begin{aligned} 5x + \quad y &= 3 \\ 2x + 0.395y &= -8 \end{aligned}$$

Solving these systems:

$$\begin{array}{llll} & 5x + \quad y = 3 & & 5x + \quad y = 3 \\ E_2 + (-\tfrac{2}{5})E_1 & -0.015y = -9.2 & & -0.005y = -9.2 \\ \text{Hence} & y = \dfrac{-9.2}{-0.015} & & y = \dfrac{-9.2}{-0.005} \\ & \quad = 613.33 & & \quad = 1840 \end{array}$$

and backward substitution gives

$$\begin{array}{llll} & 5x + 613.33 = 3 & & 5x + 1840 = 3 \\ x & = \dfrac{-610.33}{5} & x & = \dfrac{-1837}{5} \\ & = -122.07 & & = -367.4 \end{array}$$

(to five decimal digits)

But this is appalling! An adjustment of 0.005 to one of the coefficients of the system has changed the solution

from	$x = -183.4$,	$y = 920$	
to	$x = -367.4$,	$y = 1840$	(if 0.005 is added)
and to	$x = -122.07$,	$y = 613.33$	(if 0.005 is subtracted)

A system in which this kind of phenomenon occurs is said to be **ill-conditioned.** More precisely, a system is ill-conditioned if a small change in the initial data (in this case the coefficients) results in a large change in the solution. The condition of a system describes the sensitivity of the system to changes in the initial data.

This is very relevant in practical work where much of the data is obtained by measurement or is subject to roundoff error. In either case, the data will be known only approximately and the question of how a system will respond to small changes in the initial data is of real importance. The concept of condition and the problem of ill-conditioning are certainly worthy of further study†.

But now we must attempt to explain why the system

$$\begin{aligned} 5x + \quad y &= 3 \\ 2x + 0.39y &= -8 \end{aligned}$$

is so badly ill-conditioned.

The situation may be illustrated by drawing the graphs of $5x + y = 3$ and $2x + 0.39y = -8$.

† For a wider discussion of this fascinating topic, the reader is referred, once again to Conte and de Boor 1981 *Elementary numerical analysis* 3rd edn. McGraw-Hill, Chapters 4 and 6.

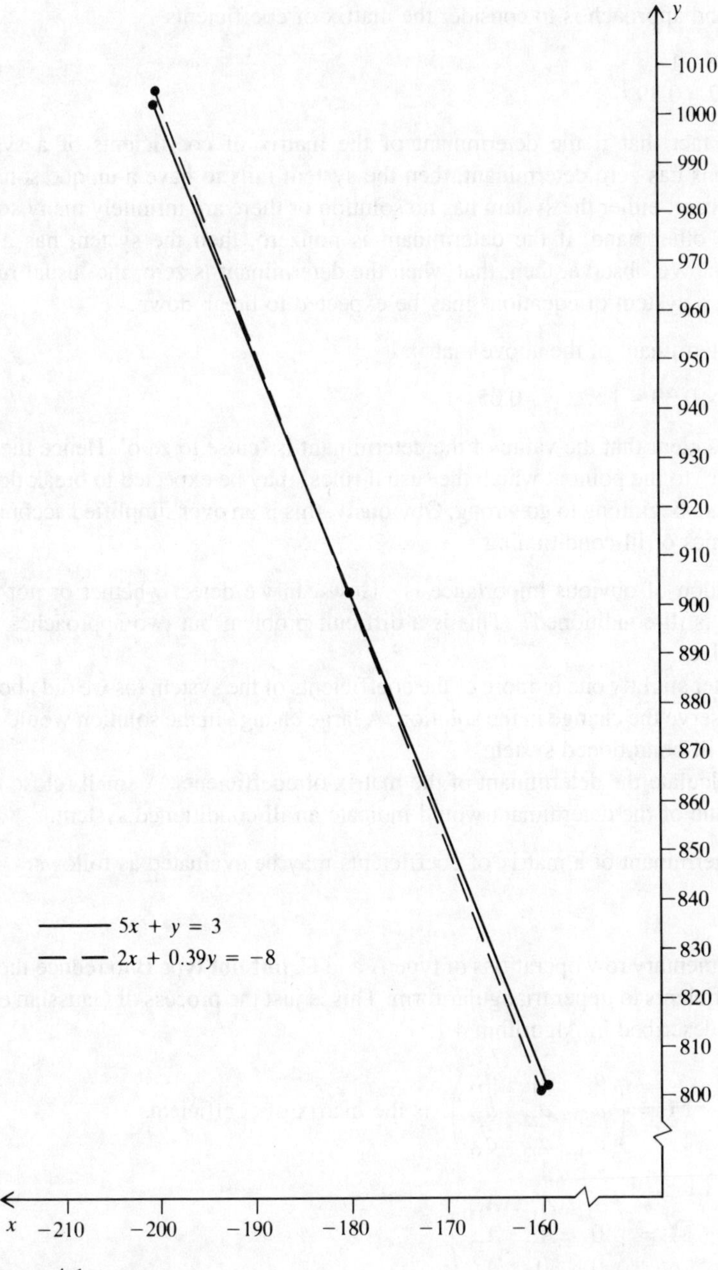

Figure 4.1

It is clear that in a neighbourhood of the point of intersection, the two lines lie very close together. Hence, a small change in the line representing the equation $2x + 0.39y = -8$ could cause a considerable shift in the point where the lines intersect. This is what was observed above.

A second approach is to consider the matrix of coefficients

$$\begin{pmatrix} 5 & 1 \\ 2 & 0.39 \end{pmatrix}$$

It is a fact that if the determinant of the matrix of coefficients of a system of equations has zero determinant, then the system fails to have a unique solution. In such a case, either the system has no solution or there are infinitely many solutions. On the other hand, if the determinant is nonzero, then the system has a unique solution. We observe, then, that when the determinant is zero, the 'usual rules' for solving a system of equations may be expected to break down.

The determinant of the above matrix is

$$5 \times 0.39 - 1 \times 2 = -0.05$$

and it is clear that the value of the determinant is 'close to zero'. Hence the system is 'close' to the point at which the 'usual rules' may be expected to break down and things are beginning to go wrong. Obviously, this is an over simplified account of the mechanics of ill-conditioning†.

A question of obvious importance is, 'How can we detect whether or not a given system is ill-conditioned?' This is a difficult problem but two approaches suggest themselves.

(a) Alter slightly one or more of the coefficients of the system (as we did above) and observe the change in the solution. A large change in the solution would suggest an ill-conditioned system.

(b) Calculate the determinant of the matrix of coefficients. A small (close to zero) value of the determinant would indicate an ill-conditioned system.

The determinant of a matrix of coefficients may be evaluated as follows:

Step 1

Use elementary row operations of type A and C, **but not** type B to reduce the matrix of coefficients to upper triangular form. This is just the process of Gaussian elimination as described in Algorithm 4.1.

So if $\mathbf{M} = \begin{pmatrix} a_{11} & a_{12} & a_{13} \\ a_{21} & a_{22} & a_{23} \\ a_{31} & a_{32} & a_{33} \end{pmatrix}$ is the matrix of coefficients

let $\hat{\mathbf{M}} = \begin{pmatrix} A_{11} & A_{12} & A_{13} \\ 0 & A_{22} & A_{23} \\ 0 & 0 & A_{33} \end{pmatrix}$

be the upper triangular matrix obtained from **M** by row operations of type A and C. Note that an elementary row operation of type B would multiply a row of a matrix by some number. This is perfectly acceptable when dealing with an augmented

† The reader is referred, once again, to Conte and de Boor 1981 *Elementary numerical analysis* 3rd edn. McGraw-Hill, for a rigorous description.

matrix and solving a system of equations: it is not permissible when evaluating the determinant of a matrix.

Step 2

Observe the numbers on the upper left-lower right diagonal of $\hat{\mathbf{M}}$ (called the leading diagonal). Then the numerical value of the determinant of $\mathbf{M}$ is given by $A_{11} \times A_{22} \times A_{33}$. (Note that $A_{11} \times A_{22} \times A_{33}$ may not give the correct sign of the determinant but, since we are interested only in whether or not the determinant is close to zero, this causes us no problems.)

Observe that if any of the numbers A_{11}, A_{22}, A_{33} are small (close to zero), then the system is likely to be ill-conditioned.

In the case of the ill-conditioned system

$$\begin{aligned} 5x + \quad y &= 3 \\ 2x + 0.39y &= -8 \end{aligned}$$

$$\mathbf{M} = \begin{pmatrix} 5 & 1 \\ 2 & 0.39 \end{pmatrix}$$

and $$\hat{\mathbf{M}} = \begin{pmatrix} 5 & 1 \\ 0 & -0.01 \end{pmatrix}$$

The numerical value of the determinant of $\mathbf{M}$ is given by $5 \times (-0.01) = -0.05$. In this case, the sign of the determinant is correct and we have the result observed above. Observe also the small number -0.01 on the leading diagonal of $\hat{\mathbf{M}}$.

Unfortunately, there is very little that can be done about an ill-conditioned system. There is no easy solution such as we found in the last section with pivoting. Ill-conditioning is a property of the system of equations and as such is quite independent of the method of solution. We can not remove ill-conditioning in a particular system; we simply have to accept that such systems do exist. Forewarned is forearmed.

Exercise 4

Section 4.1

1 Which of the following are triangular systems of equations?

(a) $$\begin{aligned} 4x - 2z &= 4 \\ 2y + z &= -1 \\ z &= 0 \end{aligned}$$

(b) $$\begin{aligned} y - 3z &= 0 \\ z + 2y &= -1 \\ 4z &= 2 \end{aligned}$$

(c) $$\begin{aligned} x_2 + x_3 &= -2 \\ x_1 - x_3 + x_2 &= 1 \\ x_3 - 2x_2 &= 4 \end{aligned}$$

(d) $$\begin{aligned} 4x_3 - 3x_2 + x_1 &= 1 \\ x_3 + 4x_2 &= 5 \\ 3x_1 - x_2 &= 4 \end{aligned}$$

(e) $$\begin{aligned} 6x_1 + 2x_2 - x_3 &= 5 \\ x_3 &= 8 \\ x_3 - 4x_2 &= -1 \end{aligned}$$

2 Solve the following triangular system of equations:

$$\begin{aligned} 4x + 2y - z &= 1 \\ 3y + z &= -8 \\ 5z &= 5 \end{aligned}$$

3 Solve the triangular system of equations:

$$\begin{aligned} 2x - y + 3z &= 9 \\ 4y - 2z &= -12 \\ 5z &= 20 \end{aligned}$$

4 Solve the triangular system of equations:

$$\begin{aligned} x + 7z &= -10 \\ 2y - 5z &= 9 \\ 6z &= -6 \end{aligned}$$

5 Solve the system of equations:

$$\begin{aligned} 5b - 3c &= -1 \\ 2a - 3b - c &= 7 \\ 7c &= -21 \end{aligned}$$

6 Solve the equations:

$$4x_2 + 5x_3 = -13, \qquad -2x_3 = 2, \qquad 6x_1 - x_3 = 13$$

7 Explain why the solution of a system of equations will be unaltered when the three elementary row operations described in Section 4.1 are applied to the system.

8 Use Gaussian elimination to solve the system of equations:

$$\begin{aligned} x - y + 4z &= 11 \\ 3x + 2y - 3z &= 8 \\ 2x - 5y + 2z &= 7 \end{aligned}$$

9 Solve the system of equations using Gaussian elimination:

$$\begin{aligned} 3x_1 + x_2 - 4x_3 &= 3 \\ 2x_1 + 3x_2 + 5x_3 &= -4 \\ x_1 - 4x_2 - 3x_3 &= -13 \end{aligned}$$

10 Find the value of a, b, and c in the following:

$$\begin{aligned} 4a - 2b \qquad &= -2 \\ 6a \qquad + 3c &= 7.5 \\ 5b - 2c &= 21 \end{aligned}$$

11 Solve the equations below for p, q and r.

$$\begin{aligned} 4p + 3q - r &= 4 \\ 5p + 2q \qquad &= 3 \\ -4q + r &= 1 \end{aligned}$$

12 Solve the system of equations shown below.

$$\begin{aligned} 3x - 2y + 3z - 4w &= 22 \\ 2x + 3y - 2z + w &= -1 \\ 5x + 3z - w &= 20 \\ 6y - z + 2w &= -11 \end{aligned}$$

13 Find a, b, c and d in the equations below.

$$\begin{aligned} 5a - 2b - d &= 12 \\ 7a + 3b - c &= 30.5 \\ 4b - 3d &= 8.5 \\ 2a + 7c + 2d &= 18.5 \end{aligned}$$

Section 4.4

14 Solve the system of equations given below. Use and display a sum check.

$$\begin{aligned} -2x + 3y - z &= -16 \\ x + 4y + 3z &= 7 \\ 3x - 2z &= 1 \end{aligned}$$

15 Using a sum check, solve the system of equations:

$$\begin{aligned} 3x_1 + 4x_3 &= -22 \\ 2x_1 - x_2 + 3x_3 &= -13 \\ -2x_1 + 4x_2 - x_3 &= -4 \end{aligned}$$

16 Solve the system of equations shown below. Use a sum check to help with the accuracy.

$$\begin{aligned} 3x - y + 2z + w &= 17 \\ 2x + 3y - z + 2w &= 3 \\ -y + 5z - 4w &= 11 \\ -x + 2z - 3w &= 0 \end{aligned}$$

Section 4.5

17 Explain what is meant by a tridiagonal system. Solve the system of equations given below.

$$\begin{aligned} 3a - 2b &= 12 \\ a + 3b - 2c &= 8 \\ 2b + 3c - d &= -7 \\ -c + 5d &= 7 \end{aligned}$$

18 Explain why a tridiagonal system of equations is easier to solve than a general system of linear equations. Solve the system of equations given below.

$$\begin{aligned} 6x_1 + x_2 &= 19 \\ -2x_1 + 3x_2 + x_3 &= -4 \\ x_2 - 2x_3 + x_4 &= 4 \\ 5x_3 - 3x_4 &= -8 \end{aligned}$$

Section 4.6

19 Solve the equations:

$$0.001x + 5.45y = 2.188$$
$$0.68x - 1.37y = 4.892$$

(a) working in four decimal digit arithmetic using chopping,
(b) working as accurately as your calculator/computer will permit,
(c) using a pivoting strategy.

Comment on your solutions.

20 Solve the system of equations:

$$0.0004a + 1.312b = 1.972$$
$$0.4561a - 1.894b = 1.72$$

(a) using Gaussian elimination without pivoting,
(b) using pivoting.

In both cases, work entirely in four decimal digit arithmetic (using chopping).

21 Solve the system of equations

$$0.001\,42x_1 + 0.040\,03x_2 = 0.011\,42$$
$$0.21x_1 - 4.909x_2 = 1.417$$

using Gaussian elimination without pivoting and working in a four decimal digit system using chopping.

Now solve the system using pivoting. Comment on your results.

Section 4.8

22 In the system of equations

$$3x + ay = 2$$
$$7x + 5y = 5$$

the value of a has to be estimated. To three decimal digits (rounded) $a = 2.15$. Investigate this system of equations by considering the limits within which the solution must lie.

23 Solve the system of equations:

$$2x - y = 2$$
$$6x - 3.06y = -4$$

The coefficient of y in the second equation (–3.06) is written correct to three decimal digits (using chopping). By considering the range of values within which the exact value of this coefficient must lie, investigate whether or not the system is ill-conditioned.

24 (a) Solve the equations:

$$5x + 6y = 3$$
$$3\tfrac{1}{3}x + 4y = -1$$

using Gaussian elimination and working in six digit chopped arithmetic. Comment on the accuracy of your solution.

(b) Now multiply the second equation by 3 and solve the resulting system. Comment on your answers.

25 By using either of the techniques suggested in Section 4.8, decide which of the following systems are ill-conditioned.

(a)
$$\begin{aligned} 7x + 4.1y &= 2.3 \\ -2.9x - 1.7y &= 1.8 \end{aligned}$$

(b)
$$\begin{aligned} 5.2x + 3.16y - 0.9z &= 4.3 \\ 21.32x + 12.97y + 2.46z &= -2.8 \\ 10.92x + 6.84y + 11.28z &= 16.7 \end{aligned}$$

(c)
$$\begin{aligned} 3a + 8.2b &= 5 \\ 1.6a + 3.05b &= 1.2 \end{aligned}$$

(d)
$$\begin{aligned} 4x + 2.1y - 5.75z &= 14.8 \\ -x + 4.7y - 5z &= 0 \\ 5.1x - y - 2.8z &= -3.2 \end{aligned}$$

(e)
$$\begin{aligned} 3x - 2.2y + 4.5z &= 16.2 \\ -2x + 3.7y + 1.98z &= -5.3 \\ x + 2.8y - 3.82z &= 28 \end{aligned}$$

26 Define what is meant by an ill-conditioned system of equations. Is the following system ill-conditioned?

$$\begin{aligned} 3.01x - 1.05y + 2.49z &= -1.61 \\ 4.33x + 0.56y - 1.78z &= 7.23 \\ -0.83x - 0.54y + 1.47z &= -3.38 \end{aligned}$$

27 Determine whether or not the following system of equations is ill-conditioned.

$$\begin{aligned} 5.16x - 0.17y + 2.49z &= 1.67 \\ 2.58x + 3.135y + 0.275z &= -0.87 \\ 7.74x + 12.625y - 0.1447z &= 2.14 \end{aligned}$$

Miscellaneous

28 Solve the equations

$$\begin{aligned} 1.1x + 0.4y + 0.3z &= 0.2 \\ 0.6x + 0.8y + 0.1z &= 1.0 \\ 1.0x + 1.0y + 2.0z &= 0.5 \end{aligned}$$

correct to four decimal places. (MEI adapted)

29 Use a pivoting strategy to solve the system of equations:

$$\begin{aligned} 4x_1 + x_2 + x_3 &= 1 \\ x_1 + 3x_2 + x_3 &= 1 \\ 2x_1 + x_2 + 3x_3 &= 2 \end{aligned}$$

Give your solution to three places of decimals. Show what checking procedure can be applied to your results.

(MEI adapted)

30 Use a pivoting strategy to solve the system of equations:

$$\begin{aligned} 6x_1 + 2x_2 + x_3 &= 1 \\ 2x_1 + 5x_2 + x_3 &= 1 \\ 3x_1 + 2x_2 + 5x_3 &= 2 \end{aligned}$$

Give your solutions to three places of decimals. (MEI adapted)

31 Use pivoting and a sum check to solve the equations:

$$\begin{aligned} 1.32x \qquad\qquad - 2.71z &= 2.43 \\ 2.11x + 1.78y \qquad\qquad &= 1.15 \\ 0.51x - 0.21y + 0.76z &= 0.83 \end{aligned}$$

(MEI)

32 Use the method of pivotal condensation, with a sum check, or any other suitable method, to find the values of x_1, x_2, x_3, from the following simultaneous equations. Tabulate the details of your working, and give your results to three significant figures. (You may use a calculator.)

$$\begin{aligned} 9.37x_1 + 3.04x_2 - 2.44x_3 &= 9.23 \\ 3.04x_1 + 6.18x_2 + 1.22x_3 &= 8.20 \\ -2.44x_1 + 1.22x_2 + 8.44x_3 &= 3.93 \end{aligned}$$

(MEI)

33 Describe a numerical method for solving a system of linear algebraic equations. Illustrate the description by solving:

$$\begin{aligned} 4.3x + 0.9y - 2.1z &= 2.72 \\ 1.8x + 5.2y - 0.8z &= -0.89 \\ -1.2x + 2.6y + 5.9z &= 2.14 \end{aligned}$$

correct to three decimal places.
Explain briefly what is meant by a system of equations being ill-conditioned.

(MEI)

34 (a) Find, using pivotal condensation with a checking procedure, the values of x_1, x_2, x_3 correct to three decimal places, where

$$\begin{aligned} 12x_1 - 2x_2 + 3x_3 &= -6 \\ 10x_1 - 80x_2 + 5x_3 &= -15 \\ 3x_1 + x_2 - 15x_3 &= 4 \end{aligned}$$

(b) By considering the solutions of the following two systems of simultaneous equations, explain what is meant by 'ill-conditioning' of a system of linear equations, and give a geometrical interpretation.

System 1:

$$\begin{aligned} x_1 + 10x_2 &= 11 \\ 10x_1 + 101x_2 &= 111 \end{aligned}$$

System 2:

$$\begin{aligned} x_1 + 10x_2 &= 11 \\ 10.1x_1 + 100x_2 &= 111 \end{aligned}$$

(MEI)

35 A solution of the following system of equations is required:

$$\begin{pmatrix} 0.60 & 0.80 & 0.10 \\ 1.10 & 0.40 & 0.30 \\ 1.71 & 1.20 & 0.41 \end{pmatrix} \begin{pmatrix} x \\ y \\ z \end{pmatrix} = \begin{pmatrix} 1.00 \\ 0.20 \\ 1.20 \end{pmatrix}$$

Find the determinant of the matrix and comment upon the result with reference to the solution of the system of equations.
Solve the system of equations correct to two decimal places. (MEI)

36 Use the method of pivotal condensation to find x_1, x_2 and x_3 in terms of b_1, b_2, and b_3, where

$$\mathbf{A}\begin{pmatrix} x_1 \\ x_2 \\ x_3 \end{pmatrix} = \begin{pmatrix} b_1 \\ b_2 \\ b_3 \end{pmatrix} \quad \text{and} \quad \mathbf{A} = \begin{pmatrix} 6.3 & -3.2 & 1.0 \\ -3.2 & 8.4 & -2.6 \\ 1.0 & -2.6 & 5.7 \end{pmatrix}$$

giving your answers to three decimal places.
Hence find an approximation to $\mathbf{A}^{-1}$.

Also show that if $\mathbf{B} = \mathbf{A} + \begin{pmatrix} 0 & 0 & 0 \\ 0 & 0 & 0 \\ 0 & 0 & 0.1 \end{pmatrix}$ then an approximate value of

$\mathbf{B}^{-1}$ is $\mathbf{A}^{-1} - \begin{pmatrix} 0 & 0 & 0 \\ 0 & 0 & 0.001 \\ 0 & 0.001 & 0.004 \end{pmatrix}$ (MEI)

37 Solve the following system of equations by any method that seems to you to be suitable. Perform all calculations in
(a) two digit arithmetic using rounding
(b) three digit arithmetic using rounding
(c) the most accurate arithmetic your computer/calculator will allow.

$$\begin{aligned} 0.24x + 0.36y + 0.12z &= 0.84 \\ 0.12x + 0.17y + 0.25z &= 0.54 \\ 0.13x + 0.23y + 0.19z &= 0.55 \end{aligned}$$

Comment on your results.

38 It is required to solve the equations

$$\begin{aligned} 4x - 3y &= -1.5 \\ 5x + 2y &= 12.5 \end{aligned}$$

using
(a) Gaussian elimination
(b) any other method

For each method of solution, count the number of multiplications/divisions and the number of additions/subtractions. Compare your methods.
Repeat the exercise for the equations

$$\begin{aligned} 3x - 2y + 4z &= 1.8 \\ 5x + 3y - 2z &= -4.2 \\ x - 5y - 3z &= -15 \end{aligned}$$

Chapter 5

Polynomials 2

5.1 Taylor polynomials (centre = 0)

In this chapter, we shall consider a further and very important application of polynomials. As mentioned in Chapter 2, polynomials are extremely well behaved functions and are, in general, easy to handle. Life would be very easy indeed if all functions had the same desirable properties as polynomials. Hence we might pose the question: 'How successful are we likely to be if we write an arbitrary function f(x) as a polynomial in x?'

To pursue this idea, let f(x) be an arbitrary function and write f(x) as (say) a cubic polynomial.

Then $\quad f(x) = a_0 + a_1x + a_2x^2 + a_3x^3 \qquad \ldots (1)$

where a_0, a_1, a_2 and a_3 are the unknown coefficients of the polynomial. To find suitable values for a_0, a_1, a_2, and a_3 we might proceed as follows.

Step 1 Put $\quad x = 0 \quad$ then $\quad f(0) = a_0 \quad$ and we have

$$\boxed{a_0 = f(0)}$$

Step 2 Differentiate both sides of Equation (1):

$$f'(x) = a_1 + 2a_2x + 3a_3x^2$$

Put $\quad x = 0 \quad$ then $\quad f'(0) = a_1 \quad$ and

$$\boxed{a_1 = f'(0)}$$

Step 3 Differentiate both sides of Equation (1) a second time:

$$f''(x) = 2a_2 + 3 \times 2a_3x$$

Put $\quad x = 0 \quad$ then $\quad f''(0) = 2a_2 \quad$ and

$$\boxed{a_2 = \tfrac{1}{2}f''(0)}$$

Step 4 Differentiate both sides of Equation (1) a third time:

$$f'''(x) = 3 \times 2a_3$$

Put $x = 0$ then $f'''(0) = 3 \times 2a_3$ and

$$\boxed{a_3 = \tfrac{1}{6}f'''(0)}$$

Putting all this together:

$$a_0 = f(0) \qquad a_1 = f'(0) \qquad a_2 = \tfrac{1}{2}f''(0) \qquad a_3 = \tfrac{1}{6}f'''(0)$$

and

$$a_0 + a_1x + a_2x^2 + a_3x^3 = f(0) + f'(0)x + \tfrac{1}{2}f''(0)x^2 + \tfrac{1}{6}f'''(0)x^3$$

This cubic polynomial is called the Taylor polynomial (of degree three) of the function f(x). In some books, the expression $f(0) + f'(0)x + \tfrac{1}{2}f''(0)x^2 + \tfrac{1}{6}f'''(0)x^3$ is called 'the first four terms of the Taylor series of f(x)'; in other books it is called 'the first four terms of the Maclaurin series'. A series is normally taken to be the sum of infinitely many terms and a Taylor (or Maclaurin) series is the sum of infinitely many terms of the type shown above. Since we shall always be considering a finite number of terms, this expression will be called a Taylor (or Maclaurin) polynomial. The distinction between a Taylor and a Maclaurin polynomial will be explained at the end of Section 5.3.

To see how well (or otherwise) the Taylor polynomial approximates the function f(x), consider a specific example.

Example 5.1

Construct the Taylor polynomial of degree two for the function $f(x) = \dfrac{1}{1 + 2x}$. Draw, on the same axes, the graphs of the Taylor polynomial and $y = \dfrac{1}{1 + 2x}$. Comment on the usefulness of the approximation.

Presentation 1

We have $f(x) = \dfrac{1}{1 + 2x} = (1 + 2x)^{-1}$

Write $(1 + 2x)^{-1} = a_0 + a_1x + a_2x^2$

Put $x = 0$ then $(1 + 2 \times 0)^{-1} = a_0$ giving $a_0 = 1$

Differentiate $(-1)(1 + 2x)^{-2} \times 2 = a_1 + 2a_2x$

Put $x = 0$ then $(-1)(1 + 2 \times 0)^{-2} \times 2 = a_1$ giving $a_1 = -2$

Differentiate $(-1)(-2)(1 + 2x)^{-3} \times 2 \times 2 = 2a_2$

Put $x = 0$ then $(-1)(-2)(1 + 2 \times 0)^{-3} \times 2 \times 2 = 2a_2$
giving $a_2 = 4$

Hence the Taylor polynomial of degree two of $\dfrac{1}{1 + 2x}$ is

$$1 - 2x + 4x^2$$

Presentation 2

Apply the formula for the terms of the Taylor polynomial directly. Write:

$$\begin{aligned} f(x) &= (1 + 2x)^{-1} &&: f(0) = (1)^{-1} = 1 \\ f'(x) &= (-1)(1 + 2x)^{-2} \times 2 &&: f'(0) = (-1)(1)^{-2} \times 2 = -2 \\ f''(x) &= (-1)(-2)(1 + 2x)^{-3} \times 2 \times 2 &&: f''(0) = (-1)(-2)(1)^{-3} \times 2 \times 2 = 8 \end{aligned}$$

The Taylor polynomial is

$$f(0) + f'(0)x + \tfrac{1}{2}f''(0)x^2 = 1 - 2x + 4x^2$$

The graphs of the two functions are illustrated in Figure 5.1

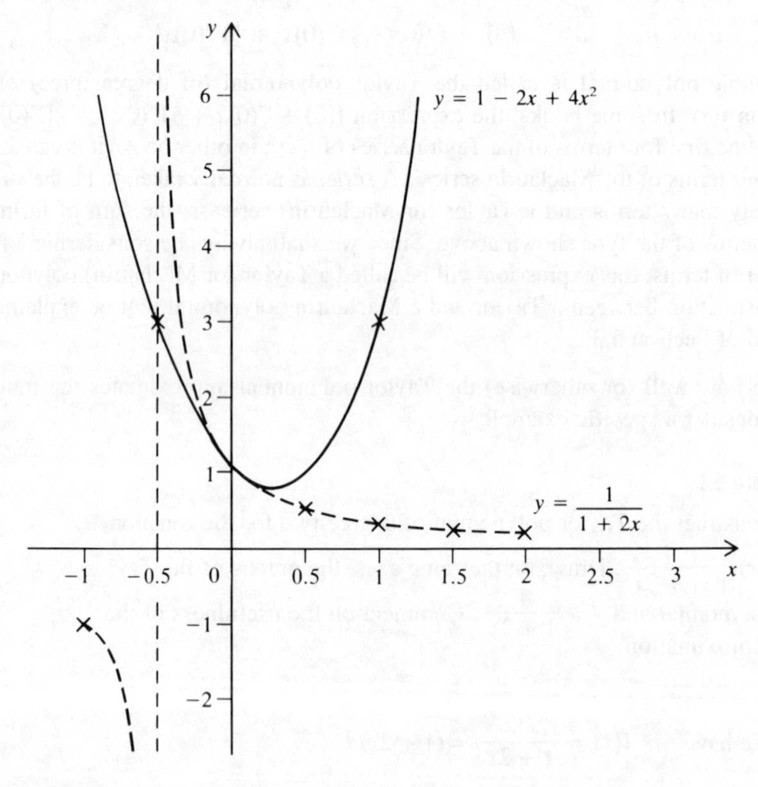

Figure 5.1

It is clear that $1 - 2x + 4x^2$ provides a good approximation to $\dfrac{1}{1 + 2x}$ for values of x in a small region surrounding $x = 0$. It is equally clear that the approximation is very poor indeed for values of x 'not close to zero'.

5.2 Taylor polynomials (centre ≠ 0)

It is reasonable to ask what happens if it is required to express $\dfrac{1}{1 + 2x}$ as a polynomial for values of x which are not close to $x = 0$. We may, for example, want to write $\frac{1}{1+2x}$ as a polynomial in some region surrounding $x = 2$. The solution is delightfully simple: we shift the centre of attention to $x = 2$ and rewrite the Taylor polynomial as a polynomial in $(x - 2)$. That is, we write (for a cubic polynomial)

$$f(x) = b_0 + b_1(x-2) + b_2(x-2)^2 + b_3(x-2)^3 \qquad \ldots (2)$$

To determine b_0, b_1, b_2 and b_3 we proceed (almost) as before.

Step 1 Put $x = 2$ then $f(2) = b_0$ and

$$\boxed{b_0 = f(2)}$$

Step 2 Differentiate Equation (2) with respect to x

$$f'(x) = b_1 + 2b_2(x-2) + 3b_3(x-2)^2$$

Put $x = 2$ then $f'(2) = b_1$ and

$$\boxed{b_1 = f'(2)}$$

Step 3 Differentiate Equation (2) a second time

$$f''(x) = 2b_2 + 3 \times 2b_3(x-2)$$

Put $x = 2$ then $f''(2) = 2b_2$ and

$$\boxed{b_2 = \tfrac{1}{2}f''(2)}$$

Step 4 Differentiate Equation (2) a third time

$$f'''(x) = 3 \times 2b_3$$

Put $x = 2$ then $f'''(2) = 3 \times 2b_3$ and

$$\boxed{b_3 = \tfrac{1}{6}f'''(2)}$$

Hence $b_0 = f(2)$ $\quad b_1 = f'(2)$ $\quad b_2 = \tfrac{1}{2}f''(2)$ $\quad b_3 = \tfrac{1}{6}f'''(2)$

and the Taylor polynomial of degree three of $f(x)$, centred at $x = 2$ is given by:

$$b_0 + b_1(x-2) + b_2(x-2)^2 + b_3(x-2)^3$$
$$= f(2) + f'(2)(x-2) + \tfrac{1}{2}f''(2)(x-2)^2 + \tfrac{1}{6}f'''(2)(x-2)^3$$

To observe the effect of centering the Taylor polynomial we return to the previous example.

Example 5.2

For the function $f(x) = \frac{1}{1+2x}$ construct the Taylor polynomial of degree two with centre $x = 2$. Draw the graphs of this polynomial and $y = \frac{1}{1+2x}$. Comment on the usefulness of the approximation.

If $\quad f(x) = \dfrac{1}{1+2x}$ $\quad$ then $\quad f(2) = \dfrac{1}{1+2\times 2} = 0.2$

$f'(x) = (-1)(1+2x)^{-2}\times 2$ $\quad$ so $\quad f'(2) = -\dfrac{2}{25} = -0.08$

$f''(x) = (-1)(-2)(1+2x)^{-3}\times 2\times 2$ $\quad$ so $\quad f''(2) = \dfrac{8}{125} = 0.064$

The required Taylor polynomial is:

$$f(2) + f'(2)(x-2) + \tfrac{1}{2}f''(2)(x-2)^2$$
$$= 0.2 - 0.08(x-2) + 0.032(x-2)^2$$

The graphs of the two functions are illustrated in Figure 5.2.

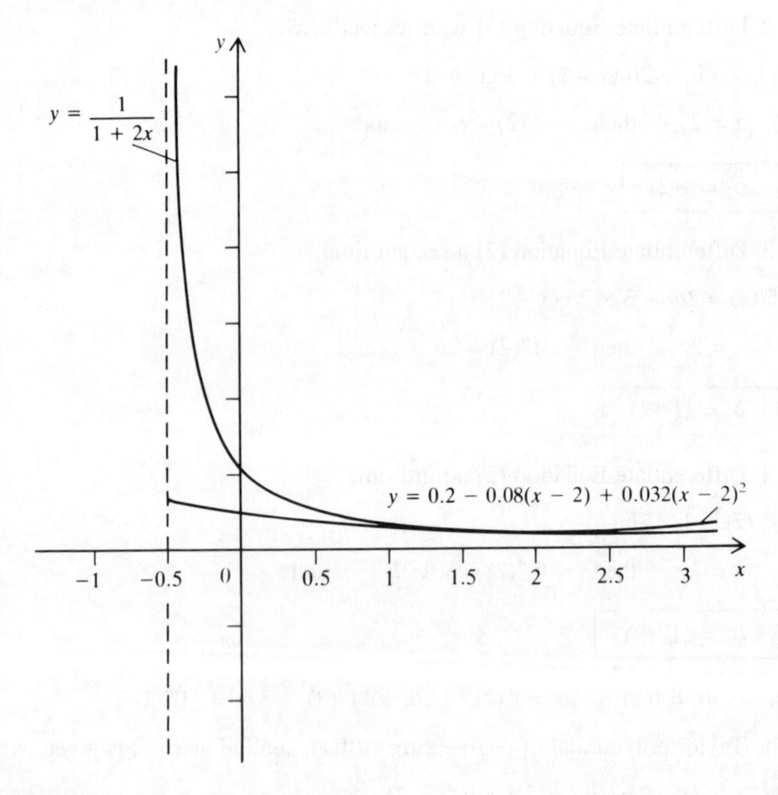

Figure 5.2

It is clear that the polynomial

$$0.2 - 0.08(x-2) + 0.032(x-2)^2$$

provides a good approximation to $f(x) = \dfrac{1}{1+2x}$ in a region surrounding $x = 2$. The approximation is clearly not so good in other regions of the x-axis and certainly, if x is less than 0.05, the approximation is very poor indeed.

As can be seen from these examples, it is extremely important when calculating a Taylor polynomial to be sure of the region in which the polynomial representation is to be used. If the Taylor polynomial is not centred on the relevant part of the x-axis, the approximation may be so poor as to be almost useless.

It may be appropriate at this point to give a general definition of the Taylor polynomial.

Definition

Let $f(x)$ be a function having derivatives $f'(x)$, $f''(x)$, ..., $f^{(n)}(x)$ and let $x = a$ represent a point on the x-axis. The Taylor polynomial (of degree n) of the function $f(x)$ is given by:

$$f(a) + f'(a)(x - a) + \frac{1}{2!}f''(a)(x - a)^2 + \frac{1}{3!}f'''(x)(x - a)^3 + \ldots$$
$$+ \frac{1}{n!}f^{(n)}(a)(x - a)^n$$

This form of the Taylor polynomial is said to be centred at $x = a$.

To indicate that the Taylor polynomial provides an approximation to the function $f(x)$, it is often written

$$f(x) \approx f(a) + f'(a)(x - a) + \frac{1}{2!}f''(a)(x - a)^2 + \ldots + \frac{1}{n!}f^{(n)}(a)(x - a)^n$$

If x is reasonably close in value to a, so that $(x - a)$ is numerically small, then high powers of $(x - a)$ will become very small indeed. Hence, if x is close to a, a Taylor polynomial of low degree may well provide a good approximation to $f(x)$. If x is not quite so close to a, then a polynomial of higher degree may be necessary. This is illustrated in the following example.

Example 5.3

The graph of $\ln x$ is sketched in Figure 5.3. Calculate the Taylor polynomials of degree one, of degree three and of degree five which will approximate the function $f(x) = \ln x$ in a region of $x = 3$.

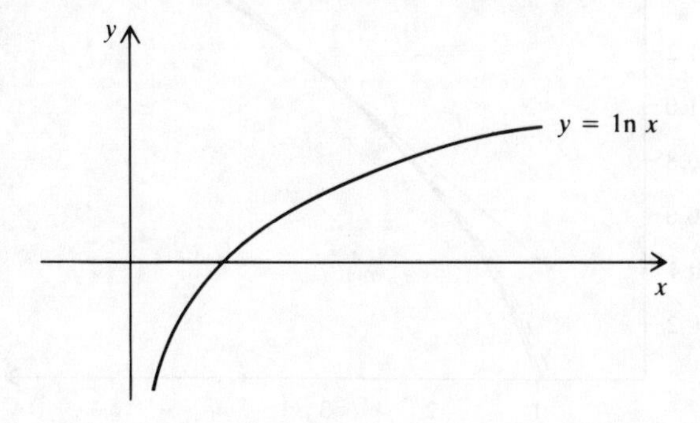

Figure 5.3

We need to calculate Taylor polynomials of degrees one, three and five with centre $x = 3$.

$$f(x) = \ln x \qquad f(3) = \ln(3) = 1.0986$$

$$f'(x) = \frac{1}{x} \qquad f'(3) = \frac{1}{3} = 0.33333$$

$$f''(x) = -\frac{1}{x^2} \qquad f''(3) = -\frac{1}{9} = -0.11111$$

$$f'''(x) = (-1)(-2) \times \frac{1}{x^3} \qquad f'''(3) = \frac{2}{27} = 0.074074$$

$$f^{(4)} = (-1)(-2)(-3) \times \frac{1}{x^4} \qquad f^{(4)}(3) = -\frac{6}{81} = -0.074074$$

$$f^{(5)} = (-1)(-2)(-3)(-4) \times \frac{1}{x^5} \qquad f^{(5)}(3) = \frac{24}{243} = 0.098765$$

The polynomial of degree one:

$$\begin{aligned} p_1(x) &= f(3) + f'(3)(x-3) \\ &= 1.0986 + 0.33333(x-3) \end{aligned}$$

The polynomial of degree three:

$$\begin{aligned} p_3(x) &= f(3) + f'(3)(x-3) + \tfrac{1}{2}f''(3)(x-3)^2 + \tfrac{1}{6}f'''(3)(x-3)^3 \\ &= 1.0986 + 0.33333(x-3) - 0.055556(x-3)^2 + 0.012346(x-3)^3 \end{aligned}$$

The polynomial of degree five:

$$\begin{aligned} p_5(x) &= f(3) + f'(3)(x-3) + \tfrac{1}{2}f''(3)(x-3)^2 + \tfrac{1}{6}f'''(3)(x-3)^3 \\ &\quad + \tfrac{1}{24}f^{(4)}(3)(x-3)^4 + \tfrac{1}{120}f^{(5)}(3)(x-3)^5 \\ &= 1.0986 + 0.33333(x-3) - 0.055556(x-3)^2 + 0.012346(x-3)^3 \\ &\quad - 0.0030864(x-3)^4 + 0.00082305(x-3)^5 \end{aligned}$$

The graphs of $y = \ln x$, $y = p_1(x)$, $y = p_3(x)$ and $y = p_5(x)$ are shown below.

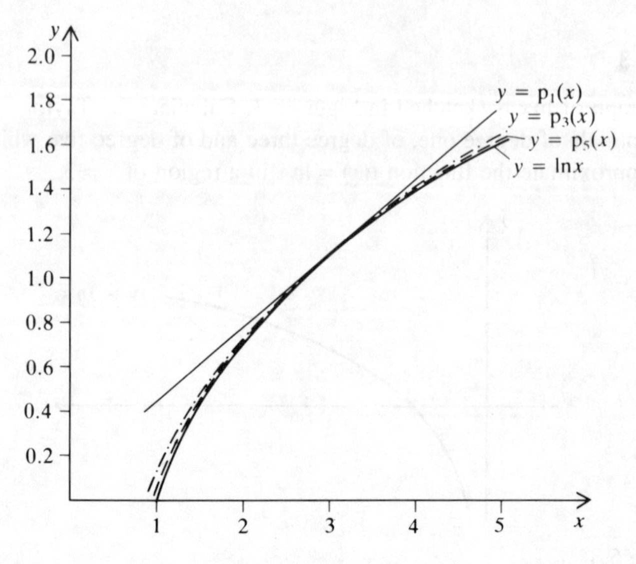

Figure 5.4

It is clear that in the interval [1.5, 4.5], $p_5(x)$ provides a very good approximation indeed to the function $f(x) = \ln x$. $p_3(x)$ provides a good approximation in the interval [2, 4] while $p_1(x)$ approximates the function well only in a small interval surrounding $x = 3$.

The Taylor polynomial may be written in an alternative form which is useful in certain situations (see in particular Chapter 7).

We have:

$$f(x) \approx f(a) + f'(a)(x-a) + \frac{1}{2!}f''(a)(x-a)^2 + ... + \frac{1}{n!}f^{(n)}(a)(x-a)^n \quad ...(3)$$

The idea now is to write $\quad x - a = h \quad$ if $\quad x > a$

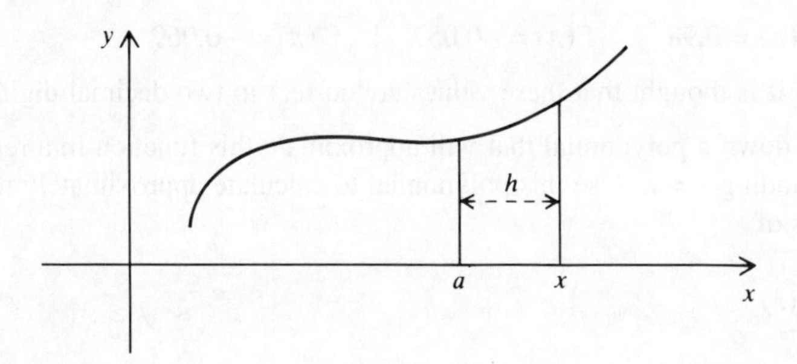

Figure 5.5

So $x = a + h$ and substitution in Equation (3) gives:

$$f(a+h) \approx f(a) + f'(a)h + \frac{1}{2!}f''(a)h^2 + ... + \frac{1}{n!}f^{(n)}(a)h^n \quad ...(4)$$

If $\quad x < a, \quad$ write $\quad x - a = -h$

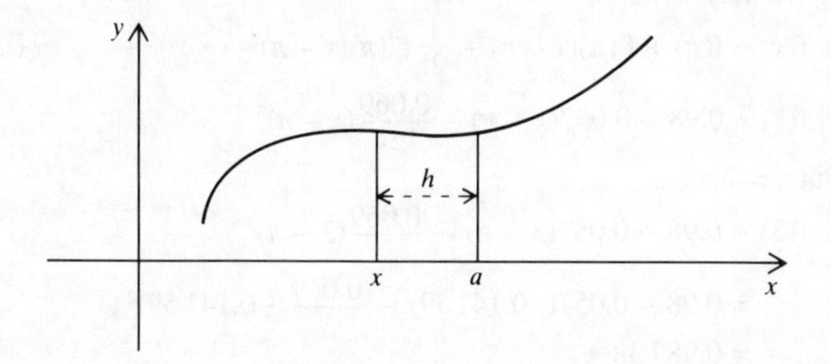

Figure 5.6

In this case, $x = a - h$ and substitution in Equation (3) gives:

$$f(a-h) \approx f(a) - f'(a)h + \frac{1}{2!}f''(a)h^2 + ... + (-1)^n\frac{1}{n!}f^{(n)}(a)h^n \quad ...(5)$$

It may be seen that the right-hand sides in both expressions are polynomials in h.

The remark that preceeded Example 5.3 is perhaps even more obvious when the polynomial is written in this form. If h is numerically small, then h^3 may well be very small indeed (depending, of course, on how small h really is). In this case, a Taylor polynomial of degree two may provide a sufficiently accurate approximation to the function in the region of interest and this polynomial may then be used to derive further information about f(x). An example will illustrate how this may be achieved.

Example 5.4

We have very limited information about a certain function f(x). It is known only that when $x = \pi$, f(x) and its derivatives take the values given below

$$f(\pi) = 0.98 \quad : \quad f'(\pi) = -0.057 \quad : \quad f''(\pi) = -0.069$$

where it is thought that these values are correct to two decimal digits.

Write down a polynomial that will approximate this function in a region surrounding $x = \pi$. Use this polynomial to calculate approximately the values of

(a) f(3)
(b) f'(3)
(c) $\displaystyle\int_{\pi}^{\frac{7\pi}{6}} f(x)\,dx$

Comment on the accuracy of your answers.

Since f'''(π) is not known, the Taylor polynomial of highest degree that can be obtained from the given information is a polynomial of degree two. This is:

$$f(\pi) + f'(\pi)(x-\pi) + \frac{1}{2!}f''(\pi)(x-\pi)^2$$

Hence we may write

$$f(x) \approx f(\pi) + f'(\pi)(x-\pi) + \frac{1}{2!}f''(\pi)(x-\pi)^2 \qquad \ldots (6)$$

or
$$f(x) \approx 0.98 - 0.057(x-\pi) - \frac{0.069}{2}(x-\pi)^2$$

(a) Put $x = 3$:

$$f(3) \approx 0.98 - 0.057(3-\pi) - \frac{0.069}{2}(3-\pi)^2$$
$$= 0.98 - 0.057(-0.141\,59) - \frac{0.069}{2}(-0.141\,59)^2$$
$$= 0.987\,38$$

(b) Differentiating both sides of Equation (6):

$$f'(x) \approx f'(\pi) + \frac{1}{2!}f''(\pi) \times 2(x-\pi)$$
$$= f'(\pi) + f''(\pi)(x-\pi)$$

When $x = 3$, we have:

$$f'(3) \approx -0.057 - 0.069(-0.1416) \qquad (\text{taking } \pi = 3.1416)$$
$$= -0.047\,23$$

(c) The calculations are simplified slightly if we use the alternative form of the Taylor polynomial. Since, for the range of integration, $x > \pi$, we use Equation (4) and write

$$f(\pi + h) \approx f(\pi) + f'(\pi)h + \frac{1}{2!}f''(\pi)h^2$$

Using the substitution $x = \pi + h$ gives

$$\int_{\pi}^{\frac{7\pi}{6}} f(x)\mathrm{d}x = \int_{0}^{\frac{\pi}{6}} f(\pi + h)\,\mathrm{d}h$$

$$\int_{0}^{\frac{\pi}{6}} (0.98 - 0.057h - 0.035h^2)\,\mathrm{d}h$$

$$= \left[0.98h - 0.057\frac{h^2}{2} - 0.035\frac{h^3}{3}\right]_0^{\frac{\pi}{6}}$$

$$= 0.50364$$

Since the given values of $f(\pi)$, $f'(\pi)$ and $f''(\pi)$ are accurate to, at most, two decimal digits, it is unreasonable to expect that any of these answers are accurate to more than two decimal digits. Without knowing more about the error in the Taylor polynomial approximation, we cannot say more than this, but experience suggests that it would be rash to expect that these answers are accurate even to two decimal digits.

This example illustrates how Taylor polynomials may be used to derive information about functions for which only limited information is available. This is an important application of Taylor polynomials.

We have seen that when using a Taylor polynomial to approximate a function $f(x)$ the error can be considerable. It would, as usual, be extremely useful to have some indication of the size of the error and in the theorem below which proves the result we have been using, an expression for the error term is given. The proof involves Rolle's theorem and is similar in structure to the proof that established the Lagrange interpolating polynomial. In this case also, the proof may appear at first sight to be difficult. In fact, if the reader manages to avoid being lost in the algebra, the basic idea can be seen to be really very simple.

5.3 Taylor's theorem

Let $f(x)$ be a function and $A(a, 0)$ and $X(x, 0)$ be two points on the x-axis.

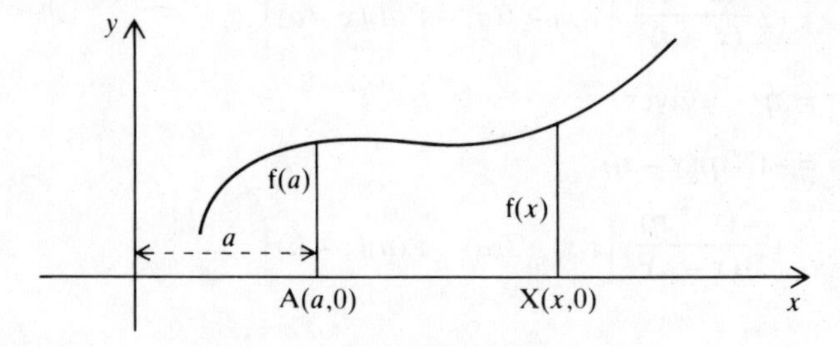

Figure 5.7

If the derivatives f'(x), f''(x), ..., $f^{(n)}(x)$ all exist at the point A and $f^{(n+1)}(x)$ exists in the interval (a, x) then

$$f(x) = f(a) + f'(a)(x-a) + \frac{1}{2!}f''(x)(x-a)^2 + \frac{1}{3!}f'''(x)(x-a)^3$$
$$+ \ldots + \frac{1}{n!}f^{(n)}(x)(x-a)^n + \frac{1}{(n+1)!}f^{(n+1)}(\eta)(x-a)^{n+1}$$

where $a < \eta < x$

and $f^{(n+1)}(\eta)$ represents the $(n+1)$th derivative of f(x) evaluated at $x = \eta$.

Hence the error in using the Taylor polynomial of degree n to approximate f(x) is

$$\frac{1}{(n+1)!}f^{(n+1)}(\eta)(x-a)^{n+1}$$

Proof

As with the Lagrange interpolating polynomial, we prove the result for $n = 1$. The general proof is left until the exercises at the end of this chapter.

The idea (as before) is to set up a subsidiary function Q(t). Again, this subsidiary function looks awful but is not as bad as it appears.

Let
$$Q(t) = f(x) - f(t) - (x-t)f'(t)$$
$$-\left(\frac{x-t}{x-a}\right)^2 \times \left[f(x) - f(a) - f'(a)(x-a)\right]$$

We regard Q(t) as being a function of t; we shall consider x to be constant.

If we put $t = a$, then Q(a) = 0 as the reader may verify.
If we put $t = x$, then Q(x) = 0.

By Rolle's theorem, for some $\eta(a < \eta < x)$ we have:

$$Q'(\eta) = 0$$

Differentiating Q(t) (as a function of t):

$$Q'(t) = -f'(t) - (f''(t)(x-t) - f'(t))$$
$$-2\left(\frac{x-t}{x-a}\right)\left(\frac{-1}{x-a}\right)\left[f(x) - f(a) - f'(a)(x-a)\right]$$

or
$$Q'(t) = -f''(t)(x-t)$$
$$+2\frac{(x-t)}{(x-a)^2}\left[f(x) - f(a) - f'(a)(x-a)\right]$$

Putting $t = \eta$ gives

$$Q'(\eta) = -f''(\eta)(x-\eta)$$
$$+2\frac{(x-\eta)}{(x-a)^2}\left[f(x) - f(a) - f'(a)(x-a)\right]$$

But $\quad Q'(\eta) = 0 \quad$ so

$$0 = -f''(\eta)(x - \eta)$$

$$+2\frac{(x-\eta)}{(x-a)^2}\left[f(x) - f(a) - f'(a)(x-a)\right]$$

This becomes:

$$f(x) - f(a) - f'(a)(x - a) = \frac{1}{2}f''(\eta)(x - a)^2$$

or

$$f(x) = f(a) + f'(a)(x - a) + \frac{1}{2}f''(\eta)(x - a)^2 \qquad \text{as required}$$

Note that in the alternative form, the theorem becomes:

$$f(a + h) = f(a) + f'(a)h + \frac{1}{2}f''(a)h^2 + \ldots + \frac{1}{n!}f^{(n)}(a)h^n + \frac{1}{(n+1)!}f^{(n+1)}(\eta)h^{n+1}$$

where $\quad a < \eta < a + h$

Example 5.5

Express the function $f(x) = e^{-x^2}$ as a quadratic polynomial in the interval [0, 1]. Calculate the greatest possible value of the error.

Since the approximation is required in the interval [0, 1], we calculate the Taylor polynomial of $f(x)$ with centre $x = 0.5$.

$$f(x) = e^{-x^2} \qquad f(0.5) = 0.778\,800\,78$$

$$f'(x) = e^{-x^2}(-2x) \qquad f'(0.5) = -0.778\,800\,78$$

$$f''(x) = (-2x)(e^{-x^2}(-2x)) + e^{-x^2}(-2)$$

$$= (4x^2 - 2)e^{-x^2} \qquad f''(0.5) = -0.778\,800\,78$$

$$f'''(x) = (4x^2 - 2)(e^{-x^2}(-2x)) + (8x)e^{-x^2}$$

$$= (-8x^3 + 12x)e^{-x^2}$$

The required polynomial is:

$$p_2(x) = f(0.5) + f'(0.5)(x - 0.5) + \frac{1}{2}f''(0.5)(x - 0.5)^2$$

$$= 0.778\,800\,78 - 0.778\,800\,78(x - 0.5) - 0.389\,400\,39(x - 0.5)^2$$

The graphs of $y = e^{-x^2}$ and $y = p_2(x)$ are shown in Figure 5.8.

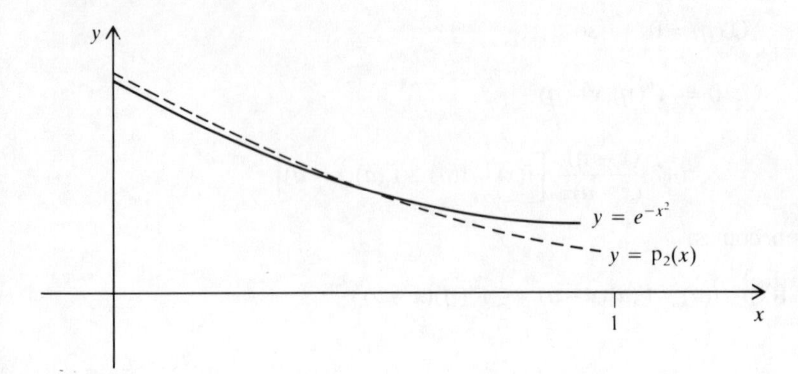

Figure 5.8

The error $e(x)$ in using $p_2(x)$ to approximate $f(x)$ is given by:

$$e(x) = \frac{1}{3!}f'''(\eta)(x - 0.5)^3 \qquad 0 < \eta < 1$$

$$= \frac{1}{6}[(-8\eta^3 + 12\eta)e^{-\eta^2}](x - 0.5)^3$$

To obtain a bound on the error $e(x)$, we maximise $(x - 0.5)^3$ and $(-8\eta^3 + 12\eta)e^{-\eta^2}$ separately.

Clearly, the greatest value that can be taken by $(x - 0.5)^3$ is $(\frac{1}{2})^3 = \frac{1}{8}$.

The greatest value that can be taken by

$$(-8\eta^3 + 12\eta)e^{-\eta^2} \qquad 0 < \eta < 1$$

may be determined by the usual maximum and minimum technique of the calculus. Differentiating gives:

$$\frac{d}{d\eta}[(-8\eta^3 + 12\eta)e^{-\eta^2}] = (-8\eta^3 + 12\eta)(e^{-\eta^2}(-2\eta)) + (-24\eta^2 + 12)e^{-\eta^2}$$

$$= (16\eta^4 - 24\eta^2)e^{-\eta^2} + (-24\eta^2 + 12)e^{-\eta^2}$$

$$= (16\eta^4 - 48\eta^2 + 12)e^{-\eta^2}$$

Now put the derivative equal to zero and solve the resulting equation for η.

To solve $\quad (16\eta^4 - 48\eta^2 + 12)e^{-\eta^2} = 0$

Divide through by 4 and observe that $e^{-\eta^2}$ is never zero.

Hence $\quad 4\eta^4 - 12\eta^2 + 3 = 0$

Let $\quad z = \eta^2 \quad$ then $4z^2 - 12z + 3 = 0$

Solving this quadratic equation:

$$z = \frac{12 \pm \sqrt{144 - 48}}{8}$$

$$= \frac{12 \pm \sqrt{96}}{8}$$

$$= \frac{12 \pm 9.7979590}{8}$$

$$= \begin{cases} 2.7247449 \\ 0.27525513 \end{cases}$$

Taking the square root gives the required value of η

$$\eta = \begin{cases} \pm 1.6506801 \\ \pm 0.52464762 \end{cases}$$

Only one of these values, $\eta = 0.52464762$, lies in the required interval (0, 1). Further, by substituting values of η, (a) less than 0.52464762 and (b) greater than 0.52464762 into the expression $(16\eta^4 - 48\eta^2 + 12)e^{-\eta^2}$, it may be shown that $\eta = 0.52464762$ produces a maximum value of $(-8\eta^3 + 12\eta)e^{-\eta^2}$. This maximum value is 3.90356614 and hence the greatest value that can be taken by the error function e(x) is

$$3.90356614 \times \frac{1}{8} = 0.48794577$$

This is clearly greater than the error that occurs in Figure 5.8. The reader should be warned that this method of calculating the maximum error provides only a rough approximation and can, on occasions, overshoot the maximum error considerably.

Note two final points.

(a) In the special case where the centre is taken to be $x = 0$, the resulting polynomial is often called the Maclaurin polynomial or a Maclaurin series. Hence the first Taylor polynomials in this chapter could well have been called Maclaurin polynomials or the first few terms of a Maclaurin series.

(b) The reader may have noticed a remarkable similarity in the error terms arising in the Lagrange interpolating polynomial and in the Taylor polynomial.

Lagrange interpolating polynomial error term	Taylor polynomial error term
$\frac{1}{(n+1)!} f^{(n+1)}(\eta)(x - x_0)(x - x_1) \ldots (x - x_n)$	$\frac{1}{(n+1)!} f^{(n+1)}(\eta)(x - a)^{n+1}$

If, in the Lagrange error term, we took

$$x_0 = x_1 = x_2 = \ldots = x_n = a$$

then we should have the Taylor polynomial error term. Why should this be so? It may be shown that as the points $x_0, x_1, x_2, \ldots, x_n$ move towards a common value a so the

Lagrange interpolating polynomial comes to resemble ever more closely the Taylor polynomial. Finally, the two polynomials become identical[†].

We have now seen how polynomials can be used to interpolate a set of data points (the Lagrange interpolating polynomial) and to approximate a given function (the Lagrange interpolating polynomial and the Taylor polynomial) and, hopefully, the reader has been persuaded that polynomials are extremely useful functions. Since polynomials are used in so many areas of mathematics it would be convenient to know an effective method of calculating the value of a polynomial for a given value of x, for example, of calculating the value of $6x^4 - 5x^3 - 3x^2 + 2x - 7$ when $x = -2$. This process is called 'evaluating the polynomial' at the given value of x, and finding an efficient algorithm for the evaluation of polynomials forms the subject matter of the remainder of this chapter.

5.4 Horner's method of polynomial evaluation

Consider the polynomial mentioned above:

$$\mathrm{p}_4(x) = 6x^4 - 5x^3 - 3x^2 + 2x - 7$$

We observe that the first four terms of the polynomial each has x as a factor, the first three terms each has x^2 as a factor and the first two terms each has x^3 as a factor. Factorising the first four terms gives:

$$\mathrm{p}_4(x) = (6x^3 - 5x^2 - 3x + 2)x - 7$$

Factorising the first three terms in the bracket $(6x^3 - 5x^2 - 3x + 2)$ gives:

$$\mathrm{p}_4(x) = (\{6x^2 - 5x - 3\}x + 2)x - 7$$

Factorising the first two terms in the { } bracket gives:

$$\mathrm{p}_4(x) = (\{[6x - 5]x - 3\}x + 2)x - 7$$

Horner's method consists of evaluating, in sequence, the brackets in this expression. To evaluate $\mathrm{p}_4(x)$ when $x = -2$ (or equivalently, to calculate $\mathrm{p}_4(-2)$), proceed as follows:

Step 1 Evaluate $[6x - 5]$ when $x = -2$ to get $6(-2) - 5 = -17$

Step 2 Evaluate $\{[-17]x - 3\}$ when $x = -2$ to get $[-17](-2) - 3 = 31$

Step 3 Evaluate $(\{31\}x + 2)$ when $x = -2$ to get $\{31\}(-2) + 2 = -60$

Step 4 Evaluate $\mathrm{p}_4(x) = (-60)x - 7$ when $x = -2$ to get $(-60)(-2) - 7 = 113$

Hence $\quad \mathrm{p}_4(-2) = 113$

We will now see how this process may be achieved without first writing down the factored expression.

[†] To see how this happens, and indeed to learn a great deal more about interpolating polynomials, the reader is referred to Chapter 2 of Conte and de Boor 1981 *Elementary numerical analysis* 3rd edn. McGraw-Hill.

Write $\quad p_4(x) = a_4x^4 + a_3x^3 + a_2x^2 + a_1x + a_0$.

Then $\quad a_4 = 6,\ a_3 = -5,\ a_2 = -3,\ a_1 = 2$ and $a_0 = -7$.

We need to have some way of storing the intermediate values of the calculation: −17, 31, −60, 113. We shall call these quantities b_3, b_2, b_1 and b_0. Also, we write:

let $\quad b_4 = a_4 \quad (= 6)$.

Now the calculation may be set out by writing the coefficients of the polynomial in a row and writing $b_4 = a_4$ (= 6) in the first column. This is illustrated below.

	coeff of x^4	coeff of x^3	coeff of x^2	coeff of x	const term
$x = -2$	$a_4 = 6$	$a_3 = -5$	$a_2 = -3$	$a_1 = 2$	$a_0 = -7$
	$b_4 = 6$				

In the column headed 'coeff of x^3', perform Step 1: $6 \times (-2) - 5 = -17$. This is achieved by multiplying b_4 by the value of x and adding the product b_4x to a_3. Call the result of this calculation b_3. This is illustrated below.

	coeff of x^4	coeff of x^3	coeff of x^2	coeff of x	const term
$x = -2$	$a_4 = 6$	$a_3 = -5$	$a_2 = -3$	$a_1 = 2$	$a_0 = -7$
		$6\times(-2) = -12\ +$			
	$\times x$ ↗				
	$b_4 = 6$	$b_3 = -17$			

In the column headed 'coeff of x^2', perform Step 2: $(-17)(-2) - 3 = 31$. To do this, multiply b_3 by the value of x and add the product b_3x to a_2. Call the result of this calculation b_2. This is illustrated below.

	coeff of x^4	coeff of x^3	coeff of x^2	coeff of x	const term
$x = -2$	$a_4 = 6$	$a_3 = -5$	$a_2 = -3$	$a_1 = 2$	$a_0 = -7$
		$6\times(-2) = -12\ +$	$(-17)\times(-2) = 34\ +$		
	$\times x$ ↗	$\times x$ ↗			
	$b_4 = 6$	$b_3 = -17$	$b_2 = 31$		

Repeat this process until b_0 has been calculated.

	coeff of x^4	coeff of x^3	coeff of x^2	coeff of x	const term
$x = -2$	$a_4 = 6$	$a_3 = -5$	$a_2 = -3$	$a_1 = 2$	$a_0 = -7$
		$6 \times (-2) = -12$	$(-17) \times (-2) = 34$	$31 \times (-2) = -62$	$-60 \times (-2) = 120$
	$\times x$ ↗	$\times x$ ↗	$\times x$ ↗	$\times x$ ↗	
	$b_4 = 6$	$b_3 = -17$	$b_2 = 31$	$b_1 = -60$	$b_0 = 113$

Then b_0 gives the value of $p_4(-2)$.

Example 5.6

Evaluate the polynomial $-2x^5 + 3x^2 - 2x - 9$ when $x = -1$.

Note that even though there are no terms involving x^4 and x^3 in the polynomial, these terms must be represented in the table, as below.

	coeff of x^5	coeff of x^4	coeff of x^3	coeff of x^2	coeff of x	const term
$x = -1$	$a_5 = -2$	$a_4 = 0$	$a_3 = 0$	$a_2 = 3$	$a_1 = -2$	$a_0 = -9$
		$(-2)(-1) = 2$	$2 \times (-1) = -2$	$(-2)(-1) = 2$	$5 \times (-1) = -5$	$(-7)(-1) = 7$
	$\times x$	$\times x$	$\times x$	$\times x$	$\times x$	
	$b_5 = -2$	$b_4 = 2$	$b_3 = -2$	$b_2 = 5$	$b_1 = -7$	$b_0 = -2$

Hence $p_5(-1) = b_0 = -2$.

This process may be illustrated, generally, as below

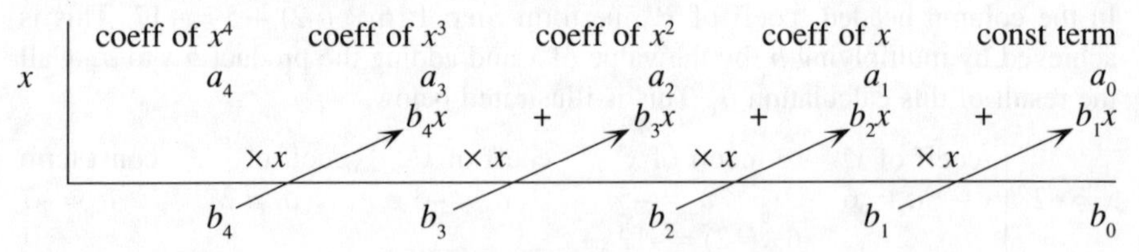

Note that it is not necessary to write *a*s and *b*s in the table; it is sufficient to record only the values taken by these quantities.

Example 5.7

What is the value of $-x^4 + 4x^2 - 3x + 1$ when $x = 3$?

	coeff of x^4	coeff of x^3	coeff of x^2	coeff of x	const term
$x = 3$	-1	0	4	-3	1
		$(-1) \times 3 = -3$	$(-3) \times 3 = -9$	$(-5)(3) = -15$	$(-18)(3) = -54$
	-1	-3	-5	-18	-53

Hence the value of the polynomial when $x = 3$ is -53.

Although the reader may now be convinced of the validity of Horner's method, the result will be stated as a theorem and proved in the case of a polynomial of degree four. The proof for a general polynomial of degree n proceeds along identical lines.

The proof will repay careful study. The importance of the role played by b_0 has been demonstrated above. The proof will also reveal a surprising and very useful interpretation of b_4, b_3, b_2 and b_1.

Theorem 5.1 Horner's method

Let $\quad p_4(x) = a_4x^4 + a_3x^3 + a_2x^2 + a_1x + a_0$

be a polynomial of degree four and let t be a given number. Let $b_4 = a_4$ and define

$$b_3 = b_4t + a_3$$
$$b_2 = b_3t + a_2$$
$$b_1 = b_2t + a_1$$
$$b_0 = b_1t + a_0$$

Then $b_0 = p_4(t)$. That is, b_0 gives the value of the polynomial $p_4(x)$ when $x = t$.

Proof

Let $\quad q_3(x) = b_4x^3 + b_3x^2 + b_2x + b_1$

We shall prove that:

$$p_4(x) = (x - t)q_3(x) + b_0$$

By substitution

$$(x - t)q_3(x) + b_0 = (x - t)(b_4x^3 + b_3x^2 + b_2x + b_1) + b_0$$
$$= b_4x^4 + b_3x^3 + b_2x^2 + b_1x - b_4tx^3 - b_3tx^2 - b_2tx - b_1t + b_0$$
$$= b_4x^4 + (b_3 - b_4t)x^3 + (b_2 - b_3t)x^2 + (b_1 - b_2t)x + (b_0 - b_1t)$$

But from the conditions of the theorem:

$$b_4 = a_4 \qquad b_3 - b_4t = a_3 \qquad b_2 - b_3t = a_2 \qquad b_1 - b_2t = a_1 \qquad b_0 - b_1t = a_0$$

Hence

$$(x - t)q_3(x) + b_0 = a_4x^4 + a_3x^3 + a_2x^2 + a_1x + a_0$$
$$= p_4(x)$$

and the assertion is proved.

By putting $x = t$ in this equation, we obtain

$0 + b_0 = p_4(t)$ and the theorem is proved.

To see how b_4, b_3, b_2, and b_1 enter the picture, consider again the equation

$$p_4(x) = (x - t)q_3(x) + b_0$$

Differentiate both sides with respect to x (regarding t as a constant). Then, using the rule for differentiating a product:

$$p_4'(x) = (x - t)q_3'(x) + q_3(x)$$

Put $x = t$:

$$\begin{aligned} p_4'(t) &= 0 + q_3(t) \\ &= q_3(t) \\ &= b_4t^3 + b_3t^2 + b_2t + b_1 \end{aligned}$$

Hence $\quad p_4'(t) = b_4t^3 + b_3t^2 + b_2t + b_1$

Because $p_4(x)$ is a polynomial of degree four, we know from calculus that $p_4'(x)$ will be a polynomial of degree three. What is surprising, however, is that the coefficients of this cubic polynomial are b_4, b_3, b_2 and b_1. Recall that b_4, b_3, b_2 and b_1 were introduced originally as a means of storing intermediate stages of the calculation. So their appearence as coefficients of the polynomial representing $p_4'(x)$ is rather surprising. We shall see in the examples that follow, and more so in Chapter 6, that this result is also very useful.

Example 5.8

Let $p(x) = 4x^3 - 2x + 1$. Evaluate (a) $p(2)$ and (b) $p'(2)$.

(a) To evaluate $p(2)$, write

	coeff of x^3	coeff of x^2	coeff of x	const term
$x = 2$	4	0	-2	1
		$4 \times 2 = 8$	$8 \times 2 = 16$	$14 \times 2 = 28$
	$b_3 = 4$	$b_2 = 8$	$b_1 = 14$	$b_0 = 29$

So $p(2) = 29$.

(b) By the above reasoning,

$$p'(t) = b_3t^2 + b_2t + b_1$$

To evaluate $p'(2)$, we use Horner's method to evaluate the polynomial $b_3t^3 + b_2t + b_1$ at $t = 2$. This is achieved most efficiently by writing the stages of the calculation underneath the original calculation i.e. underneath b_3, b_2 and b_1 in the previous table. We shall use c_3, c_2 and c_1 to store the intermediate stages of the calculation. Observe that because b_1 is the constant term in the polynomial being evaluated, the process stops when c_1 has been calculated. c_1 will then hold the value $p'(2)$.

	coeff of x^3	coeff of x^2	coeff of x	const term
$x = 2$	4	0	-2	1
$x = 2$	$b_3 = 4$	$b_2 = 8$	$b_1 = 14$	$b_0 = 29$
		$4 \times 2 = 8$	$16 \times 2 = 32$	
	$c_3 = 4$	$c_2 = 16$	$c_1 = 46$	

We see then that $p'(2) = c_1 = 46$.

In the next example, the calculation is performed without further comment.

Example 5.9

The polynomial f(x) is given by

$$f(x) = 4x^3 - 2x^2 + 9$$

Evaluate (a) f(–2) (b) f'(–2).

	coeff of x^3	coeff of x^2	coeff of x	const term
$x = -2$	4	–2	0	9
		$4 \times (-2) = -8$	$-10 \times (-2) = 20$	$20 \times (-2) = -40$
$x = -2$	4	–10	20	–31
		$4 \times (-2) = -8$	$(-18)(-2) = 36$	
	$c_3 = 4$	$c_2 = -18$	$c_1 = 56$	

Hence f(–2) = –31 and the value of f'(–2) is given by $c_1 = 56$. The reader may verify by direct differentiation that this is indeed the case.

Example 5.10

If $f(x) = 5x^5 - 4x^3 + 2x - 7$, find the value of f'(–5).

We perform Horner's method twice, eliminating b_0 from the second calculation.

	coeff of x^5	coeff of x^4	coeff of x^3	coeff of x^2	coeff of x	const term
$x = -5$	5	0	–4	0	2	–7
		$5 \times (-5) = -25$	$(-25) \times (-5) = 125$	$121 \times (-5) = -605$	$-605 \times (-5) = 3025$	$3027 \times (-5) = -15135$
$x = -5$	$b_5 = 5$	$b_4 = -25$	$b_3 = 121$	$b_2 = -605$	$b_1 = 3027$	$b_0 = -15142$
		$5 \times (-5) = -25$	$(-50) \times (-5) = 250$	$371 \times (-5) = -1855$	$(-2460) \times (-5) = 12300$	
	$c_5 = 5$	$c_4 = -50$	$c_3 = 371$	$c_2 = -2460$	$c_1 = 15327$	

Hence f'(–5) = 15327.

So far, in this section, each polynomial has been centred at $x = 0$. Horner's method may be applied immediately, however, to a polynomial centred at $x = c$ where $c \neq 0$. Consider the polynomial

$$a_4(x - c)^4 + a_3(x - c)^3 + a_2(x - c)^2 + a_1(x - c) + a_0$$

By writing $y = x - c$, this polynomial may be written

$$a_4y^4 + a_3y^3 + a_2y^2 + a_1y + a_0$$

and with suitable choice of y, the method proceeds as before. An example should make this clear.

Example 5.11

The polynomial p(x) is centred at $x = -1$ and

$$p(x) = 5(x + 1)^3 - 2(x + 1)^2 + 3(x + 1) - 4$$

Evaluate (a) p(0.5) (b) p'(0.5).

Write $y = x + 1$; then the polynomial becomes

$$5y^3 - 2y^2 + 3y - 4$$

Since the polynomial is to be evaluated at $x = 0.5$, write $y = x + 1 = 1.5$ and use Horner's method as before to evaluate this polynomial and its derivative at $y = 1.5$.

$y = 1.5$	5	-2	3	-4
		$5 \times 1.5 = 7.5$	$5.5 \times 1.5 = 8.25$	$11.25 \times 1.5 = 16.875$
$y = 1.5$	$b_3 = 5$	$b_2 = 5.5$	$b_1 = 11.25$	$b_0 = 12.875$
		$5 \times 1.5 = 7.5$	$13 \times 1.5 = 19.5$	
	$c_3 = 5$	$c_2 = 13$	$c_1 = 30.75$	

Hence p(0.5) = b_0 = 12.875 and p'(0.5) = c_1 = 30.75.

We shall study further applications of this technique in the exercises and in Chapter 6.

There is a second interpretation of the equation

$$p_4(x) = (x - t)q_3(x) + b_0$$

This is that if $p_4(x)$ is divided by $x - t$ then $q_3(x)$ is the quotient polynomial and b_0 the remainder. Dividing both sides of the above equation by $x - t$ gives:

$$\frac{p_4(x)}{x - t} = q_3(x) + \frac{b_0}{x - t}$$

and it may be seen that the assertion holds. The usefulness of this result is illustrated in Chapter 6.

As mentioned earlier, Horner's method provides a very powerful and efficient algorithm for evaluating polynomials. The algorithm is efficient in three ways.

(a) It is straightforward to implement on a computer.
(b) In comparison with the usual method of polynomial evaluation, Horner's method provides a quite dramatic reduction in the number of arithmetic operations required.
(c) By using Horner's method, the propagation of rounding errors in polynomial evaluation is kept almost to a minimum.

Opposite is an algorithm for Horner's method, satisfying (a), followed by an example to illustrate (b). Further questions on (b) will be found in the exercises at the end of the chapter. Point (c) lies beyond the scope of this book[†].

Algorithm 5.1 Horner's method

To evaluate the polynomial $p_n(x) = a_n x^n + a_{n-1}x^{n-1} + ... + a_1 x + a_0$ and the derivative $p_n'(x)$ at $x = t$.

input: the degree of the polynomial n;
the coefficients $a_n, a_{n-1}, a_{n-2}, ..., a_0$;
the number t

$b_n := a_n$
$c_n := b_n$
for $k = n - 1, n - 2, ..., 1$
  $b_k := b_{k+1}t + a_k$
  $c_k := c_{k+1}t + b_k$
endloop
$b_0 := b_1 t + a_0$
output: $b_0 = p_n(t)$, $c_1 = p_n'(t)$, $b_1, b_2, ..., b_n$

To give some idea of the efficiency of Horner's method, we shall compare this method of polynomial evaluation with the usual method. For purposes of comparison, we shall evaluate the polynomial $2x^4 - 3x^3 + 8x^2 - 5x + 7$ at $x = 3$.

In the usual method, we would evaluate

(a) $2(3)^4 = 162$ involving four multiplications
$3(3)^3 = 81$ involving three multiplications
$8(3)^2 = 72$ involving two multiplications
$5(3) = 15$ involving one multiplication
(b) $162 - 81 + 72 - 15 + 7$ involving four additions/subtractions

In total: $4 + 3 + 2 + 1 =$ ten multiplications
four additions/subtractions

In Horner's method, we would

let $b_4 = 2$

$b_3 = 2 \times 3 - 3 = 3$ involving one multiplication and one addition/subtraction

$b_2 = 3 \times 3 + 8 = 17$ involving one multiplication and one addition/subtraction

$b_1 = 17 \times 3 - 5 = 46$ involving one multiplication and one addition/subtraction

$b_0 = 46 \times 3 + 7 = 145$ involving one multiplication and one addition/subtraction

[†] A full discussion of this fascinating topic may be found in Chapter 1 of Golub (ed.) *Studies in numerical analysis*, Mathematical Association of America. This chapter was written by JH Wilkinson, one of the great pioneers of numerical analysis in this century and this chapter, in particular, cannot be recommended too highly. It conveys some of the excitement of a young and developing subject and it contains some extraordinary horror stories of what can go wrong when calculations are performed without any consideration of errors. Definitely recommended reading!

In total: four multiplications and four additions/subtractions

Hence, there is a 60% reduction in the number of multiplications required. This is an impressive reduction in what is a comparatively simple calculation and this illustrates something of the power and usefulness of numerical analysis. Futher questions on this topic are given in the exercises.

Exercise 5

Section 5.1

1 Write the function $f(x) = \sin x$ in the form $p_3(x) = a_0 + a_1x + a_2x^2 + a_3x^3$ where a_0, a_1, a_2, and a_3 are to be determined.
Using (a) your calculator and (b) your polynomial $p_3(x)$ find the value of
(a) $\sin(0.1)$
(b) $\sin(0.5)$
(c) $\sin(1)$ (all in radians)
Comment on your results.

2 Express e^x in the form $a_0 + a_1x + a_2x^2 + a_3x^3$ where a_0, a_1, a_2 and a_3 are to be determined. Use your polynomial to estimate the values of (a) $e^{0.2}$ (b) $e^{0.4}$ (c) e^{-1} (d) e^4.
Comment on the accuracy of your approximations.

3 Write the function $f(x) = \ln(1 + x)$ in the form $a_0 + a_1x + a_2x^2 + a_3x^3$. Draw the graphs of $y = \ln(1 + x)$ and of your polynomial. For which range of values of x do you think the polynomial might be used as an approximation to $\ln(1 + x)$? Repeat this question for the function $f(x) = \ln(1 - x)$.

4 Find $p_3(x)$, the Taylor polynomial of degree three centred at the origin, of the function $f(x) = \dfrac{1}{\sqrt{(1 + x)}}$.
Plot, on the same axes, the graphs of $y = \dfrac{1}{\sqrt{(1 + x)}}$ and $y = p_3(x)$. Use your graphs to estimate the error $\left| f(x) - p_3(x) \right|$
(a) when $x = 0.5$
(b) when $x = 1$
(c) when $x = 1.5$
Comment.

5 Calculate the Taylor polynomial of degree two centred at the origin for the function $f(x) = xe^{-x}$.On the same axes draw the graphs of $y = xe^{-x}$ and of your polynomial.

For $0 \leqslant x \leqslant 2$, find the maximum error in using your polynomial to approximate xe^{-x}.

6 It is known that, for a certain function f(x)

$$f(0) = 1.6 \qquad f'(0) = 0.52 \qquad f''(0) = 0.35$$

Write down the Taylor polynomial of degree two for the function f(x). Use your polynomial to estimate the value of f(x) when
(a) $x = 0.15$
(b) $x = 0.25$
(c) $x = 0.35$
Which of parts (a), (b), or (c) is likely to be the most accurate?

7 For a quantity S, it is believed that when $x = 0$

$$S = 5.3 \qquad \frac{dS}{dx} = 1.6 \quad \text{and} \quad \frac{d^2S}{dx^2} = 0.4$$

Write down a polynomial whose values will approximate those of S. Use your polynomial to estimate
(a) the value of S when (i) $x = 0.2$ (ii) $x = 0.25$
(b) the value of $\frac{dS}{dx}$ when (i) $x = 0.15$ (ii) $x = 0.2$

8 Write down the Taylor polynomials of degree five centred at $x = 0$ of the following functions:
(a) $f(x) = \sin x$
(b) $f(x) = \cos x$
(c) $f(x) = \tan x$

Use your polynomials to find the value of the following functions as $x \to 0$

(d) $\dfrac{\sin x}{x}$

(e) $\dfrac{\cos x - 1}{x^2}$

(f) $\dfrac{\tan x - x}{x - \sin x}$

9 Find the Taylor polynomial of degree four centred at $x = 0$ of the function $f(x) = \ln(1 + \sin x)$.

Hence determine the value of the function $\dfrac{\ln(1 + \sin x) - x}{x^2}$ as $x \to 0$.

Section 5.2

10 Write down for the function $f(x) = \dfrac{1}{1 + x^2}$, the Taylor polynomial of degree three
(a) centred at $x = 0$ and
(b) centred at $x = 1$.

On the same axes, plot the graphs of $f(x) = \dfrac{1}{1 + x^2}$ and the two polynomials.
Describe carefully the circumstances under which you would use each of the polynomials to approximate the function $\dfrac{1}{1 + x^2}$.

11 For the function $f(x) = e^{\sqrt{x}}$ calculate the Taylor polynomials
(a) of degree one
(b) of degree two
(c) of degree three
all centred at $x = 2$.

For $0 \leqslant x \leqslant 4$ draw the graphs of $y = e^{\sqrt{x}}$ and of your three polynomials. For each of (a), (b) and (c) write down the value of x in the interval [0, 4] at which the error $|f(x) - p(x)|$ is a maximum.

12 Calculate the Taylor polynomial of degree two centred at $x = 0.1$ (in radians) of the function $\dfrac{\sin x}{x}$. Use your polynomial to calculate the value of this function as $x \to 0$.
Compare your answer with that given in Question 8.

13 The function f(x) and its derivatives take the following values when $x = 2$:

$$f(2) = 20.5 \qquad f'(2) = 1.67 \qquad f''(2) = 0.78$$

Write down the Taylor polynomial of degree two centred at $x = 2$ for the function f(x). Use your polynomial to estimate the values of
(a) f(2.1)
(b) f(2.2)
(c) f(1.97)

14 It is known that when $x = 1$ the function f(x) and it derivatives take the following values:

$$f(1) = 1 \qquad f'(1) = 0.8 \qquad f''(1) = 0.1 \qquad f'''(1) = 0.07$$

Write down the Taylor polynomial of f(x) of degree three (centred at $x = 1$). Use your polynomial to calculate approximately the values of
(a) f(0.8)
(b) f(1.1)
(c) f'(1.05)
Using all the information you possess, draw a graph to approximate the function $y = f(x)$. Indicate a region in which you think your approximation will be reasonably accurate.

15 It is given that, for the function f(x), when $x = 10$

$$f(10) = 41.15 \qquad f'(10) = 4.05 \qquad f''(10) = -0.005 \qquad f'''(10) = 0.001$$

(a) Write down a polynomial approximation to f(x). Use your polynomial to estimate
  (i) f(10.5) and f'(10.5)
  (ii) f(8.5) and f'(8.5)
(b) Which of (i) or (ii) is likely to be the more accurate?
(c) In fact, $f(x) = 4x + \ln(\sqrt{x})$. Is your answer to (b) correct?

16 It is required to approximate the function $f(x) = \ln(1 + \sqrt{x})$ by a polynomial of degree two so as to evaluate (approximately) the integral $\int_1^2 \ln(1 + \sqrt{x})\,dx$. About which point should the polynomial be centred? Find a suitable polynomial and use this polynomial to approximate the integral.

Section 5.3

17 Write down, for the function $f(x) = e^{-x}$ the Taylor polynomial of degree two with centre $x = -1.5$. For values of x lying in the interval $-2 \leqslant x \leqslant -1$, calculate the greatest value that can be taken by the error term in the Taylor polynomial.

18 $f(x) = \dfrac{1}{1 + x}$. Write down $p(x)$, the Taylor polynomial of $f(x)$ of degree three with centre $x = 2$. If $1 \leqslant x \leqslant 3$, calculate the greatest value that can be taken by the error term $\left| f(x) - p(x) \right|$. Draw the graphs of both functions and compare the theoretical maximum error with the observed maximum error.

Section 5.4

19 Use Horner's method to evaluate the polynomial $3x^2 - 5x + 2$
(a) when $x = 1$
(b) when $x = 2$

20 Evaluate the polynomial $4x^3 - 5x^2 - 2x - 3$
(a) when $x = -1$
(b) when $x = 1$
(c) when $x = 2.5$

21 The polynomial $-x^4 + 3x^3 - 2x + 5$ is to be evaluated
(a) when $x = -2$
(b) when $x = 3$
Evaluate the polynomial at these values of x using Horner's method.

22 If $P(x) = 7x^3 + 103x^2 + 101$, use Horner's method to calculate
(a) $P(-4)$ (b) $P(-1.5)$ (c) $P(0.5)$ (d) $P(3.2)$

23 Write a program that will evaluate the polynomial $4x^3 - 2x + 3$ at the points $x = -1, x = -0.99, x = -0.98, \ldots, x = 0.99, x = 1$.

Section 5.5

24 If $p(x) = -2x^3 + 4x - 3$, evaluate
(a) $p(-1.5)$, $p'(-1.5)$
(b) $p(2.5)$, $p'(2.5)$

25 For the polynomial $p(x) = 5x^5 - 3x^3 + 2x$, evaluate
(a) $p'(-1)$
(b) $p'(0)$
(c) $p'(2)$

26 If $y = \frac{1}{2}x^4 - 4x^2 + 2x - 3$, find the value of $\frac{dy}{dx}$
(a) when $x = 1$
(b) when $x = 4.5$
(c) when $x = -2$

27 It is given that an object moves so that its distance s (in metres) from a fixed point is related to time t (in seconds) by the formula

$$s = -3t^3 + 4t^2 - 6$$

Calculate the speed of the body after
(a) 1s (b) 2 s (c) 3 s

28 If $f(x) = \frac{5}{3}x^3 - 2x^2 - 12x + 1$ evaluate
(a) $f'(1)$
(b) $f'(2)$
What do your answers tell you about the position of a stationary value of $y = f(x)$? Find the position of the stationary value(s) and check your answer.

29 Propose an extension of Horner's method that will evaluate $p''(3)$ when $p(x) = 4x^3 - 3x^2 + 2x - 1$. Test your proposal. Write the proposal in the form of an algorithm and write a program that will evaluate $p''(t)$ for given polynomial $p(x)$ and given x-value t.

30 Let $p(x) = a_3x^3 + a_2x^2 + a_1x + a_0$ and let b_3, b_2, b_1, b_0 and c_3, c_2, c_1 be as in the proof of Horner's theorem. Write $d_3 = c_3$, $d_2 = d_3t + c_2$. Let $q_1(x) = c_3x + c_2$.
Show that
(a) $b_3x^2 + b_2x + b_1 = (x - t)q_1(x) + c_1$
(b) $p(x) = (x - t)^2q_1(x) + c_1(x - t) + b_0$
(c) $p'(t) = c_1$
(d) $p''(t) = 2(c_3t + c_2) = 2d_2$

31 Use Horner's method to evaluate the polynomial

$$2(x + 3)^2 - 4(x + 3) - 8$$

(a) when $x = 1$
(b) when $x = -1$
(c) when $x = -4$

32 If $p(x) = 2(x - 2)^3 + 5(x - 2)^2 - 3(x - 2) + 2$ use Horner's method to evaluate
(a) $p(3)$ (b) $p(-1)$ (c) $p(-2)$

33 For the polynomial $p(x) = 3(x - \frac{1}{2})^3 + 5(x - \frac{1}{2}) + 2$ evaluate
(a) $p(1.5)$ and $p'(1.5)$
(b) $p(3)$ and $p'(3)$
(c) $p(-1.7)$ and $p'(-1.7)$

34 If $y = 2(x + 5)^6 - 3(x + 5)^3 - 2(x + 5) + 1$ find
(a) the value of y
(b) the value of $\frac{dy}{dx}$ when $x = -3$

35 The polynomial $p(x) = 5x^2 - 3x + 2$ is to be evaluated at several values of x. Calculate the number of multiplications and the number of additions/subtractions that will be required if the evaluation is made using Horner's method. Repeat the exercise in the case where the evaluation is made by 'the standard method'.

36 Adapt Algorithm 5.1 so that the polynomial

$$p(x) = a_n(x-c)^n + a_{n-1}(x-c)^{n-1} + \ldots + a_1(x-c) + a_0$$

and its derivative may be evaluated at $x = t$.

Miscellaneous

37 Given that $f(x) = \tan x$, write expressions for $f'(x)$, $f''(x)$ and $f'''(x)$. Deduce a Taylor approximation for $\tan\alpha$, when α is small, in the form

$$\tan\alpha \approx a + b\alpha + c\alpha^2 + d\alpha^3$$

where a, b, c, d are constants. (SMP)

38 The expression

$$1 - \tfrac{3}{2}x^2 + \tfrac{5}{2}x^3$$

is the Taylor approximation to $f(x)$ for small values of x.
Write down the values of $f'(0)$, $f''(0)$, $f'''(0)$.
Sketch the graph of f near $x = 0$. (SMP)

39 Calculate Taylor polynomials of degree three with their error terms for the following polynomials. (The centres of the polynomials are as indicated in the question.)

(a) $f(x) = e^{-x}$ centre $x = 1$

(b) $f(x) = \dfrac{1}{\sqrt{x+1}}$ centre $x = 3$

(c) $f(x) = \sin^2 x$ centre $x = \dfrac{\pi}{2}$ (in radians)

40 For the function $f(x) = xe^{-x}$, $0 \leqslant x \leqslant 1$

(a) Calculate $p_2(x)$ the Lagrange polynomial of degree two that interpolates $f(x)$ at $x = 0$, $x = 0.5$, $x = 1$.

(b) Calculate $T_2(x)$, the Taylor polynomial of degree two of $f(x)$ centred at $x = 0$.

Draw on the same axes, the graphs $y = xe^{-x}$, $y = p_2(x)$, $y = T_2(x)$.
Comment.

41 If

$$\begin{aligned} f(5) &= 0.333\,333 \\ f'(5) &= -0.055\,555\,6 \\ f''(5) &= 0.018\,518\,5 \\ f'''(5) &= -0.009\,259\,26 \\ f^{iv}(5) &= 0.006\,172\,84 \end{aligned}$$

(a) Write down $p_3(x)$, the Taylor polynomial of degree three for the function $f(x)$.

(b) Use your polynomial to calcuate approximately the values of (i) f(5.1) (ii) f(5.5).

(c) Write down an expression for the error term in $p_3(x)$.

In fact, $f(x) = \dfrac{2}{1+x}$.

(d) Calculate the maximum value that can be taken by the error term in each of your approximations in (b).

(e) Calculate f(5.1) and f(5.5) exactly and check that the errors in your approximations lie within the bounds calculated in (d).

42 Let $f(x)$ be a function and let $p_3(x)$ be the Taylor polynomial of degree three of $f(x)$ centred at $x = 2$. Show that if the error in using $p_3(x)$ to approximate $f(x)$ in the interval [2, 2.4] is to be less than 0.0005, then

$$\left|f^{iv}(x)\right| \leqslant 0.468\,75 \qquad \text{for all } x \text{ satisfying} \qquad 2 \leqslant x \leqslant 2.4$$

Suggest a function $f(x)$ that will satisfy this requirement and one that will not.

43 (a) Find the Taylor polynomial of degree four, centred at $x = 0$ for the function $f(x) = e^x$.

If $r(x)$ denotes the error in using this polynomial to approximate e^x write down an expression for $r(x)$.

If $0 < x < 1$, prove that $\left|r(x)\right| < \dfrac{e}{5!}$.

(b) Write down $p_n(x)$, the Taylor polynomial of degree n, centred at $x = 0$ for $f(x) = e^x$.

Calculate the degree n of the Taylor polynomial that will ensure that if $0 < x < 1$ then

$$\left|e^x - p_n(x)\right| < 0.000\,05$$

44 Given that $\dfrac{dy}{dx} = \cos x + y \sin x$, subject to $y = 0$ when $x = 0$, find the Maclaurin expansion for y in terms of x as far as the term involving x^5. Hence find the value of y when $x = 0.1$, correct to six significant figures.

Justify by means of a sketch the approximation

$$\left(\frac{dy}{dx}\right)_0 \approx \frac{1}{2h}(y_1 - y_{-1})$$

and apply it to find the approximate value of y when $x = 0.2$, correct to four significant figures.

Check the accuracy of this value by using the Maclaurin expansion to evaluate y when $x = 0.2$.

(MEI)

45 Prove Taylor's theorem (page 183) for general n.

46 Throughout this question, y_n denotes the value of y at $x = nh$ with $h = 0.1$. Furthermore, y satisfies the differential equation

$$y'' = xy$$

with $y_0 = 2$ and $y_0' = -3$. By repeated differentiation of the differential equation, obtain the values of y_0'', y_0''', $y_0^{(iv)}$, $y_0^{(v)}$.
Use the Maclaurin series of y to obtain the values of y_1 correct to four decimal places, explaining why your are confident about the accuracy of your answer.
Find an approximate value of y_1'', and by considering y''', $y^{(iv)}$ and $y^{(v)}$, estimate y_2 and y_3.

(MEI adapted)

47 Given that $f(x) = 2x^4 + x^3 - 11x^2 - 4x + 12$, use nested multiplication to check your evaluation of $f(-4)$.How many elementary arithmetical operations (addition, subtraction, multiplication or division of two numbers) were needed to find $f(-4)$?
Suppose $g(x)$ is any polynomial of degree n, and x_0 is any value of x. Estimate the number of elementary arithmetical operations needed to evaluate $g(x_0)$ by (a) nested multiplication, and (b) evaluating each term of $g(x)$ and adding. Comment on your results.

(MEI part)

48 Explain, by reference to a general cubic polynomial, the method of 'nested multiplication' for the evaluation of a polynomial.
Show that, for the polynomial

$$A_n x^n + A_{n-1} x^{n-1} + \dots + A_0$$

where n is a positive integer and $A_n, A_{n-1}, \dots, A_0$ are real numbers, evaluation by the method of nested multiplication requires a smaller number of multiplication operations than evaluation term-by-term.

(MEI)

Chapter 6

Solution of equations 2

In this chapter, we consider a fourth method of obtaining numerical solutions of an equation. This method is called The Newton-Raphson method after the great Isaac Newton (1642–1727) and the slightly less great Joseph Raphson (c. 1715). This method, like those in Chapter 3, is an iterative method and will consist of calculating a sequence of approximations x_0, x_1, x_2, ... which will (hopefully) converge to the required solution. The great appeal of this method, as we shall see in Section 6.4, lies in the speed of its convergence.

6.1 The Newton-Raphson method

Let $f(x) = 0$ be the equation to be solved and let λ represent the required solution. Let x_0 represent a first approximation to λ,

then $\lambda = x_0 + e_0$ where e_0 is the error in x_0

Now $f(\lambda) = 0$ since λ is a solution.

Therefore, $f(x_0 + e_0) = 0$ . . . (1)

By Taylor's theorem:

$$f(x_0 + e_0) = f(x_0) + e_0\, f'(\eta) \qquad x_0 < \eta < x_0 + e_0$$

and if $\left| e_0 \right|$ is small, we may write

$$f(x_0 + e_0) \approx f(x_0) + e_0 f'(x_0) \qquad \ldots (2)$$

Putting Equations (1) and (2) together:

$$0 \approx f(x_0) + e_0 f'(x_0)$$

and $$e_0 \approx \frac{-f(x_0)}{f'(x_0)}$$

Since $\lambda = x_0 + e_0$ it makes sense to take as the next approximation

$$x_1 = x_0 - \frac{f(x_0)}{f'(x_0)}$$

Using the same reasoning, we write next:

$$x_2 = x_1 - \frac{f(x_1)}{f'(x_1)}$$

$$x_3 = x_2 - \frac{f(x_2)}{f'(x_2)}$$

and, in general, when x_n is known, x_{n+1} is calculated from the equation:

$$x_{n+1} = x_n - \frac{f(x_n)}{f'(x_n)} \qquad n \geqslant 0 \qquad \ldots (3)$$

The Newton-Raphson method for solving the equation $f(x) = 0$ consists of generating the sequence $x_0, x_1, x_2, \ldots$ using Equation (3) and stopping either when a satisfactory solution has been obtained, or when it is clear that the sequence is not going to converge to λ.

Algorithm 6.1 The Newton-Raphson method

To obtain a solution of the equation $f(x) = 0$

input: x_0 (a starting value)
for $n = 0, 1, \ldots,$ until satisfied
$b := f(x_n)$
$c := f'(x_n)$
$x_{n+1} := x_n - \frac{b}{c}$
endloop
output: $x_0, x_1, x_2, \ldots$: a sequence of approximations to the required solution

Example 6.1

Use the Newton-Raphson method to calculate to three decimal digits the positive solution of the equation $x^2 = e^{-x}$.

Written in the form $f(x) = 0$ the equation takes the form $x^2 - e^{-x} = 0$.

So $\quad f(x) = x^2 - e^{-x}$
$\quad f'(x) = 2x + e^{-x}$

and the iterative equation is

$$x_{n+1} = x_n - \frac{x_n^2 - e^{-x_n}}{2x_n + e^{-x_n}}$$

A graph of $y = x^2 - e^{-x}$ appears in Figure 6.1.

From the graph, it is clear that a sensible starting value is given by $x_0 = 1$.

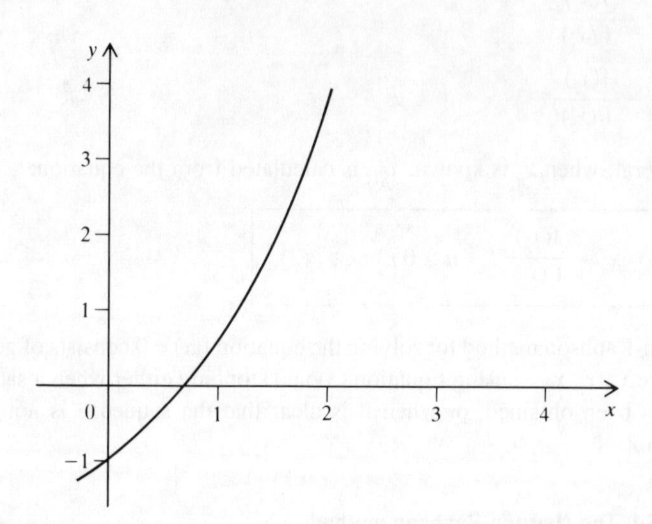

Figure 6.1

Since $$x_1 = x_0 - \frac{x_0^2 - e^{-x_0}}{2x_0 + e^{-x_0}}$$

we have $$x_1 = 1 - \frac{1 - e^{-1}}{2 + e^{-1}}$$

$$= 0.73304$$

Also $$x_2 = x_1 - \frac{x_1^2 - e^{-x_1}}{2x_1 + e^{-x_1}}$$

$$= 0.73304 - \frac{(0.73304)^2 - e^{-0.73304}}{2 \times 0.73304 + e^{-0.73304}}$$

$$= 0.70381$$

Also $$x_3 = x_2 - \frac{x_2^2 - e^{-x_2}}{2x_2 + e^{-x_2}}$$

$$= 0.70381 - \frac{(0.70381)^2 - e^{-0.70381}}{2 \times 0.70381 + e^{-0.70381}}$$

$$= 0.70347$$

Similary, we obtain

$$x_4 = 0.70347$$

and we deduce that the required solution is

$$x = 0.703 \quad \text{to three decimal digits.}$$

Example 6.2

Use
(a) the fixed point iterative method
(b) the secant method
(c) the Newton-Raphson method
to obtain, correct to four decimal digits, the solution of $2x - 3\cos\frac{x}{2} = 0$ lying in the interval [1,2]. Observe that x is measured in radians. Comment on the convergence properties of the three methods.

(a) The fixed point iterative method
Write the equation in the form $x = g(x)$.

We write $\quad x = \frac{3}{2}\cos\frac{x}{2}$

Taking $x_0 = 1$ and writing answers to six decimal digits

$$n = 1 \qquad x_1 = \frac{3}{2}\cos\frac{x_0}{2} = \frac{3}{2}\cos\frac{1}{2} = 1.31637$$

$$n = 2 \qquad x_2 = \frac{3}{2}\cos\frac{x_1}{2} = \frac{3}{2}\cos\frac{1.31637}{2} = 1.18665$$

$$n = 3 \qquad x_3 = \frac{3}{2}\cos\frac{1.18665}{2} = 1.24363$$

$$n = 4 \qquad x_4 = \frac{3}{2}\cos\frac{1.24363}{2} = 1.21923$$

and so on until

$$x_{10} = 1.22657$$
$$x_{11} = 1.22664$$
$$x_{12} = 1.22661$$

From which it is safe to judge that

$$x = 1.227 \qquad \text{to four decimal digits}$$

(b) The secant rule
For the equation $f(x) = 0$, the iterative equation for the secant method is:

$$x_{n+1} = \frac{x_{n-1}f(x_n) - x_n f(x_{n-1})}{f(x_n) - f(x_{n-1})}$$

Since $f(x) = 2x - 3\cos\left(\frac{x}{2}\right)$, the iterative equation becomes

$$x_{n+1} = \frac{x_{n-1}\left(2x_n - 3\cos\frac{x_n}{2}\right) - x_n\left(2x_{n-1} - 3\cos\frac{x_{n-1}}{2}\right)}{\left(2x_n - 3\cos\frac{x_n}{2}\right) - \left(2x_{n-1} - 3\cos\frac{x_{n-1}}{2}\right)}$$

Taking $x_0 = 1$, $x_1 = 2$, we get:

$$x_2 = \frac{x_0\left(2x_1 - 3\cos\frac{x_1}{2}\right) - x_1\left(2x_0 - 3\cos\frac{x_0}{2}\right)}{\left(2x_1 - 3\cos\frac{x_1}{2}\right) - \left(2x_0 - 3\cos\frac{x_0}{2}\right)}$$

$$= \frac{1(2 \times 2 - 3\cos 1) - 2(2 \times 1 - 3\cos\frac{1}{2})}{(2 \times 2 - 3\cos 1) - (2 \times 1 - 3\cos\frac{1}{2})}$$

$$= 1.21009 \qquad \text{to six decimal digits}$$

$$x_3 = \frac{x_1\left(2x_2 - 3\cos\frac{x_2}{2}\right) - x_2\left(2x_1 - 3\cos\frac{x_1}{2}\right)}{\left(2x_2 - 3\cos\frac{x_2}{2}\right) - \left(2x_1 - 3\cos\frac{x_1}{2}\right)}$$

$$= 1.22547 \qquad \text{to six decimal digits}$$

Similarly,

$$x_4 = 1.22662$$
$$x_5 = 1.22662$$

Hence the required solution is

$$x = 1.227 \qquad \text{to four decimal digits}$$

(c) The Newton-Raphson method

We have $\quad f(x) = 2x - 3\cos\frac{x}{2}$

$$f'(x) = 2 + \frac{3}{2}\sin\frac{x}{2}$$

and the iterative equation is:

$$x_{n+1} = x_n - \frac{f(x_n)}{f'(x_n)}$$

which becomes

$$x_{n+1} = x_n - \frac{2x_n - 3\cos\frac{x_n}{2}}{2 + \frac{3}{2}\sin\frac{x_n}{2}}$$

Taking $x_0 = 1$

$$n = 1 \quad x_1 = x_0 - \frac{2x_0 - 3\cos\frac{x_0}{2}}{2 + \frac{3}{2}\sin\frac{x_0}{2}}$$

$$= 1 - \frac{2 - 3\cos\frac{1}{2}}{2 + \frac{3}{2}\sin\frac{1}{2}}$$

$$= 1.23270 \quad \text{to six decimal digits}$$

$$n = 2 \quad x_2 = x_1 - \frac{2x_1 - 3\cos\frac{x_1}{2}}{2 + \frac{3}{2}\sin\frac{x_1}{2}}$$

$$= 1.23270 - \frac{2 \times 1.23270 - 3\cos\frac{1.23270}{2}}{2 + \frac{3}{2}\sin\frac{1.23270}{2}}$$

$$= 1.22662$$

$$n = 3 \quad x_3 = x_2 - \frac{2x_2 - 3\cos\frac{x_2}{2}}{2 + \frac{3}{2}\sin\frac{x_2}{2}}$$

$$= 1.22662 - \frac{2 \times 1.22662 - 3\cos\frac{1.22662}{2}}{2 + \frac{3}{2}\sin\frac{1.22662}{2}}$$

$$= 1.22662$$

and we take $x = 1.227$ as the required solution.

It is clear from the calculations that the fixed point iterative method was the slowest to converge of the three methods. The secant method converged very much more quickly, but not quite as fast as the Newton-Raphson method. The secant method, in convergence terms, was seen to be closer to the Newton-Raphson method than to the fixed point iterative method. We shall learn more about this in Section 6.4.

Example 6.3

The cubic polynomial $x^3 + 3x^2 + 2 = 0$ has one real solution. It is known that this solution is negative. Use the Newton-Raphson method with $x_0 = -1$ to find this solution.

We have $\quad f(x) = x^3 + 3x^2 + 2$

and $\quad f'(x) = 3x^2 + 6x$

The iterative equation is:

$$x_{n+1} = x_n - \frac{f(x_n)}{f'(x_n)}$$

$$= x_n - \frac{x_n^3 + 3x_n^2 + 2}{3x_n^2 + 6x_n}$$

Taking $x_0 = -1$:

$$n = 0 \quad x_1 = -1 - \frac{(-1)^3 + 3(-1)^2 + 2}{3(-1)^2 + 6(-1)}$$

$$= 0.33333 \qquad \text{to five decimals digits}$$

$$n = 1 \quad x_2 = 0.33333 - \frac{(0.33333)^3 + 3(0.33333)^2 + 2}{3(0.33333)^2 + 6(0.33333)}$$

$$= -0.68255$$

$$n = 2 \quad x_3 = -0.68255 - \frac{(-0.68255)^3 + 3(-0.68255)^2 + 2}{3(-0.68255)^2 + 6(-0.68255)}$$

$$= 0.45904$$

In the same way, we obtain

$$x_4 = -0.34680$$
$$x_5 = 1.0015$$
$$x_6 = 0.33467$$

and it is clear that the sequence $x_0, x_1, x_2, \ldots$ is not going to converge.

To see what went wrong here, consider a graphical interpretation of the Newton-Raphson method.

6.2 Graphical representation of the Newton-Raphson method

To solve the equation $f(x) = 0$, let the starting value be x_0. Then x_1 is given by:

$$x_1 = x_0 - \frac{f(x_0)}{f'(x_0)}$$

Rearranging:

$$f'(x_0)(x_1 - x_0) = -f(x_0)$$

$$f'(x_0) = \frac{-f(x_0)}{(x_1 - x_0)}$$

giving

$$f'(x_0) = \frac{f(x_0)}{x_0 - x_1} \qquad \ldots (4)$$

But $f'(x_0)$ gives the gradient of the tangent to the curve $y = f(x)$ at the point where $x = x_0$. The graph of $y = f(x)$ is illustrated in Figure 6.2. A is the point on the x-axis where $x = x_0$; P is the point on the curve with x coordinate x_0 and PT is the tangent to the curve at P. T lies on the x-axis and has x coordinate X.

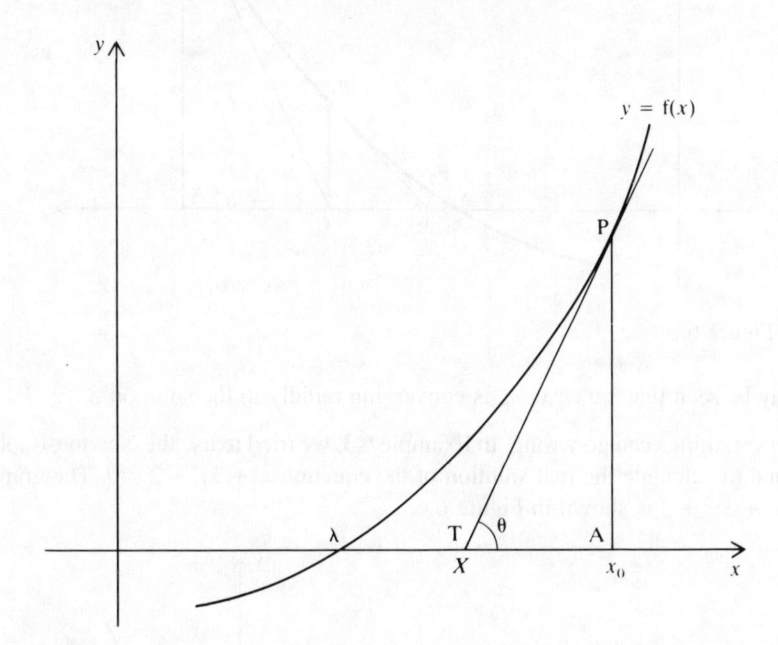

Figure 6.2

Then, by the above, $\quad f'(x_0) = \tan\theta$

$$= \frac{AP}{TA}$$

$$= \frac{f(x_0)}{x_0 - X}$$

But equation (4) gives $\quad f'(x_0) = \dfrac{f(x_0)}{x_0 - x_1}$

Therefore $\quad X = x_1$

and geometrically, x_1 is given by the point at which the tangent from $P(x_0, f(x_0))$ cuts the x-axis.

After several iterations of the Newton-Raphson method, we have a diagram similar to Figure 6.3.

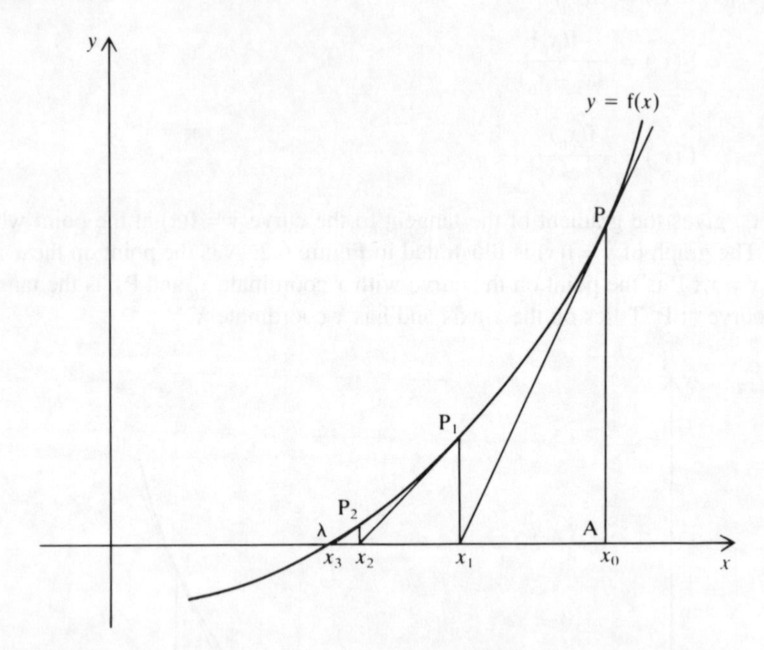

Figure 6.3

It may be seen that x_0, x_1, x_2, ... is converging rapidly to the solution λ.

However, things can go wrong. In Example 6.3, we tried to use the Newton-Raphson method to calculate the real solution of the equation $x^3 + 3x^2 + 2 = 0$. The graph of $y = x^3 + 3x^2 + 2$ is shown in Figure 6.4.

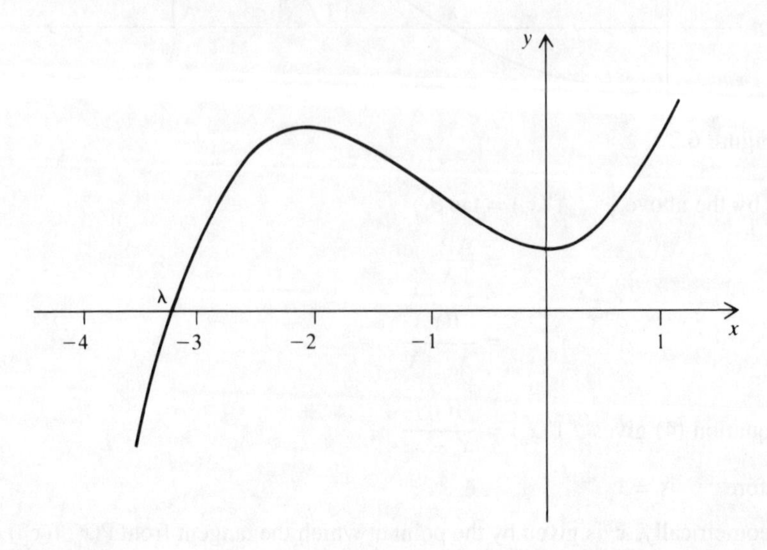

Figure 6.4

With $x_0 = -1$ we construct the sequence $x_1, x_2, x_3, \ldots$ as shown in Figure 6.5.

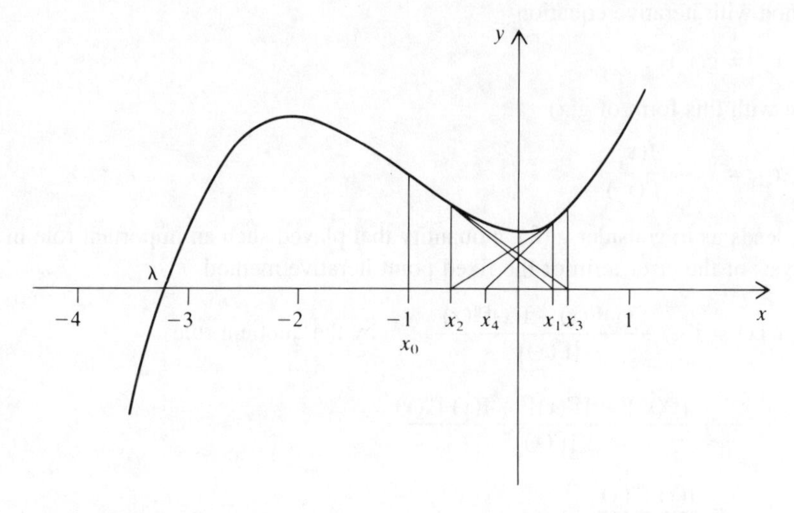

Figure 6.5

It is clear that the sequence is not going to converge quickly: and it may not converge at all.

6.3 Convergence

Precise conditions under which the sequence $x_0, x_1, x_2, \ldots$ will converge to the required solution will not be stated†. We shall, however, make the more general statement:

> If the curve representing $y = f(x)$ possesses turning points or points of inflection between the starting value x_0 and the exact solution λ or between x_0 and x_1, the sequence $x_0, x_1, x_2, \ldots$ may not converge to λ, although it may converge to some other solution.

For the curve given in Example 6.3, when $x_0 = -1$ is chosen as the starting value, there is a turning point lying between x_0 and λ. (There is also a turning point lying between x_0 and x_1.) Hence the statement above predicts correctly that the Newton-Raphson method with $x_0 = -1$ will run into difficulties.

6.4 Errors and convergence

By writing

$$g(x) = x - \frac{f(x)}{f'(x)}$$

† The interested reader is referred to Conte and de Boor 1981 *Elementary numerical analysis* 3rd edn. McGraw-Hill or to Burden and Faires 1985 *Numerical analysis* 3rd edn. Prindle, Weber and Schmidt for the details.

we see that the Newton-Raphson method is a special case of the fixed point iterative method with iterative equation

$$x_{n+1} = g(x_n)$$

since with this form of $g(x)$

$$x_{n+1} = x_n - \frac{f(x_n)}{f'(x_n)}$$

This leads us to consider $g'(\lambda)$, a quantity that played such an important role in the analysis of the error term of the fixed point iterative method.

$$g'(x) = 1 - \frac{f'(x)f'(x) - f(x)f''(x)}{[f'(x)]^2} \quad \text{by the quotient rule}$$

$$= \frac{[f'(x)]^2 - [f'(x)]^2 + f(x)\, f''(x)}{[f'(x)]^2}$$

$$= \frac{f(x)\, f''(x)}{[f'(x)]^2}$$

If $x = \lambda$ this gives

$$g'(\lambda) = \frac{f(\lambda)f''(\lambda)}{[f'(\lambda)]^2} \quad \text{if } f'(\lambda) \neq 0$$

$$= 0 \quad \text{since } f(\lambda) = 0$$

This is an encouraging result. We concluded in Section 3.12 that the speed of convergence of the fixed point iterative method $x_{n+1} = g(x_n)$ depended on the smallness of $|g'(\lambda)|$. In the Newton-Raphson method, $|g'(\lambda)|$ has achieved its smallest possible value.

Continuing the error analysis described in Section 3.12, write:

$$\lambda = x_n - e_n \qquad \lambda = x_{n+1} - e_{n+1}$$

Then $\qquad x_{n+1} = g(x_n)$

becomes $\qquad \lambda + e_{n+1} = g(\lambda + e_n)$

$$= g(\lambda) + e_n\, g'(\lambda) + \frac{1}{2!} e_n^2\, g''(\lambda) + \ldots$$

using Taylor's theorem.

But $\qquad \lambda = g(\lambda) \qquad$ since λ is a solution

and $\qquad g'(\lambda) = 0$

Hence $\qquad e_{n+1} \approx \frac{1}{2!} g''(\lambda) e_n^2$

and we have the important result that

$$|e_{n+1}| \approx C\, |e_n|^2 \quad \text{where} \quad C = \frac{1}{2!} g''(\lambda)$$

Referring to the convergence criteria in Section 3.10, we see that for the Newton-Raphson method, $p = 2$ and that, in consequence, we would expect the Newton-Raphson method to have better convergence properties than the secant method and much better convergence properties than the fixed point iterative method. Example 6.2 tends to confirm this.

A 'convergence league table' for the three methods is:

Newton-Raphson method	$p = 2$
secant method	$p \approx 1.68$
fixed point iterative method	$p = 1$

and we see that the Newton-Raphson method is the most powerful equation-solving technique so far described.

One objection to the Newton-Raphson method is that it is often difficult or inconvenient to evaluate $f'(x)$. Because of this, it is interesting to see what happens when $f'(x)$ is replaced by a suitable approximation. If x_n, x_{n-1} are the most recently calculated approximations to the solution, it is sensible to write (see in Chapter 7):

$$f'(x_n) \approx \frac{f(x_n) - f(x_{n-1})}{x_n - x_{n-1}}$$

Then, the Newton-Raphson iterative equation

$$x_{n+1} = x_n - \frac{f(x_n)}{f'(x_n)}$$

becomes

$$x_{n+1} = x_n - \frac{f(x_n)}{\dfrac{f(x_n) - f(x_{n-1})}{x_n - x_{n-1}}}$$

This equation may be written:

$$\begin{aligned} x_{n+1} &= x_n - \frac{(x_n - x_{n-1})\, f(x_n)}{f(x_n) - f(x_{n-1})} \\ &= \frac{x_n(f(x_n) - f(x_{n-1})) - (x_n - x_{n-1})f(x_n)}{f(x_n) - f(x_{n-1})} \\ &= \frac{x_n f(x_n) - x_n f(x_{n-1}) - x_n f(x_n) + x_{n-1} f(x_n)}{f(x_n) - f(x_{n-1})} \\ &= \frac{x_{n-1} f(x_n) - x_n f(x_{n-1})}{f(x_n) - f(x_{n-1})} \end{aligned}$$

giving finally

$$x_{n+1} = \frac{x_n f(x_{n-1}) - x_{n-1} f(x_n)}{f(x_{n-1}) - f(x_n)}$$

which is the iterative equation associated with the secant method.

We may, then, think of the Newton-Raphson method as the special case of the secant method that occurs when x_{n-1} and x_n coincide,

since $\quad f'(x_n) = \lim_{x_{n-1} \to x_n} \left\{ \dfrac{f(x_n) - f(x_{n-1})}{x_n - x_{n-1}} \right\}$

In this case, the chord joining $(x_{n-1}, f(x_{n-1}))$ to $(x_n, f(x_n))$ in the secant method becomes the tangent to the curve at $(x_n, f(x_n))$ in the Newton-Raphson method.

Geometrically:

Secant method	**Newton-Raphson method**
	The chord AB has become the tangent to the curve at B

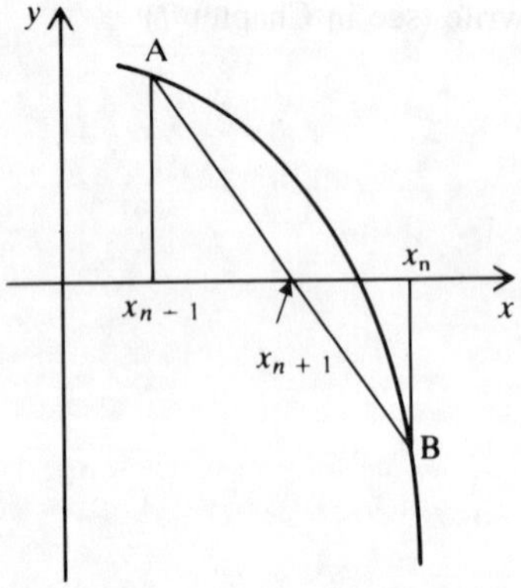

Figure 6.6

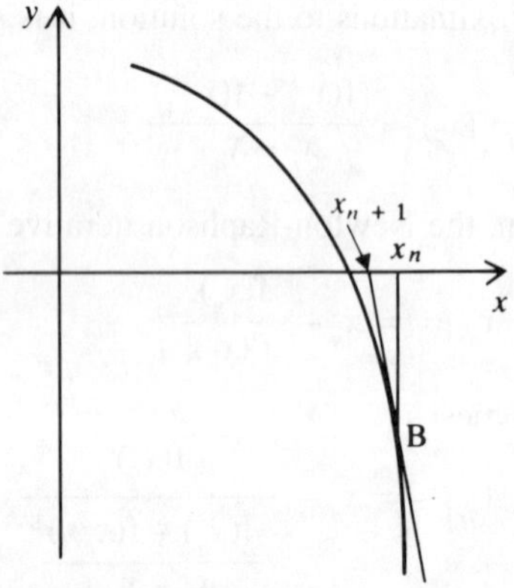

Figure 6.7

Finally, bringing together ideas from another area of numerical analysis, we shall see how the Newton-Raphson method may be used to obtain all real solutions of a polynomial.

6.5 Finding the real solutions of a polynomial

We shall use the Newton-Raphson method together with Horner's method for the evaluation of a polynomial (see Section 5.4). First recap briefly some of the essential ideas from that section.

Let $f(x) = a_0 + a_1x + a_2x^2 + ... + a_nx^n$ be the polynomial of degree n whose real solutions we are attempting to find. Algorithm 5.1 provides an efficient procedure for evaluating, for any number t, $f(t)$ and $f'(t)$. For convenience, the algorithm is restated here in a slightly different form.

Algorithm 5.1 Horner's method

To evaluate the polynomial $f(x) = a_0 + a_1x + a_2x^2 + ... + a_nx^n$ and the derivative $f'(x)$ at $x = t$.

input: the degree of the polynomial n;
the coefficients $a_0, a_1, a_2 ..., a_n$;
the number t

$b_n := a_n$
$c_n := b_n$
for $k = n - 1, n - 2, ..., 1$
$b_k := b_{k+1}t + a_k$
$c_k := c_{k+1}t + b_k$
endloop
$b_0 := b_1t + a_0$
output: $b_0 = f(t)$, $c_1 = f'(t)$, $b_1, b_2, ..., b_n$

Observe that if $f(x)$ is to be evaluated at $x = z$, then the value of $f(z)$ is stored in b_0 and the value of $f'(z)$ is stored in c_1.

Recall that $b_1, b_2, b_3, ..., b_n$ are also of considerable interest and that we can write

$$f(x) = (x - z)\,(b_1 + b_2x + b_3x^2 + ... + b_nx^{n-1}) + b_0$$

Since $b_0 = f(z)$, we have:

$$f(x) = (x - z)(b_1 + b_2x + b_3x^2 + ... + b_nx^{n-1}) + f(z)$$

If z is a solution of $f(x) = 0$, then $f(z) = 0$ and

$$f(x) = (x - z)\,(b_1 + b_2x + b_3x^2 + ... + b_nx^{n-1}) \quad ...(5)$$

This equation tells us that $x = z$ is a solution of the equation $f(x) = 0$ (which we knew) and that the remaining solutions of $f(x) = 0$ are precisely the solutions of

$$b_1 + b_2x + b_3x^2 + ... + b_nx^{n-1} = 0$$

This fact is extremely important and is fundamental to the algorithm that follows.

The different stages of the procedure for obtaining all the real solutions of the polynomial equation

$$a_0 + a_1x + a_2x^2 + ... + a_nx^n = 0$$

can now be described.

Procedure

Step 1 Find out how many real solutions the equation possesses and obtain rough estimates of these solutions. This may be achieved using a graph plotter or by considering sign changes in $y = f(x)$.

Step 2 Select the solution with smallest absolute value. Call this solution λ_1. Let x_0 represent the first approximation to λ_1. Let $z = x_0$.

Step 3 Use Algorithm 5.1 to evaluate $b_0 = f(z)$ and $c_1 = f'(z)$.

Step 4 Use the Newton-Raphson method to calculate an improved approximation to λ_1. This iterative equation would normally be written

$$x_{m+1} = x_m - \frac{f(x_m)}{f'(x_m)}$$

We shall write, using the above results and 'computer algebra'

$$z = z - \frac{b_0}{c_1}$$

and the latest value of z will give the improved approximation.

Step 5 Repeat steps 3 and 4 until satisfactory accuracy has been achieved. In the algorithm that follows, we use the criterion abs(f(z)) < ε for some acceptably small number ε. This is written abs(b_0) < ε.

Step 6 Take z to be our approximation to λ_1.

Step 7 Shift attention to the polynomial

$$b_1 + b_2x + b_3x^2 + \ldots + b_nx^{n-1}$$

and obtain all real solutions of the equation

$$b_1 + b_2x + b_3x^2 + \ldots + b_nx^{n-1} = 0 \qquad \ldots (6)$$

To do this, rewrite Equation (6) in the form

$$a_0 + a_1x + a_2x^2 + \ldots + a_nx^n = 0$$

and then use the procedure described in Steps 2 to 6. Hence we have:

Step 8 Decrease by 1 the degree of the polynomial. So write:

Let $\quad n = n - 1$

With this new value of n, write

Let $\quad a_0 = b_1$
$\quad a_1 = b_2$
$\quad \ldots$
$\quad a_n = b_{n+1}$

Now the equation to be solved (Equation(6)) is written

$$a_0 + a_1x + a_2x^2 + \ldots + a_nx^n = 0$$

and Steps 2 to 8 are repeated (with slight, and obvious modifications) until all the real solutions have been calculated.

When all this is written as an algorithm, we have:

Algorithm 6.2

To calculate all real solutions of the polynomial equation

$$a_0 + a_1x + a_2x^2 + ... + a_nx^n = 0$$

input: the degree n of the polynomial;
the coefficients of the polynomial, $a_0, a_1, ..., a_n$
the number r of real solutions

Step 1
for $k = 1, 2, ..., r$
input: x_0 (an approximation to one of the real solutions)
$z := x_0$
$b_n := a_n$
$c_n := b_n$

Step 2
for $i = n - 1, n - 2, ..., 1$
$b_i := a_i + z \times b_{i+1}$
$c_i := b_i + z \times c_{i+1}$
endloop
$b_0 := a_0 + z \times b_1$

Step 3
if abs(b_0) is sufficiently small then
goto Step 4
otherwise

$$z := z - \frac{b_0}{c_1}$$

and repeat Steps 2 and 3
endif

Step 4
$\lambda_k := z$

Step 5
$n := n - 1$
for $i = 0, 1, ..., n$
$a_i := b_{i+1}$
endloop
endloop
output: $\lambda_1, \lambda_2, ..., \lambda_r$; the real solutions of the equation

One of the problems with this method is that the errors can become large. This is because, at each stage, $b_1, b_2, ..., b_n$ are approximate values of the coefficients of the polynomial and when we are repeatedly obtaining approximate solutions of inexact polynomials, it is reasonable to expect that the errors are going to increase. The problem is eased somewhat by calculating the solutions in order of magnitude. Calculate the solution of smallest magnitude first, the solution of second smallest magnitude next and so on. This explains the instruction in Step 2 of the procedure to begin with the solution of smallest absolute value. This instruction was not repeated in the algorithm, but it is good advice and we recommend that the reader follows this advice when using Algorithm 6.2. An example requiring the solution of $3x^3 - 10x + 1 = 0$ will hopefully make this discussion more real.

It is expected that this technique will be implemented on a computer and a program listing follows which is written in what is hoped is a reasonably compatible version of BASIC. Hence we shall not perform the calculations; we shall let the computer do that and we shall limit discussion to commenting on the interpretation of the output. The results that follow were obtained on a Research Machines Nimbus.

Program to calculate the real solutions of the polynomial equation $a_0 + a_1x + a_2x^2 + ... + a_nx^n = 0$

```
10 REM find real solutions of polynomial
20 PRINT: INPUT "Degree of polynomial"; n
30 m = n
40 DIM a(n), b(n), c(n)
50 PRINT: INPUT " Number of real solutions "; r
60 DIM x(r)
70 FOR i = 0 TO n
80 PRINT: PRINT " Enter a("; i;") ";
90 INPUT a(i)
100 NEXT i
110 REM finding the r solutions begins here
120 FOR k = 1 TO r
130 PRINT: INPUT " Enter approximate solution "; z
140 b(m) = a(m) : c(m) = b(m)
150 REM n – r method loop: 160 – 220
160 FOR i = m – 1 TO 1 STEP –1
170 b(i) = a(i) + z*b(i + 1)
180 c(i) = b(i) + z*c(i + 1)
190 NEXT i
200 b(0) = a(0) + z*b(1)
210 z = z – b(0)/c(1)
220 IF ABS(b(0)) < 0.00001 THEN 230 ELSE 160
230 x(k) = z
240 PRINT : PRINT "x(";k;") = ";x(k)
250 PRINT
260 FOR i = 1 TO m
270 PRINT "b(";i;") = ";b(i)
```

```
280 NEXT i
290 REM reduce polynomial
300 m = m – 1
310 FOR i = 0 TO m
320 a(i) = b(i + 1)
330 NEXT i
340 NEXT k
350 END
```

Example 6.4

Calculate the three real solutions of the equation $3x^3 - 10x + 1 = 0$.

A sketch of the graph of $y = 3x^3 - 10x + 1$ is given in Figure 6.8.

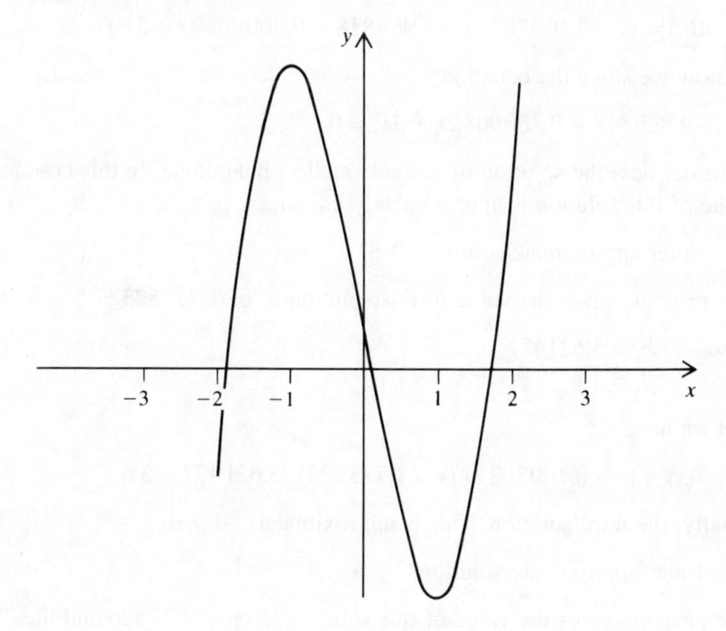

Figure 6.8

There are three real solutions: $x \approx -1.9$, $x \approx 0.2$, $x \approx 1.8$. Written in the form

$$a_0 + a_1x + a_2x^2 + a_3x^3 = 0$$

the equation becomes

$$1 - 10x + 3x^3 = 0$$

Hence $a_0 = 1$ $a_1 = -10$ $a_2 = 0$ $a_3 = 3$

Using the program:

Degree of polynomial is ? 3
Number of real solutions is ? 3

Enter a(0)? 1
Enter a(1)? –10
Enter a(2)? 0
Enter a(3)? 3

We begin with the solution of smallest magnitude. This is approximately 0.2 so:

Enter approximate solution? 0.2

The program gives the value of this solution to be 0.1003027.

At this point $b_1 = -9.969818$
$b_2 = 0.3009082$
$b_3 = 3$

and we have, from Equation (5):

$$f(x) = (x - 0.1003027)(-9.969818 + 0.3009082x + 3x^2)$$

So now we solve the equation

$$-9.969818 + 0.3009082x + 3x^2 = 0$$

Consider next the solution of second smallest magnitude. In this case the value of this solution is approximately 1.8 so:

Enter approximate solution? 1.8

The program gives the value of this solution to be 1.773523.

Also $b_1 = 5.621477$
$b_2 = 3$

and we have

$$f(x) = (x - 0.1003027)(x - 1.773523)(5.621477 + 3x)$$

Finally, the third solution. This is approximately –1.9 so:

Enter approximate solution? –1.9

The program gives the value of this solution to be –1.873826 and the three solutions are:

–1.873826 0.1003027 1.773523

Several points should be noted.

(a) Although the three solutions given above are accurate to the number of digits written, this technique will not always achieve such a high level of accuracy. One way of improving the accuracy is to use the solutions obtained as the starting values x_0 in straightforward applications of the Newton-Raphson method. This will almost always avoid the pitfalls that can occur in the Newton-Raphson method and high levels of accuracy can be obtained by the use of this modification.
(b) The concerns expressed in (a) are unlikely to be important when the degree of the polynomial is low (less than or equal to three).

(c) It is possible to use this technique to obtain **all** solutions (real and complex) of a cubic equation. Once the first solution has been found (let us say that this solution is $x = \lambda_1$) then

$$f(x) = (x - \lambda_1)(b_1 + b_2x + b_3x^2)$$

The remaining two solutions may then be found by using the formula for the solution of a quadratic equation and solving

$$b_1 + b_2x + b_3x^2 = 0$$

If the solutions are complex, then this formula will reveal their values.

(d) The discussion in (c) could be applied to any polynomial having just two complex solutions and a last example is provided to illustrate this.

Example 6.5

Solve completely the equation

$$x^4 - 3x^3 + 2x - 1 = 0$$

The graph of $y = x^4 - 3x^3 + 2x - 1$, sketched in Figure 6.9, reveals that the equation

$$x^4 - 3x^3 + 2x - 1 = 0$$

has two real solutions and that these are given approximately by

$$x \approx -0.9 \qquad \text{and} \qquad x \approx 2.8$$

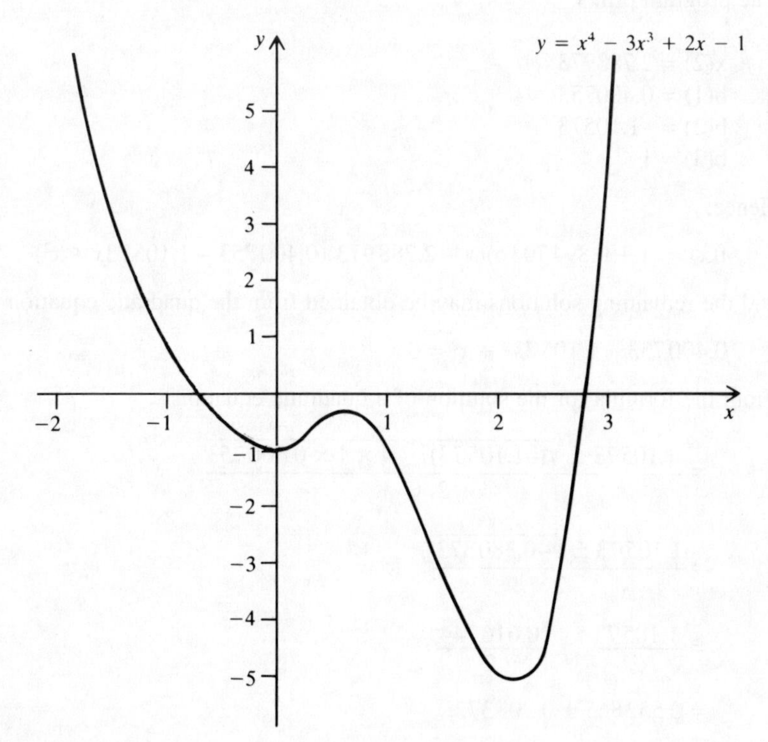

Figure 6.9

Write the equation in the form

$$a_0 + a_1x + a_2x^2 + a_3x^3 + a_4x^4 = 0$$

so $\quad a_0 = -1 \quad a_1 = 2 \quad a_2 = 0 \quad a_3 = -3 \quad a_4 = 1$

Running the program:

Degree of polynomial? 4
Number of real solutions? 2
Enter a(0)? –1
Enter a(1)? 2
Enter a(2)? 0
Enter a(3)? –3
Enter a(4)? 1
Enter approximate solution? –0.9

The program prints:

x(1) = –0.8947036
b(1) = –1.117689
b(2) = 3.484605
b(3) = –3.894704
b(4) = 1
Enter approximate solution? 2.8

The program prints:

x(2) = 2.788973
b(1) = 0.400753
b(2) = –1.10573
b(3) = 1

Hence:

$$f(x) = (x + 0.8947036)(x - 2.788973)(0.400753 - 1.10573x + x^2)$$

and the remaining solutions may be obtained from the quadratic equation

$$0.400753 - 1.10573x + x^2 = 0$$

From the formula for the solution of a quadratic equation

$$x = \frac{1.10573 \pm \sqrt{(-1.10573)^2 - 4 \times 1 \times 0.400753}}{2}$$

$$= \frac{1.10573 \pm \sqrt{-0.380373}}{2}$$

$$= \frac{1.10573}{2} \pm \frac{0.616744}{2}i$$

$$= 0.552865 \pm 0.308372i$$

Hence the solutions of the equation

$$x^4 - 3x^3 + 2x - 1 = 0$$

are -0.8947036, 2.788973, $0.552865 + 0.308372i$ and $0.552865 - 0.308372i$.

Exercise 6

There may be slight differences between your answers and those given in the book. These differences could be due to rounding error or, perhaps, to taking different starting values.

Section 6.1

1 Use the Newton-Raphson method to obtain, correct to four decimal digits, the solution lying in each of the given intervals for the following equations

(a) $6x^2 + \sqrt{x} - 10 = 0$ $[0, 1]$
(b) $x^2 + e^x - 5 = 0$ $[-3, -2]$, $[1, 2]$
(c) $x^3 + 5x^2 - 3 = 0$ $[-5, -4]$, $[-1, 0]$, $[0,1]$

2 Use the Newton-Raphson method to calculate successive approximations $x_0, x_1, x_2, \ldots$ to the solution, lying in each of the given intervals, of the following equations. Take as your approximation to the solution the first term x_r for which $|f(x_r)| < 0.0005$.

(a) $5x + 3\tan x - 2 = 0$ $[0,1]$ $[1,2]$ (in radians)
(b) $x = 3\sin x$ $[-3, -2]$, $[2, 3]$ (in radians)

3 Using the Newton-Raphson method, calculate a sequence of approximations $x_0, x_1, x_2 \ldots$ to the solution, lying in each of the given intervals, of the equations written below. Let the calculated solution be the first term of the sequence x_{r+1} for which $|x_{r+1} - x_r| < 0.0001$

(a) $\ln x + x^3 - 7 = 0$ $[1, 2]$
(b) $5(x + 1) = (x + 2)^3$ $[-5, -4]$
(c) $xe^x - 1 - x = 0$ $[0, 1]$

Section 6.2

4 To solve the equation $x^3 - 4x + 3 = 0$, use the Newton-Raphson method (with $x_0 = -2$) to calculate successive approximations $x_1, x_2, x_3, \ldots$ to the solution lying in the interval $[-3, -2]$. Plot the positions of x_0, x_1, x_2 on a graph of $y = x^3 - 4x + 3$ and illustrate geometrically how the Newton-Raphson method works.

5 It is required to solve the equation $x^3 + x^2 + 2 = 0$. Take $x_0 = 1$ and use the Newton-Raphson method to calculate successive approximations x_1, x_2, x_3. Plot the positions of x_0, x_1, x_2, x_3 on a graph of $y = x^3 + x^2 + 2$ and illustrate the Newton-Raphson method graphically. Do you think the sequence is going to converge? Repeat the procedure with $x_0 = -1$. Do you think that the new sequence $x_0, x_1, x_2, \ldots$ is going to converge?

Section 6.3

6 Find the solution of the equation $x = \tan x$ that lies closest to 100 (working in radians) using the Newton-Raphson method. (Unless x is chosen very carefully, the Newton-Raphson method produces a divergent sequence.)

7 Use
(a) the fixed point iterative method
(b) the secant method
(c) the Newton-Raphson method
to find, correct to four decimal digits, (in radians) the solution of the equation $x^2 - 5\sin x = 0$ that lies in the interval [2, 3]. Comment on the convergence properties of the three methods.

8 Solve the equation

$$5x^3 = 3\ln x + 6$$

using
(a) the k technique described in Section 3.13
(b) the Newton-Raphson method
Describe carefully which of the two methods you prefer.

9 Solve the equation $3x + 2\cos x = 0$ using
(a) the accelerated convergence algorithm
(b) the Newton-Raphson method
Which method has the fastest convergence? Comment on your answer. (Note that x is measured in radians).

10 Use the fixed point iterative method to obtain the solution lying in the interval [1, 2] of the equation

$$\sqrt{x} + x + x^2 = 4$$

Choose, as your approximation to the solution, the first term x_r for which $|x_r - x_{r-1}| < 0.005$. Choose the form of the iterative function carefully and try to achieve the fastest convergence that the method will allow.
Now use the Newton-Raphson method to obtain this solution.
Which of the two techniques is the more efficient?

11 Draw the graph of the function $y = 5 - 2x^2 - 6x^3$. It is required to use the Newton-Raphson method to obtain the real solution of this equation. Write down a value of x_0 which will cause the sequence $x_0, x_1, x_2, \ldots$ to converge to this solution. Check your answer by calculating $x_0, x_1, x_2, \ldots$. Find also a value of x_0 which will cause the Newton-Raphson method to diverge. Again, check your answer.

Section 6.5

12 Use the Newton-Raphson method in conjunction with Horner's method to solve completely the following polynomial equations. Give your answers correct to

five decimal digits (rounded).
(a) $x^3 - 4x^2 + 8 = 0$
(b) $4x^3 - 7x + 2 = 0$
(c) $2x^3 - 11x + 1 = 0$
(d) $5x^4 + 2x^3 - 9x^2 - 4x + 1 = 0$
(e) $5x^3 + x^2 - 7x - 3 = 0$
(f) $7x^3 + 2x^2 - 9x + 1 = 0$

13 Use the Newton-Raphson method, together with Horner's method to obtain **all** solutions of the following polynomial equations.
(a) $3x^4 + 3x^3 + 5x - 1 = 0$
(b) $5x^4 - 4x^2 + 3x + 2 = 0$
(c) $4x^3 + 7x - 8 = 0$
(d) $x^4 + 3x^3 - 5x + 1 = 0$
(e) $4x^4 - 9x^3 + 3x^2 - 4x - 2 = 0$
(f) $4x^3 - 5x^2 + 4x + 2 = 0$

Miscellaneous

14 Find both roots of the quadratic equation

$$x^2 - 10.2x + 0.125 = 0$$

each correct to three significant figures. (MEI)

15 Show that the equation $x^3 - x^2 - 1 = 0$ has a real root between $x = 1$ and $x = 2$. Taking $x = 2$ as your first approximation, use Newton's method to find this root to two significant figures.
(MEI)

16 Establish the Newton-Raphson formula for obtaining an approximation to a root of the equation $f(x) = 0$.
Show that the equation $x^3 + 3x - 3 = 0$ has only one real root and find this root, to three significant figures, using Newton's method.
(MEI)

17 By applying the Newton-Raphson method to the function f defined by

$$f(x) = 1 - \frac{7}{x^2}$$

develop an iterative formula for calculating $\sqrt{7}$.

Hence, using 2 as a first approximation to $\sqrt{7}$, calculate $\sqrt{7}$ correct to two places of decimals.

Show that if x_n, the nth approximation to $\sqrt{7}$, has a small error e_n, then the next approximation, x_{n+1}, has an error of magnitude about $0.6e_n^2$.
(MEI)

18 Show graphically that the equation

$$x^2 = 7 \log_{10}x + 2.347$$

has two real positive roots.

Taking $x = 2.2$ as an initial approximation to the larger of these roots, obtain a second approximation
(a) by the Newton-Raphson method
(b) by writing the equation in the form

$$x = \sqrt{(7 \log_{10}x + 2.347)}$$

and using an iterative method.

Work to three decimal places and give your answers to two decimal places.

(MEI)

19 Draw the graph of $y = 2x^3 - 3x^2 - 2x + 1$ over an appropriate range of x to find approximate solutions to the equation

$$2x^3 - 3x^2 - 2x + 1 = 0$$

By rewriting this equation in two different ways, find two convergent iterative formulae of the form $x_{n+1} = \mathrm{F}(x_n)$ to find the middle solution. Use the formula with the faster convergence to find this middle solution correct to four decimal places. How might this iterative method be adapted to improve the rate of convergence?

Use the Newton-Raphson formula to find the remaining solutions correct to four decimal places.

(MEI)

20 The Newton-Raphson iterative formula for the solution of the equation $\mathrm{f}(x) = 0$ may be written

$$x_{r+1} = \phi(x_r)$$

where $$\phi(x) = x - \frac{\mathrm{f}(x)}{\mathrm{f}'(x)}$$

Show that if X is a root of $\mathrm{f}(x) = 0$ then $\phi'(X) = 0$. Show further that, in general $\phi''(x) \neq 0$.

Hence show that the errors e_r and e_{r+1} in the iterates x_r and x_{r+1} are such that

$$e_{r+1} \approx \mathrm{k}e_r^2$$

for some constant k. Explain briefly what this relationship indicates about the convergence of the Newton-Raphson process.

(MEI)

21 The equation $x^3 - 5x + 3 = 0$ has a root λ between 0 and 1. Show that Newton's method leads to the iteration formula

$$x_{n+1} = x_n - \frac{(x_n^3 - 5x_n + 3)}{(3x_n^2 - 5)}$$

connecting two successive approximations x_n and x_{n+1} to λ. Suppose $x_n = \lambda + \delta_n$ and $x_{n+1} = \lambda + \delta_{n+1}$, where $|\delta_n|$ and $|\delta_{n+1}|$ are both much less than 1. Use the binomial expansion and the formula above to show that $\delta_{n+1} \approx K\delta_n^2$, where K is independent of n.

A second iteration formula connecting two successive approximations y_n and y_{n+1} to λ is

$$y_{n+1} = \tfrac{1}{5}(y_n^3 + 3)$$

Obtain an approximate relation between ε_{n+1} and ε_n, where $y_n = \lambda + \varepsilon_n$ and $y_{n+1} = \lambda + \varepsilon_{n+1}$, and $|\varepsilon_n|$ and $|\varepsilon_{n+1}|$ are both much less than 1.

Comment, in the light of your results, on the relative effectiveness of these two formulae.

(MEI)

22 A man in a rowing boat on still water is at a point P which is 1 mile from the nearest point A on a straight shore line. He wishes to reach a point Q which is 1 mile directly inland from a point B also on the shore line. The distance between A and B is 2 miles. The man decides to row directly to a point X on the shore line between A and B which is at a distance of x miles from A. He then walks directly to Q.

Find an expression for the total time taken in terms of x and hence show that in order to reach Q in the minimum time the point X must be such that

$$\frac{\sin\theta_r}{\sin\theta_w} = \frac{v_r}{v_w}$$

where θ_r is the acute angle between PX and the normal to AB, θ_w is the acute angle between QX and the normal, v_r is the constant rowing speed and v_w the constant walking speed.

If the man can row at 2 mph and walk at 4 mph show that x satisfies the equation

$$3x^4 - 12x^3 + 15x^2 + 4x - 4 = 0$$

Use the Newton-Raphson iterative formula to find x correct to two significant figures.

(MEI)

23 Use the Newton-Raphson method to find

(a) the solutions

(b) the stationary values

of the function $y = x^4 - 2x^3 - 5x^2 + 12x - 5$.

24 Show that the equation

$$x^4 + 2x^3 - 7x^2 + 3 = 0$$

has two real solutions. Use the Newton-Raphson method to calculate these solutions correct to five decimal digits (using rounding).

If these solutions are denoted by a and b, use Horner's method to write the above equation in the form

$$(x - a)(x - b)Q(x)$$

where $Q(x)$ is a quadratic expression in x.

Hence obtain all four solutions of the equation. Comment on the accuracy of your four solutions.

25 Show that the equation $x^3 + x^2 = 100$ has only one real solution. Evaluate this solution correct to two decimal digits (rounded). Hence find the complex solutions, giving both real and imaginary parts correct to one decimal digit.

26 The equation $x^3 - 34x + 1 = 0$ has a root α in the interval [5.80, 5.88]. It is given that $f(5.80) < 0$ and $f(5.88) > 0$, where $f(x) = x^3 - 34x + 1$.

(a) Apply the interval bisection method to find an interval $[a, a + 0.01]$ within which α lies.

(b) Use the Newton-Raphson method to determine α to six significant figures, starting with the value of a as an approximation to α.

(c) Given that the remaining roots of the equation $f(x) = 0$ are the roots of $x^2 + px + q = 0$, express p and q in terms of α.

(d) Using the value obtained in (b) for α, find the roots of the quadratic equation in (c) to four significant figures, using a method to minimise any possible loss of accuracy.

(e) Obtain the remaining roots of $f(x) = 0$ to six significant figures.

(C)

Chapter 7

Numerical differentiation and integration

Introduction

The distances covered in each second by an Olympic class athlete running 100 m are given in the table below.

Time (s)	0	1	2	3	4	5	6	7	8	9	10
Distance (m)	0	2.5	10.1	20	30.2	41.9	53.6	65.4	77.1	88.7	100.1

Several questions may be asked. 'What was the speed of the runner after 1 second?' 'What was the speed of the runner as he approached the tape?' (i.e., 'What was the speed of the runner after 9 seconds?') 'What was the acceleration of the runner after 1 second and after 9 seconds?'

Speed and acceleration are normally calculated using calculus where, if y represents the distance travelled in x seconds, then

$$\text{speed} = \frac{dy}{dx}$$

$$\text{acceleration} = \frac{d^2y}{dx^2}$$

In this case, there is no equation giving y in terms of x and so differentiation, in the sense of the calculus, is impossible. What is possible, however, is to use the data to calculate approximate values of $\frac{dy}{dx}$ and $\frac{d^2y}{dx^2}$. Techniques which perform such calculations are known as numerical differentiation.

One of the principal applications of numerical differentiation lies in the numerical solution of differential equations. This is an extremely important area of mathematics and will be discussed in some detail in Chapter 8. The topic of numerical differentiation forms the first part of the present chapter.

The second part of the chapter is concerned with the evaluation of definite integrals: that is, with integrals of the form $\int_a^b f(x)\,dx$.

$$\int_0^1 \sqrt{x}\,dx \qquad \int_0^{\frac{\pi}{2}} \sin x\,dx \qquad \text{and} \qquad \int_0^{\infty} e^{-\frac{1}{2}x}\,dx$$

are three examples of definite integrals.

The reader should be aware that a definite integral represents a number and is not an expression in x. The techniques that we shall describe seek to calculate (approximately) the value of this number.

There are several occasions when such a technique might be useful.

(a) When attempting to evaluate $\int_a^b f(x)\,dx$ it may be impossible to integrate the function $f(x)$ using the usual rules of calculus. This would be the case, for example, if we were attempting to evaluate $\int_0^1 e^{-x^2}\,dx$.

(b) It may simply be rather difficult to integrate $f(x)$ and using a numerical technique might well be very much faster than performing the integration. This would be the case with $\int_0^{\frac{\pi}{4}} \sqrt{\tan x}\,dx$ for example.

(c) The value of $f(x)$ may be known only at a finite number of points. This would be the case if, for example, $f(x)$ were given as a set of data. In this situation, the usual methods of integration simply cannot be applied.

The techniques for calculating approximately the value of a definite integral are known collectively as numerical integration.

7.1 Numerical differentiation

The aim of this section is to start with a given set of values taken by a function $y = f(x)$

x	x_0	x_1	x_2	x_3	x_i
$y = f(x)$	$y_0 = f(x_0)$	y_1	y_2	y_3	$y_i = f(x_i)$

and see how to calculate approximately the value of $\frac{dy}{dx}$ and $\frac{d^2y}{dx^2}$ for a given value of x. Expressions for the errors in these approximations will also be found. But first, some notation.

Notation

Assume that $x_0, x_1, x_2, \ldots$ are equally spaced along the x-axis and that successive values of x are h units apart.

Thus $\quad x_1 = x_0 + h$

$\qquad\quad x_2 = x_0 + 2h \qquad$ and so on.

Then we have the situation shown in Figure 7.1

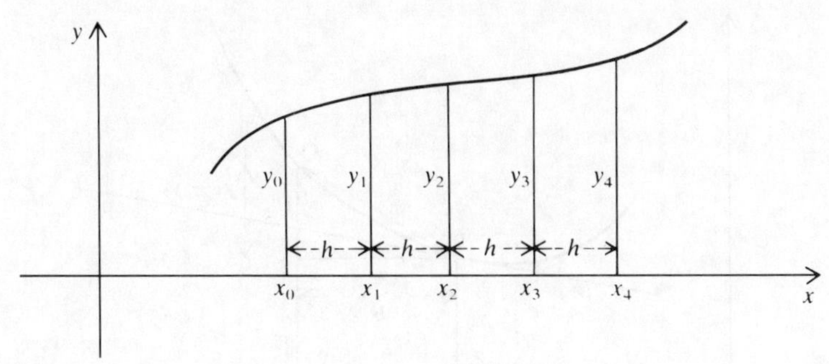

Figure 7.1

The following notation will also be used in connection with the expression $y = f(x)$:

$f'(x_2)$ is the value of $\frac{dy}{dx}$ when $x = x_2$

$f''(x_n)$ is the value of $\frac{d^2y}{dx^2}$ when $x = x_n$

This notation is particularly useful if it is required to illustrate simultaneously both the function $f(x)$ and the value of x at which $\frac{dy}{dx}$ or $\frac{d^2y}{dx^2}$ is being evaluated.

7.2 To calculate approximately the value of $\frac{dy}{dx}$

Perhaps the easiest way to proceed is to consider the definition of $\frac{dy}{dx}$:

$$\frac{dy}{dx} = \lim_{\delta x \to 0} \frac{f(x + \delta x) - f(x)}{\delta x}$$

Stated fairly crudely, this says that

$$\frac{dy}{dx} = \frac{\text{increase in } y}{\text{increase in } x}$$

as the increase in x becomes very small.

Using this, a reasonable approximation to $\frac{dy}{dx}$ when $x = x_2$ is provided by

$$f'(x_2) \approx \frac{y_3 - y_2}{h}$$

Similarly, an approximation to $\frac{dy}{dx}$ when $x = x_0$ is given by

$$f'(x_0) \approx \frac{y_1 - y_0}{h}$$

Figure 7.2 shows an interpretation of this. Geometrically, the chord PQ is being used to approximate the tangent PR.

This method is called the forward difference formula and overleaf we have:

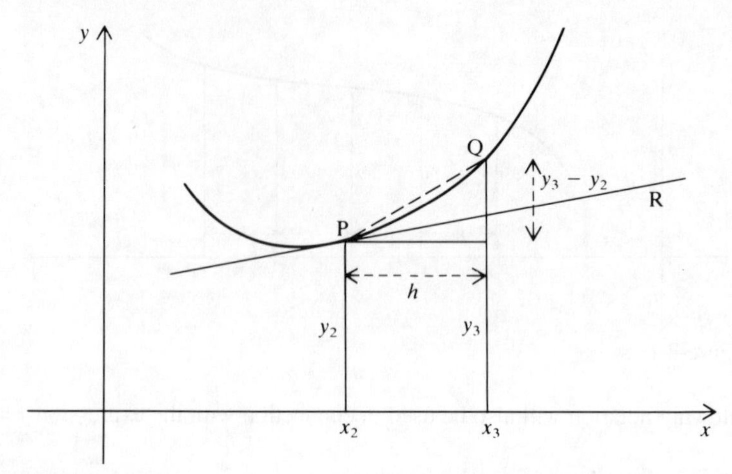

Figure 7.2

The forward difference formula

The value of $\frac{dy}{dx}$ when $x = x_n$ is given approximately by

$$f'(x_n) \approx \frac{y_{n+1} - y_n}{h}$$

Example 7.1

Using the data in the table at the beginning of this chapter, estimate the speed of the runner when (a) $x = 1$ and (b) $x = 9$.

The speed is given by $\frac{dy}{dx}$. For the data in the table, $h = 1$; hence approximations to the speed are

(a) $x = 1$ $\quad f'(1) \approx \frac{y_2 - y_1}{1} = \frac{10.1 - 2.5}{1} = 7.6\,\text{m s}^{-1}$

(b) $x = 9$ $\quad f'(9) \approx \frac{y_{10} - y_9}{1} = \frac{100.1 - 88.7}{1} = 11.4\,\text{m s}^{-1}$

One way of improving the approximation is to use more information. If y_1 and y_3 are both known, we might base the approximation on a larger interval.

An approximation to $\frac{dy}{dx}$ when $x = x_2$ is provided by

$$f'(x_2) \approx \frac{y_3 - y_1}{2h}$$

Geometrically, the chord LN is being used to approximate the tangent PR. For the curve illustrated in Figure 7.3, it is clear that in this case, the second method provides the more accurate approximation.

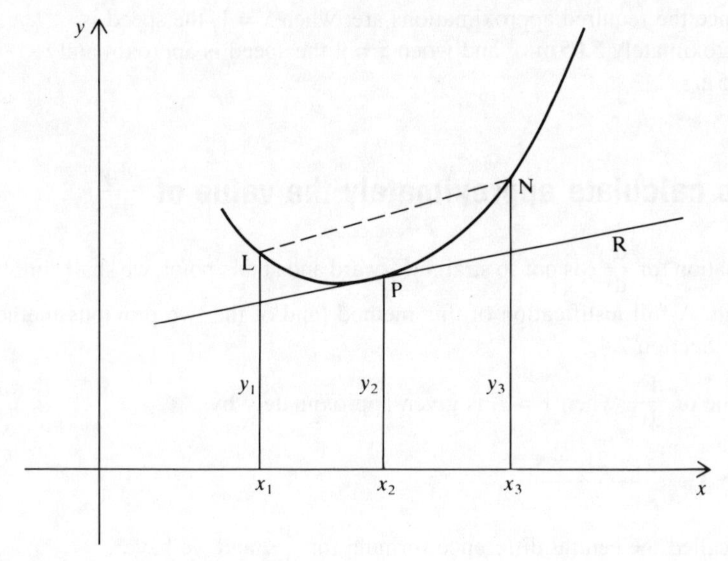

Figure 7.3

If we attempted to use the second method at $x = x_0$ there would be a problem. Unless the value of the function is known when $x = x_0 - h$, the second method cannot be used. If, however, $f(x_0 - h)$ is known, and in this case it is usual to write $f(x_0 - h) = y_{-1}$, then we have

$$f'(x_0) \approx \frac{y_1 - y_{-1}}{2h}$$

This method is called the central difference formula and we have:

The central difference formula

The value of $\frac{dy}{dx}$ when $x = x_n$ is given approximately by

$$f'(x_n) \approx \frac{y_{n+1} - y_{n-1}}{2h}$$

Example 7.2

Use the central difference formula to estimate the speed of the runner in the table at the start of this chapter, when (a) $x = 1$ and (b) $x = 9$.

By the definition and by the table:

(a) $f'(1) \approx \frac{y_2 - y_0}{2h} = \frac{10.1 - 0}{2} = 5.05\,\text{m s}^{-1}$

(b) $f'(9) \approx \frac{y_{10} - y_8}{2h} = \frac{100.1 - 77.1}{2} = 11.5\,\text{m s}^{-1}$

Hence the required approximations are: when $x = 1$, the speed is approximately $5.05\,\mathrm{m\,s^{-1}}$ and when $x = 9$ the speed is approximately $11.5\,\mathrm{m\,s^{-1}}$.

7.3 To calculate approximately the value of $\frac{d^2y}{dx^2}$

The situation for $\frac{d^2y}{dx^2}$ is not so straightforward and at this point, we shall simply state the result. A full justification of this method (and of the two previous methods) is given in Section 7.4.

The value of $\frac{d^2y}{dx^2}$ when $x = x_2$ is given approximately by

$$f''(x_2) \approx \frac{y_3 - 2y_2 + y_1}{h^2}$$

This is called the central difference formula for $\frac{d^2y}{dx^2}$ and we have:

The central difference formula for $\frac{d^2y}{dx^2}$

The value of $\frac{d^2y}{dx^2}$ when $x = x_n$ is given approximately by

$$f''(x_n) \approx \frac{y_{n+1} - 2y_n + y_{n-1}}{h^2}$$

Example 7.3

Use the central difference formula to estimate the acceleration of the runner when (a) $x = 1$ and (b) when $x = 9$.

Acceleration is given by $\frac{d^2y}{dx^2}$. Hence we calculate f''(1) and f''(9).

(a) $$f''(1) \approx \frac{y_2 - 2y_1 + y_0}{h^2} = \frac{10.1 - 2 \times 2.5 + 0}{1} = 5.1\,\mathrm{m\,s^{-2}}$$

and the acceleration when $x = 1$ is approximately $5.1\,\mathrm{m\,s^{-2}}$.

(b) $$f''(9) \approx \frac{y_{10} - 2y_9 + y_8}{h^2} = \frac{100.1 - 2 \times 88.7 + 77.1}{1} = -0.2\,\mathrm{m\,s^{-2}}$$

and the acceleration when $x = 9$ is approximately $-0.2\,\mathrm{m\,s^{-2}}$.

An obvious question to ask is, 'Which will provide the better approximation, the forward difference formula or the central difference formula?' To make progress with this question, the errors involved in the approximations will be considered.

7.4 The error term in numerical differentiation

We must now put the theory on a firm mathematical basis. Such a basis is provided by Taylor polynomials. We shall consider the problem of calculating approximate values of $\frac{dy}{dx}$ and $\frac{d^2y}{dx^2}$ when $x = a$. In this case, it will be useful to have the Taylor polynomials centred at $x = a$.

The error term in the forward difference formula

From Chapter 5, the Taylor polynomial of degree one is

$$f(a+h) = f(a) + hf'(a) + \frac{h^2}{2!}f''(\eta) \qquad \text{where} \qquad a < \eta < a + h$$

Since the aim is to calculate approximately the value of $f'(a)$, we solve this equation for $f'(a)$.

$$hf'(a) = f(a+h) - f(a) - \frac{h^2}{2!}f''(\eta)$$

and $$f'(a) = \frac{f(a+h) - f(a)}{h} - \frac{h}{2}f''(\eta)$$

In the notation used above: if $a = x_n$ this becomes

$$f'(x_n) = \frac{y_{n+1} - y_n}{h} - \frac{h}{2}f''(\eta)$$

and we have established the forward difference formula **and** the associated error term. Written in functional notation:

The forward difference formula

$$f'(a) \approx \frac{f(a+h) - f(a)}{h}$$

with the error term $\quad -\frac{h}{2}f''(\eta) \qquad a < \eta < a + h$

The error term in the central difference formula

Now consider the Taylor polynomial of degree two:

$$f(a+h) = f(a) + hf'(a) + \frac{h^2}{2!}f''(a) + \frac{h^3}{3!}f'''(\eta_1) \qquad a < \eta_1 < a + h$$

Replacing h with $-h$ gives

$$f(a-h) = f(a) - hf'(a) + \frac{h^2}{2!}f''(a) - \frac{h^3}{3!}f'''(\eta_2) \qquad a-h < \eta_2 < a$$

Subtracting these equations:

$$f(a+h) - f(a-h) = 2hf'(a) + \frac{h^3}{3!}f'''(\eta_1) + \frac{h^3}{3!}f'''(\eta_2)$$

Solving for f'(a) gives

$$2hf'(a) = f(a+h) - f(a-h) - \frac{h^3}{3!}[f'''(\eta_1) + f'''(\eta_2)]$$

and $$f'(a) = \frac{f(a+h) - f(a-h)}{2h} - \frac{h^2}{12}[f'''(\eta_1) + f'''(\eta_2)]$$

The expression $\frac{h^2}{12}[f'''(\eta_1) + f'''(\eta_2)]$ may be simplified by appealing to the intermediate value theorem. Observe that if a and b are any two numbers, then the average of a and b must lie between a and b. So if $a < b$ we have the inequality $a < \frac{a+b}{2} < b$. Applying this to the numbers $a = f'''(\eta_1)$, $b = f'''(\eta_2)$ (and assuming, for ease of expression, that $f'''(\eta_1) < f'''(\eta_2)$) gives the inequality

$$f'''(\eta_1) < \frac{f'''(\eta_1) + f'''(\eta_2)}{2} < f'''(\eta_2)$$

If f'''(x) is continuous in the interval $[a-h, a+h]$, then the conditions for the intermediate value theorem are satisfied. Applying this theorem to f'''(x) gives:

There exists a number η lying in the interval $(a-h, a+h)$ such that

$$\frac{f'''(\eta_1) + f'''(\eta_2)}{2} = f'''(\eta)$$

Hence $$\frac{h^2}{12}[f'''(\eta_1) + f'''(\eta_2)] = 2 \times \frac{h^2}{12}f'''(\eta)$$

and we may write:

$$f'(a) = \frac{f(a+h) - f(a-h)}{2h} - 2 \times \frac{h^2}{12}f'''(\eta)$$

or $$f'(a) = \frac{f(a+h) - f(a-h)}{2h} - \frac{h^2}{6}f'''(\eta)$$

The expression $-\frac{h^2}{6}f'''(\eta)$ thus becomes the error term when $\frac{f(a+h) - f(a-h)}{2h}$ is used to approximate f'(a).

In the notation used earlier, if $a = x_n$, this becomes

$$f'(x_n) = \frac{y_{n+1} - y_{n-1}}{2h} - \frac{h^2}{6}[f'''(\eta)]$$

and we have:

The central difference formula

$$f'(a) \approx \frac{f(a+h) - f(a-h)}{2h}$$

with error term $\quad -\dfrac{h^2}{6} f'''(\eta) \qquad a - h < \eta < a + h$

As discussed in Chapter 5, the error terms cannot, in general, be evaluated. We have no method for determining the value of η and, if f(x) is known only as a set of data, we do not know the form of $f'''(x)$. However, the error terms are of considerable theoretical importance.

One way in which the error terms can be of use is to observe that in the forward difference formula the error is proportional to h while in the central difference formula the error is proportional to h^2. Hence, for $h < 1$, we would expect the central difference formula to be more accurate than the forward difference formula. This is not always the case, however, as we shall now see.

Approximate values of $f'(a)$ are to be found for the functions illustrated in Figures 7.4a – c.

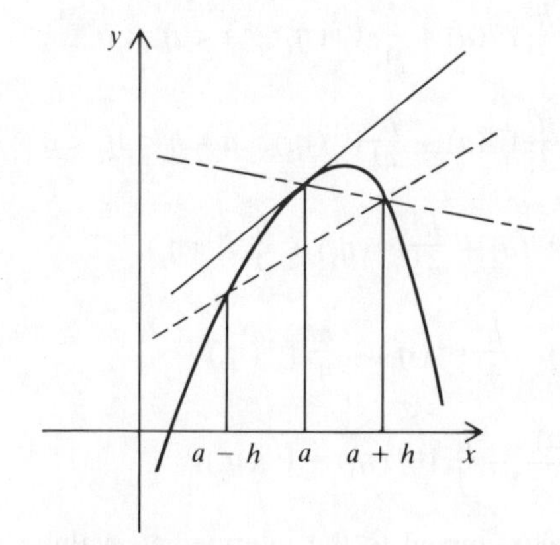

Figure 7.4a

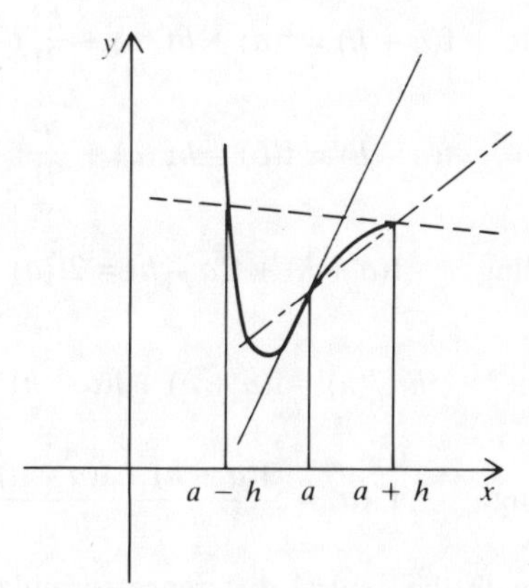

Figure 7.4b

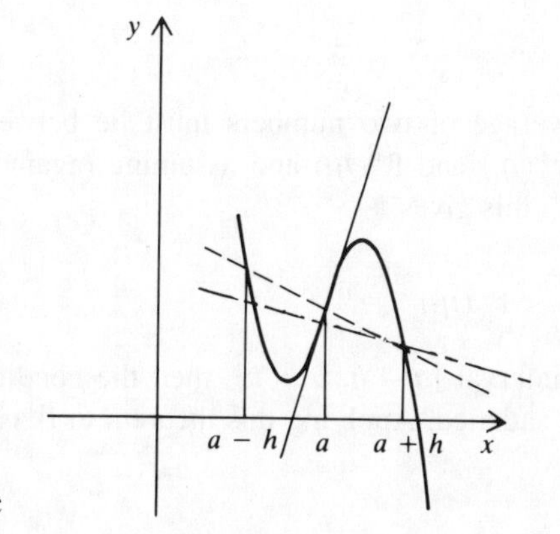

Figure 7.4c

By observing the tangent at $x = a$, it is clear that in each case, $f'(a) > 0$. However, it is also found that:
in Figure 7.4a the forward difference formula returns a negative value while the central difference formula provides a good approximation to $f'(a)$,
in Figure 7.4b, the forward difference formula will be reasonably accurate while the central difference formula returns a negative value,
in Figure 7.4c, both methods provide a negative approximation to $f'(a)$.

The problem in this case is, of course, that h is too large. But reducing the size of h will not always solve all the problems. Both methods involve the subtraction of two numbers which may, for certain values of h, be almost equal. As we saw in Chapter 1 this is a highly dangerous operation, which often leads to serious accuracy problems. Unfortunately, in the case of numerical differentiation, there is no easy way out of this difficulty.

Task T investigates this problem further. For the moment, however, we simply record that the reader must always be aware that numerical differentiation is a very unstable process.

The error term in $\frac{d^2y}{dx^2}$

Write $f(a + h) = f(a) + hf'(a) + \frac{h^2}{2!}f''(a) + \frac{h^3}{3!}f'''(a) + \frac{h^4}{4!}f^{(4)}(\eta_1) \quad a < \eta_1 < a + h$

and $f(a - h) = f(a) - hf'(a) + \frac{h^2}{2!}f''(a) - \frac{h^3}{3!}f'''(a) + \frac{h^4}{4!}f^{(4)}(\eta_2) \quad a - h < \eta_2 < a$

Adding $f(a + h) + f(a - h) = 2f(a) + \frac{2h^2}{2!}f''(a) + \frac{h^4}{4!}f^{(4)}(\eta_1) + \frac{h^4}{4!}f^{(4)}(\eta_2)$

hence $h^2f''(a) = f(a + h) + f(a - h) - 2f(a) - \frac{h^4}{4!}f^{(4)}(\eta_1) - \frac{h^4}{4!}f^{(4)}(\eta_2)$

giving $f''(a) = \frac{f(a + h) + f(a - h) - 2f(a)}{h^2} - \frac{h^2}{4!}[f^{(4)}(\eta_1) + f^{(4)}(\eta_2)]$

As with the central difference formula, we may appeal to the intermediate value theorem to simplify the expression

$$\frac{h^2}{4!}[f^{(4)}(\eta_1) + f^{(4)}(\eta_2)]$$

Again, we observe that the average of two numbers must lie between the two numbers. If the numbers are $f^{(4)}(\eta_1)$ and $f^{(4)}(\eta_2)$ and assuming (again for ease of expression) that $f^{(4)}(\eta_1) < f^{(4)}(\eta_2)$ this gives

$$f^{(4)}(\eta_1) < \frac{f^{(4)}(\eta_1) + f^{(4)}(\eta_2)}{2} < f^{(4)}(\eta_2)$$

If $f^{(4)}(x)$ is continuous in the interval $[a - h, a + h]$, then the conditions of the intermediate value theorem are satisfied. Applying this theorem to $f^{(4)}(x)$:

there exists a number η where $a - h < \eta < a + h$ such that

$$\frac{f^{(4)}(\eta_1) + f^{(4)}(\eta_2)}{2} = f^{(4)}(\eta)$$

Hence $\quad \frac{h^2}{4!}[f^{(4)}(\eta_1) + f^{(4)}(\eta_2)] = 2 \times \frac{h^2}{4!} f^{(4)}(\eta)$

and $\quad f''(a) = \frac{f(a + h) - 2f(a) + f(a - h)}{h^2} - \frac{h^2}{12} f^{(4)}(\eta)$

It may be seen, therefore, that $-\frac{h^2}{12} f^{(4)}(\eta)$ represents the error when $\frac{f(a + h) - 2f(a) + f(a - h)}{h^2}$ is used to approximate $f''(a)$.

In the notation used earlier, if $a = x_n$ this equation becomes

$$f''(x_n) = \frac{y_{n+1} - 2y_n + y_{n-1}}{h^2} - \frac{h^2}{12}[f^{(4)}(\eta)]$$

In general we have:

The central difference formula for $\frac{d^2y}{dx^2}$

$$f''(a) \approx \frac{f(a + h) - 2f(a) + f(a - h)}{h^2}$$

with error term $\quad -\frac{h^2}{12} f^{(4)}(\eta) \qquad a - h < \eta < a + h$

Observe that the error term is proportional to h^2 and that the danger of losing accuracy from the subtraction of almost equal numbers remains. The background to an alternative approach to numerical differentiation will be presented in the exercises at the end of the chapter. But it is proper to end this section with a warning. Numerical differentiation is basically an unstable process and may be damaging to the health of your calculation.

7.5 Numerical integration

The aim of numerical integration is to provide effective techniques for the approximate evaluation of definite integrals. In practice, this means that a given set of values from a function $f(x)$

x	a	x_1	x_2	x_3	...	b
$f(x)$	$f(a)$	$f(x_1)$	$f(x_2)$	$f(x_3)$	...	$f(b)$

is the data used to calculate approximately the value of the integral $\int_a^b f(x)\,dx$.

The principal idea behind all the techniques we shall consider may be stated very simply.

(i) Replace the function to be integrated, $f(x)$, by a polynomial $p(x)$. (Figure 7.5.)

(ii) Integrate $p(x)$ and take $\int_a^b p(x)\,dx$ to be the required approximation to $\int_a^b f(x)\,dx$.

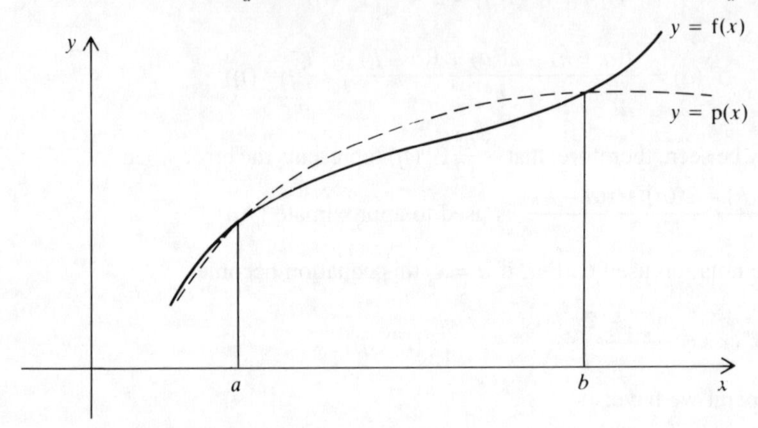

Figure 7.5

We shall use, for this purpose, both the Lagrange interpolating polynomial and Taylor polynomials. We shall use the error terms associated with the Taylor polynomials to derive expressions for the error in the approximations to $\int_a^b f(x)\,dx$.

7.6 The trapezium rule

We begin with the simplest case and replace $f(x)$ by the Lagrange interpolating polynomial of degree one: $p_1(x)$. (Figure 7.6.)

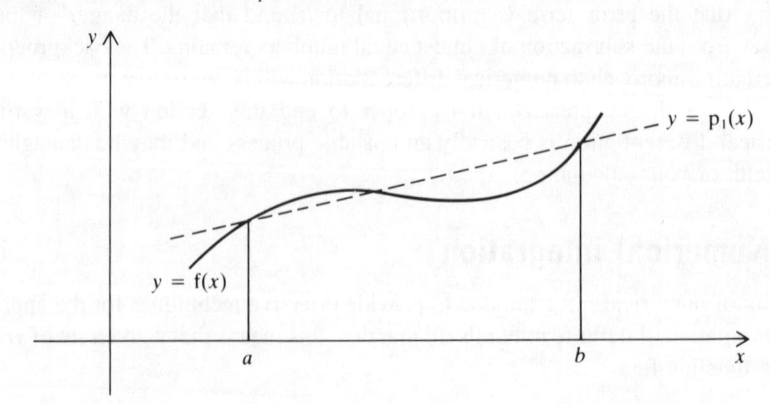

Figure 7.6

Write $x_0 = a$ and $x_1 = b$. Then, if $p_1(x)$ interpolates $f(x)$ at the points $(x_0, f(x_0))$, $(x_1, f(x_1))$, we have

$$p_1(x) = \frac{x - x_1}{x_0 - x_1} f(x_0) + \frac{x - x_0}{x_1 - x_0} f(x_1)$$

Writing $a = x_0$, $b = x_1$ gives

$$p_1(x) = \frac{x-b}{a-b}\,f(a) + \frac{x-a}{b-a}\,f(b)$$

Hence
$$\int_a^b p_1(x)\,dx = \frac{f(a)}{a-b}\int_a^b (x-b)\,dx + \frac{f(b)}{b-a}\int_a^b (x-a)\,dx$$

$$= \frac{f(a)}{a-b}\left[\frac{(x-b)^2}{2}\right]_a^b + \frac{f(b)}{b-a}\left[\frac{(x-a)^2}{2}\right]_a^b$$

$$= \frac{f(a)}{a-b}\left[0-\frac{(a-b)^2}{2}\right] + \frac{f(b)}{b-a}\left[\frac{(b-a)^2}{2}-0\right]$$

$$= -f(a)\frac{(a-b)}{2} + f(b)\frac{(b-a)}{2}$$

$$= f(a)\left(\frac{b-a}{2}\right) + f(b)\left(\frac{b-a}{2}\right)$$

$$= \frac{(b-a)}{2}\,[f(a)+f(b)]$$

and we take, as our approximation

$$\int_a^b f(x)\,dx \approx \int_a^b p_1(x)\,dx$$

Hence,
$$\int_a^b f(x)\,dx \approx \frac{b-a}{2}\,[f(a)+f(b)]$$

Geometrically, it may be seen that in Figure 7.7:

$$\int_a^b p_1(x)\,dx = \text{area of trapezium ABCD}$$

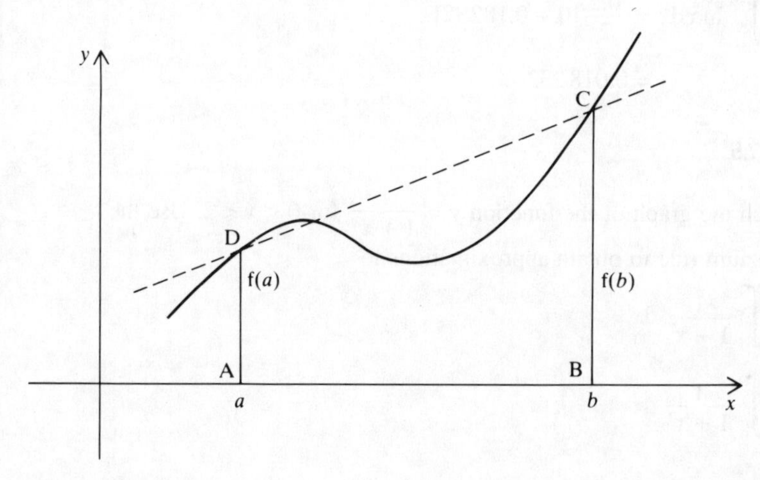

Figure 7.7

For this reason, this rule is usually described as the trapezium rule (or trapezoidal rule). If we write $h = b - a$, the trapezium rule may be written in the more familiar form:

Trapezium rule

$$\int_a^b f(x)\,dx \approx \frac{h}{2}[f(a) + f(b)]$$

Note that if $f(x)$ is itself a polynomial of degree one, then the trapezium rule will give the exact value of $\int_a^b f(x)\,dx$. This point will be discussed again when the error term for the trapezium rule is derived, later in the chapter.

Example 7.4

Use the trapezium rule to calculate approximately the value of $\int_1^{1.2} \ln x\,dx$.

Here $f(x) = \ln x$, $a = 1$, $b = 1.2$, $h = 0.2$
and we have

x	1	1.2
$f(x)$	0	0.182 32

$$\int_a^b f(x)\,dx \approx \frac{h}{2}[f(a) + f(b)]$$

becomes

$$\int_1^{1.2} \ln x\,dx \approx \frac{0.2}{2}[0 + 0.182\,32]$$

$$= 0.018\,232$$

Example 7.5

Sketch the graph of the function $y = \dfrac{1}{1 + x^2}$ for $0 \leqslant x \leqslant 2$. Use the trapezium rule to obtain approximations to

(a) $\displaystyle\int_0^2 \frac{1}{1 + x^2}\,dx$

(b) $\displaystyle\int_0^1 \frac{1}{1 + x^2}\,dx$

(c) $\displaystyle\int_1^2 \frac{1}{1 + x^2}\,dx$.

Show that

$$\int_0^2 \frac{1}{1+x^2}\,dx = \int_0^1 \frac{1}{1+x^2}\,dx + \int_1^2 \frac{1}{1+x^2}\,dx.$$

Using this result, obtain two approximations to

$$\int_0^2 \frac{1}{1+x^2}\,dx.$$

Which of these results is the more accurate? Suggest a reason why this should be so.

The graph of $y = \dfrac{1}{1+x^2}$ is shown in Figure 7.8.

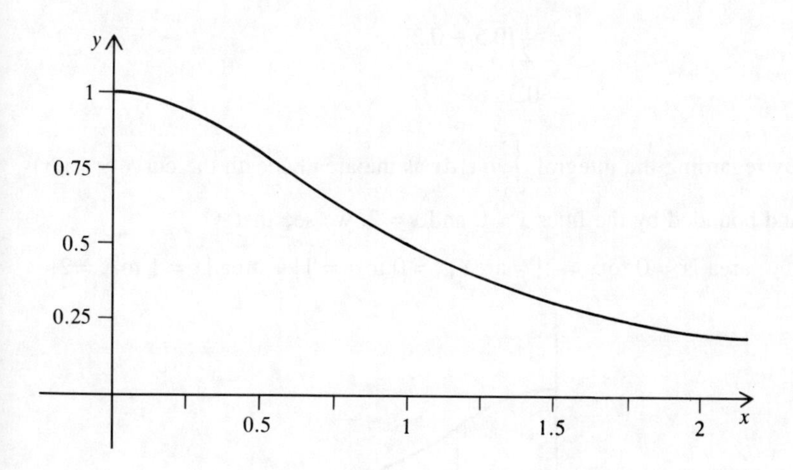

Figure 7.8

In this example, $f(x) = \dfrac{1}{1+x^2}$ and we have

x	0	1	2
$f(x)$	1	0.5	0.2

(a) The trapezium rule approximation with $h = 2$ is given by

$$\begin{aligned}\int_0^2 \frac{1}{1+x^2}\,dx &\approx \frac{h}{2}[f(0) + f(2)]\\ &= \frac{2}{2}[1 + 0.2]\\ &= 1.2\end{aligned}$$

(b) For $\int_0^1 \frac{1}{1+x^2}\,dx$ take $h = 1$ and write

$$\int_0^1 \frac{1}{1+x^2}\,dx \approx \frac{h}{2}[f(0) + f(1)]$$
$$= \frac{1}{2}[1 + 0.5]$$
$$= 0.75$$

(c) For $\int_1^2 \frac{1}{1+x^2}\,dx$ take $h = 1$ and write

$$\int_1^2 \frac{1}{1+x^2}\,dx \approx \frac{h}{2}[f(1) + f(2)]$$
$$= \frac{1}{2}[0.5 + 0.2]$$
$$= 0.35$$

By regarding the integral $\int_0^2 f(x)\,dx$ as the area beneath the curve $y = f(x)$ and bounded by the lines $x = 0$ and $x = 2$, we see that:

area [$x = 0$ to $x = 2$] = area [$x = 0$ to $x = 1$] + area [$x = 1$ to $x = 2$]

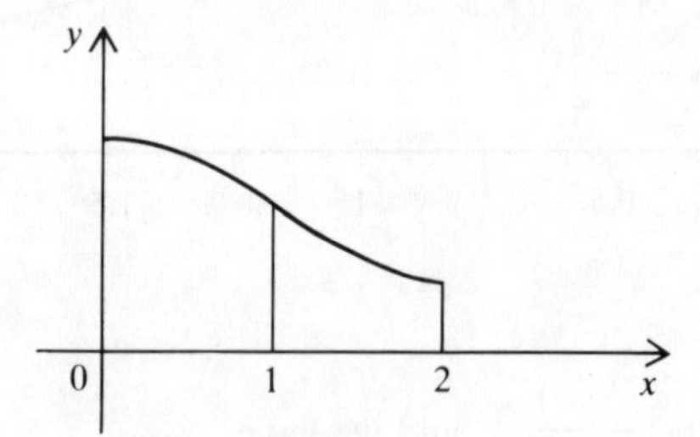

Figure 7.9

giving

$$\int_0^2 f(x)\,dx = \int_0^1 f(x)\,dx + \int_1^2 f(x)\,dx$$

Using this result:

$$\int_0^2 \frac{1}{1+x^2}\,dx = \int_0^1 \frac{1}{1+x^2}\,dx + \int_1^2 \frac{1}{1+x^2}\,dx$$

and $$\int_0^2 \frac{1}{1+x^2}\,dx \approx 0.75 + 0.35$$
$$= 1.10$$

Hence the two approximate values obtained for $\int_0^2 \frac{1}{1+x^2}\,dx$ are: from part (a) 1.2 and from the above 1.10.

Since $\int_0^2 \frac{1}{1 + x^2} \, dx = \left[\tan^{-1} x \right]_0^2$ the exact value of the integral is:

$\tan^{-1} 2 = 1.1071$ (to five decimal digits) and the second approximation is the more accurate.

One explanation of why this should be so is the following. To obtain the first approximation, the curve $f(x) = \frac{1}{1 + x^2}$ was replaced by a single straight line, whereas to obtain the second approximation, the curve was replaced by a pair of straight lines. From Figure 7.10 it is clear that the second procedure will produce the more accurate answer.

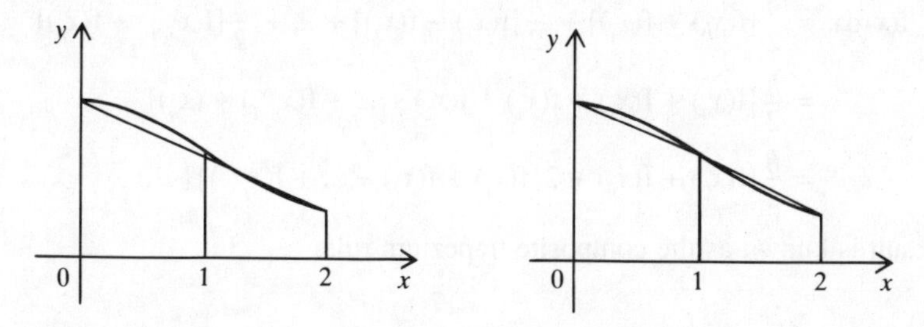

Figure 7.10

Intuitively, it is clear that the trapezium rule will be more accurate on curves which are almost straight lines than on curves which are 'curvy'. We cannot alter the shape of the curve that is being integrated, but Example 7.5 suggests a way in which the accuracy of the trapezium rule can be improved. Divide the range of integration into a sequence of smaller intervals and apply the trapezium rule in each interval separately. It is reasonable to assume that if the curve is replaced by a sequence of straight lines, we shall incur a smaller error than if the curve is replaced by a single straight line (Figure 7.11).

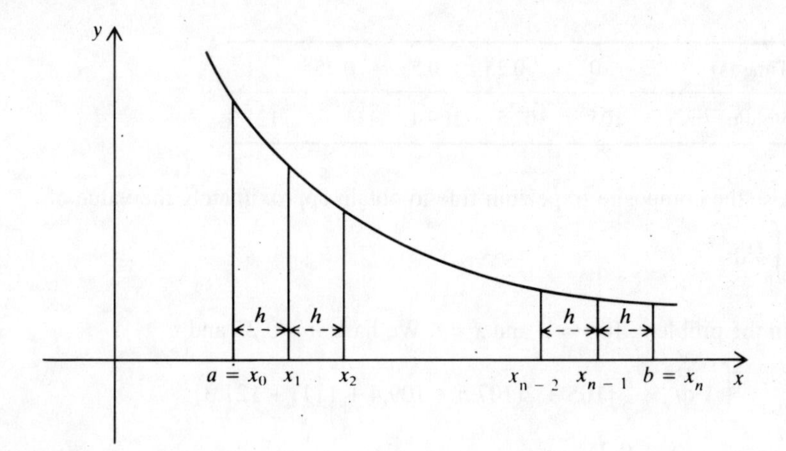

Figure 7.11

We shall assume that $x_0, x_1, \ldots x_n$ are equally spaced in the interval $[a, b]$ and that the distance between adjacent points is h. Now, using the result

$$\int_a^b f(x)\,dx = \int_{x_0}^{x_n} f(x)\,dx$$

$$= \int_{x_0}^{x_1} f(x)\,dx + \int_{x_1}^{x_2} f(x)\,dx + \ldots + \int_{x_{n-1}}^{x_n} f(x)\,dx$$

we may apply the trapezium rule to each integral on the right-hand side separately and get

$$\int_a^b f(x)\,dx \approx \frac{h}{2}[f(x_0) + f(x_1)] + \frac{h}{2}[f(x_1) + f(x_2)] + \ldots + \frac{h}{2}[f(x_{n-1}) + f(x_n)]$$

$$= \frac{h}{2}[f(x_0) + f(x_1) + f(x_1) + f(x_2) + \ldots + f(x_{n-1}) + (x_n)]$$

$$= \frac{h}{2}[f(x_0) + f(x_n) + 2\{f(x_1) + f(x_2) + \ldots + f(x_{n-1})\}]$$

This result is known as the composite trapezium rule.

Composite trapezium rule

$$\int_a^b f(x)\,dx \approx \frac{h}{2}[f(x_0) + f(x_n) + 2\{f(x_1) + f(x_2) + \ldots + f(x_{n-1})\}]$$

Example 7.6

The volume of a certain gas is measured at intervals of 0.25 seconds and the results are given in the table.

Time (s)	0	0.25	0.5	0.75	1
Volume (m^3)	105	107.5	109.4	111	121.3

Use the composite trapezium rule to obtain approximately the value of $\int_0^1 V\,dt$.

In the problem, $f(x) = V$ and $x = t$. We have $h = 0.25$ and

$$\int_0^1 V\,dt \approx \frac{h}{2}[105 + 2(107.5 + 109.4 + 111) + 121.3]$$

$$= \frac{0.25}{2}(882.1)$$

$$= 110.2625$$

Example 7.7

Use the composite trapezium rule to calculate approximately the value of $\int_1^3 \frac{1}{\sqrt{1+x}}\,dx$ using (a) two intervals (b) four intervals and (c) six intervals.
Evaluate the integral exactly and comment on the accuracy.

$$f(x) = \frac{1}{\sqrt{1+x}}$$

(a) With two intervals, $h = 1$ and we have:

x	1	2	3
f(x)	0.707 11	0.577 35	0.5

From the trapezium rule

$$\int_1^3 \frac{1}{\sqrt{1+x}}\,dx \approx \frac{1}{2}[0.707\,11 + 2(0.577\,35) + 0.5]$$
$$= 1.1809$$

(b) With four intervals, $h = 0.5$ and we have

x	1	1.5	2	2.5	3
f(x)	0.707 11	0.632 46	0.577 35	0.534 52	0.5

$$\int_1^3 \frac{1}{\sqrt{1+x}}\,dx \approx \frac{0.5}{2}[0.707\,11 + 2(0.632\,46 + 0.577\,35 + 0.534\,52) + 0.5]$$
$$= 1.1739$$

(c) With six intervals, $h = \frac{1}{3}$ and we have

x	1	$1\frac{1}{3}$	$1\frac{2}{3}$	2	$2\frac{1}{3}$	$2\frac{2}{3}$	3
f(x)	0.707 11	0.654 65	0.612 37	0.577 35	0.547 72	0.522 23	0.5

$$\int_1^3 \frac{1}{\sqrt{1+x}}\,dx \approx \frac{\frac{1}{3}}{2}[0.707\,11 + 2(0.654\,65 + 0.612\,37 + 0.577\,35 + 0.547\,72 + 0.522\,23) + 0.5]$$
$$= 1.1726$$

$$\text{But} \int_1^3 \frac{1}{\sqrt{1+x}}\,dx = \left[(2(1+x)^{\frac{1}{2}}\right]_1^3$$
$$= 2 \times 4^{\frac{1}{2}} - 2 \times 2^{\frac{1}{2}}$$
$$= 1.1716$$

We observe that as h decreases, so the accuracy of the approximation improves. This could have been predicted on intuitive grounds. However, this result will be examined further in Section 7.8.

Algorithm 7.1 The trapezium rule

To approximate the integral $\int_a^b f(x)\,dx$

input: end points a, b;
number of intervals, m

$h := \dfrac{b-a}{m}$

$ye := f(a) + f(b)$

$y := 0$

for $i = 1,2, \ldots, m-1$

$x := a + ih$

$y := y + f(x)$

endloop

$I := \dfrac{h}{2}(ye + 2y)$

output: I $\left(\approx \int_a^b f(x)\,dx\right)$

7.7 Simpson's rule

A second and more accurate method of numerical integration is obtained by replacing f(x) by $p_2(x)$, an interpolating polynomial of degree two, and taking $\int_a^b p_2(x)\,dx$ to be an approximation to $\int_a^b f(x)\,dx$.

To construct $p_2(x)$ we need three points which lie on the curve $y = f(x)$. We shall write these, as usual, in the form $(x_0, f(x_0))$, $(x_1, f(x_1))$ and $(x_2, f(x_2))$. Then, as in Chapter 2:

$$p_2(x) = \frac{(x-x_1)(x-x_2)}{(x_0-x_1)(x_0-x_2)} f(x_0) + \frac{(x-x_0)(x-x_2)}{(x_1-x_0)(x_1-x_2)} f(x_1) + \frac{(x-x_0)(x-x_1)}{(x_2-x_0)(x_2-x_1)} f(x_2)$$

Write $x_0 = a$, $x_2 = b$ and let x_1 be the mid-point of the interval $[a, b]$. Hence $x_1 = \dfrac{a+b}{2}$. Further, write $x_1 - x_0 = h$. Then $x_2 - x_1 = h$ and $x_2 - x_0 = 2h$ (Figure 7.12).

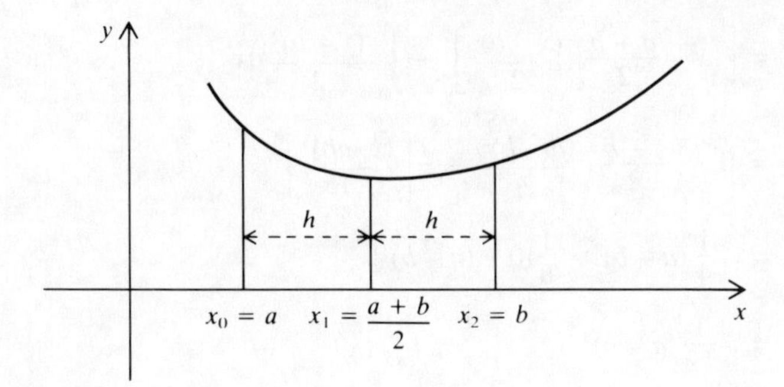

Figure 7.12

With the above notation, $p_2(x)$ may be written

$$p_2(x) = \frac{f(a)}{(-h)(-2h)}(x - x_1)(x - b) + \frac{f(x_1)}{(h)(-h)}(x - a)(x - b)$$

$$+ \frac{f(b)}{(2h)(h)}(x - a)(x - x_1)$$

Our aim now is to obtain a simple expression for $\int_a^b p_2(x)\,dx$.

$$\int_a^b p_2(x)\,dx = \int_a^b \frac{f(a)}{2h^2}(x - x_1)(x - b)\,dx + \int_a^b \frac{f(x_1)}{-h^2}(x - a)(x - b)\,dx$$

$$+ \int_a^b \frac{f(b)}{2h^2}(x - a)(x - x_1)\,dx$$

From the first of these integrals, observe that

$$\int_a^b \frac{f(a)}{2h^2}(x - x_1)(x - b)\,dx = \frac{f(a)}{2h^2}\int_a^b \left(x - \frac{a+b}{2}\right)(x - b)\,dx$$

since $\quad x_1 = \dfrac{a+b}{2}$

Write $\quad I = \displaystyle\int_a^b \left(x - \frac{a+b}{2}\right)(x - b)\,dx$

It is possible to evaluate this integral by multiplying out the brackets and integrating the resulting quadratic expression. However, it is more efficient to use integration by parts:

$$\int u\frac{dv}{dx}\,dx = [uv] - \int v\frac{du}{dx}\,dx$$

Let $\quad u = \left(x - \dfrac{a+b}{2}\right) \quad$ and $\quad \dfrac{dv}{dx} = x - b$

then $\quad \dfrac{du}{dx} = 1 \quad$ and $\quad v = \dfrac{(x-b)^2}{2}$

Then $$I = \left[\left(x - \frac{a+b}{2}\right)\frac{(x-b)^2}{2}\right]_a^b - \int_a^b \frac{(x-b)^2}{2}\,\mathrm{d}x$$

$$= 0 - \left(\frac{a-b}{2}\right)\frac{(a-b)^2}{2} - \frac{1}{2}\left[\frac{(x-b)^3}{3}\right]_a^b$$

$$= -\frac{1}{4}(a-b)^3 - \frac{1}{6}[0-(a-b)^3]$$

$$= -\frac{1}{12}(a-b)^3$$

$$= -\frac{1}{12}(-2h)^3 \qquad \text{since } 2h = b - a$$

$$= \frac{2}{3}h^3$$

Therefore

$$\int_a^b \frac{\mathrm{f}(a)}{2h^2}(x-x_1)(x-b)\,\mathrm{d}x = \frac{\mathrm{f}(a)}{2h^2}I$$

$$= \frac{\mathrm{f}(a)}{2h^2} \times \frac{2}{3}h^3$$

$$= \frac{h}{3}\mathrm{f}(a)$$

In a similar manner, it may be shown that:

the second integral is $\frac{4h}{3}\mathrm{f}\left(\frac{a+b}{2}\right)$

and the third integral is $\frac{h}{3}\mathrm{f}(b)$

Hence $$\int_a^b \mathrm{p}_2(x)\,\mathrm{d}x = \frac{h}{3}\mathrm{f}(a) + \frac{4h}{3}\mathrm{f}\left(\frac{a+b}{2}\right) + \frac{h}{3}\mathrm{f}(b)$$

$$= \frac{h}{3}\left[\mathrm{f}(a) + 4\mathrm{f}\left(\frac{a+b}{2}\right) + \mathrm{f}(b)\right]$$

This justly famous method of numerical integration is called Simpson's rule.

Simpson's rule to calculate approximately the value of $\int_a^b \mathrm{f}(x)\,\mathrm{d}x$

Let $h = \frac{b-a}{2}$ then

$$\int_a^b \mathrm{f}(x)\,\mathrm{d}x \approx \frac{h}{3}\left[\mathrm{f}(a) + 4\mathrm{f}\left(\frac{a+b}{2}\right) + \mathrm{f}(b)\right]$$

Example 7.8

Use Simpson's rule to calculate approximately the value of $\int_0^2 \frac{1}{1+x^2}\,dx$.

$$f(x) = \frac{1}{1+x^2}$$

$$a = 0 \qquad b = 2 \qquad \text{and} \qquad h = \frac{(b-a)}{2} = 1$$

$$\int_0^2 f(x)\,dx \approx \frac{h}{3}[f(0) + 4f(1) + f(2)]$$

Using

x	0	1	2
f(x)	1	0.5	0.2

gives

$$\int_0^2 \frac{1}{1+x^2}\,dx \approx \frac{1}{3}[1 + 4 \times 0.5 + 0.2]$$

$$= 1.0667 \qquad \text{to five decimal digits}$$

As observed in Example 7.5, if the integral is evaluated exactly, the value of the integral (to five decimal digits) is 1.1071.

Note that if f(x) is itself a polynomial of degree two then Simpson's rule will return the exact value of $\int_a^b f(x)\,dx$. In fact, and surprisingly, Simpson's rule will also return the exact value of the integral when f(x) is a polynomial of degree three. We have, then, the important rule: if f(x) is a polynomial of degree less than or equal to three, then Simpson's rule will return the exact value of $\int_a^b f(x)\,dx$.

This result is far from obvious. The example below illustrates the result in the case of a particular cubic polynomial. A proof of this result is given in Section 7.8.

Example 7.9

Apply Simpson's rule to the integral

$$\int_1^3 (4x^3 - 3x^2 + x - 5)\,dx.$$

Evaluate the integral exactly and record the error in the value obtained from Simpson's rule.

$$f(x) = 4x^3 - 3x^2 + x - 5 \qquad a = 1 \qquad b = 3 \qquad \text{and} \qquad h = 1$$

Using

x	1	2	3
f(x)	–3	17	79

we have

$$\int_1^3 f(x)\,dx \approx \frac{h}{3}[f(1) + 4f(2) + f(3)]$$
$$= \frac{1}{3}(-3 + 4 \times 17 + 79)$$
$$= 48$$

But

$$\int_1^3 (4x^3 - 3x^2 + x - 5)\,dx = \left[\frac{4x^4}{4} - \frac{3x^3}{3} + \frac{x^2}{2} - 5x\right]_1^3$$
$$= 43.5 - (-4.5)$$
$$= 48$$

The error in the approximation is zero and Simpson's rule, when applied to a cubic polynomial, is exact.

Observe also that since Simpson's rule is exact for polynomials of degree less than or equal to three, while the trapezium rule is exact for polynomials of degree less than or equal to one, there are grounds for believing that Simpson's rule is the more accurate of the two methods of numerical integration.

To improve the accuracy of Simpson's rule, we might consider dividing the range of integration into a number of smaller intervals and applying Simpson's rule over the narrower intervals. Intuitively, it is reasonable to suppose that the smaller the value of h, the smaller the error in using $p_2(x)$ to approximate $f(x)$ and hence, the smaller the error in using $\int_a^b p_2(x)\,dx$ to approximate $\int_a^b f(x)\,dx$. We shall see in Section 7.8 that this is indeed the case.

Observe, however, that in each single application of Simpson's rule, two intervals are required, $\left[a, \frac{a+b}{2}\right]$ and $\left[\frac{a+b}{2}, b\right]$ (Figure 7.13).

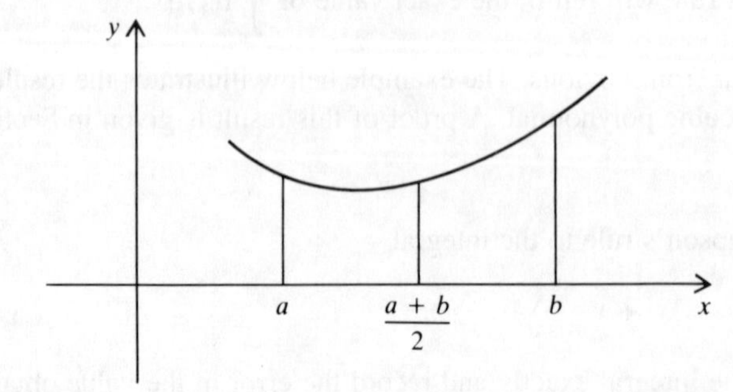

Figure 7.13

So if we are to divide the range of integration into a number of smaller intervals, we must be sure that an even number of intervals are created. For example, to improve the approximation to $\int_0^2 \frac{1}{1+x^2}\,dx$ we might divide the interval [0, 2] in one of the ways shown in Figure 7.14.

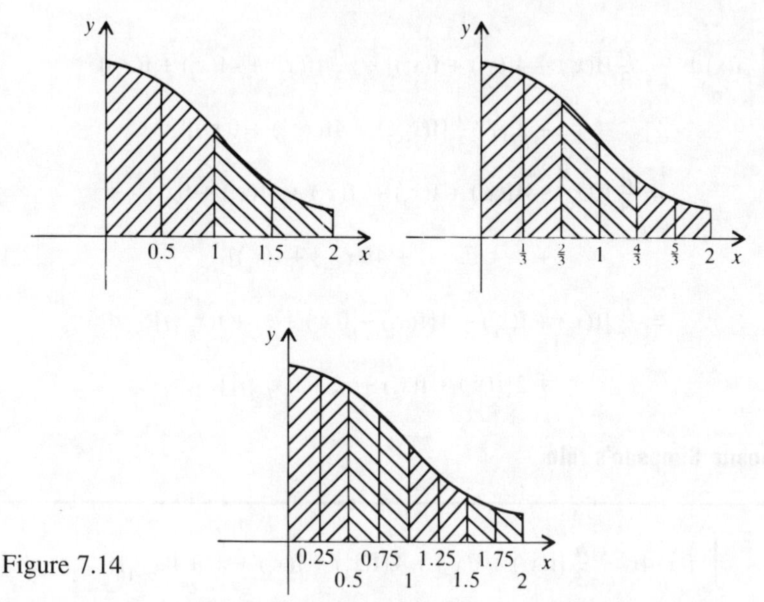

Figure 7.14

Then Simpson's rule could be applied to each adjacent pair of intervals.

This may be made more precise by considering the composite Simpson's rule.

The composite Simpson's rule

To calculate approximately the value of $\int_a^b f(x)\,dx$, let $x_0 = a$, $x_n = b$ and suppose that $x_0, x_1, ..., x_n$ divide the interval $[a, b]$ into n equal parts, where n is an even number (Figure 7.15).

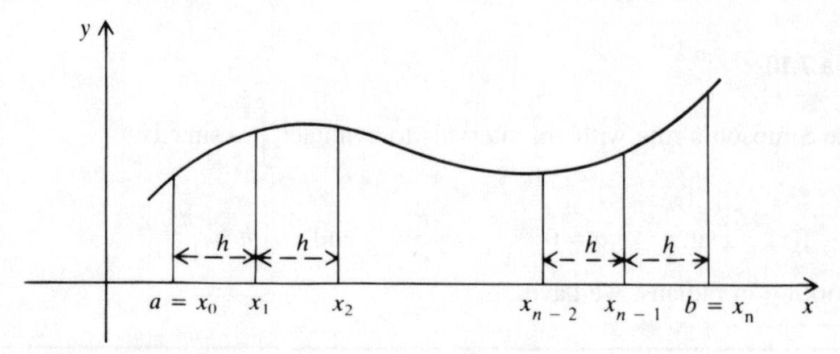

Figure 7.15

Let the distance between adjacent points on the x-axis be h. Then

$$\int_a^b f(x)\,dx = \int_{x_0}^{x_n} f(x)\,dx$$

$$= \int_{x_0}^{x_2} f(x)\,dx + \int_{x_2}^{x_4} f(x)\,dx + ... + \int_{x_{n-2}}^{x_n} f(x)\,dx$$

Because this division has created an even number of intervals, each of these integrals may now be replaced by its appropriate value under Simpson's rule. Hence:

$$\int_a^b f(x)\,dx \approx \frac{h}{3}[f(x_0) + 4f(x_1) + f(x_2)] + \frac{h}{3}[f(x_2) + 4f(x_3) + f(x_4)]$$
$$+ \ldots + \frac{h}{3}[f(x_{n-2}) + 4f(x_{n-1}) + f(x_n)]$$
$$= \frac{h}{3}[f(x_0) + 4f(x_1) + f(x_2) + f(x_2) + 4f(x_3) + f(x_4)$$
$$+ \ldots + f(x_{n-2}) + 4f(x_{n-1}) + f(x_n)]$$
$$= \frac{h}{3}[f(x_0) + f(x_n) + 4\{f(x_1) + f(x_3) + \ldots + f(x_{n-1})\}$$
$$+ 2\{f(x_2) + f(x_4) + \ldots + f(x_{n-2})\}]$$

Composite Simpson's rule

$$\int_a^b f(x)\,dx \approx \frac{h}{3}[f(x_0) + f(x_n) + 4\{f(x_1) + f(x_3) + \ldots + f(x_{n-1})\}$$
$$+ 2\{f(x_2) + f(x_4) + \ldots + f(x_{n-2})\}]$$

The composite Simpson's rule may be remembered by writing

$$\int_a^b f(x)\,dx \approx \frac{h}{3}(\text{first} + \text{last} + 4 \times \text{odds} + 2 \times \text{evens})$$

where 'odds' and 'evens' refer to the subscript on x in the expressions $f(x)$.

But beware! This rule applies only if we write $x_0 = a$: if we write $x_1 = a$, then 'odds' become 'evens' and 'evens' become 'odds'.

Example 7.10

Use Simpson's rule with six intervals to evaluate $\int_0^{\frac{\pi}{2}} x\sin x\,dx$.

$f(x) = x\sin x \quad a = 0 \quad b = \frac{\pi}{2} \quad$ and $\quad h = \frac{\pi}{12}$

Working in radians, we have

x	0	$\frac{\pi}{12}$	$\frac{\pi}{6}$	$\frac{\pi}{4}$	$\frac{\pi}{3}$	$\frac{5\pi}{12}$	$\frac{\pi}{2}$
$f(x)$	0	0.067759	0.26180	0.55536	0.90690	1.2644	1.5708

Hence, $$\int_0^{\frac{\pi}{2}} x\sin x\,dx \approx \frac{h}{3}\left[f(0) + 4\left\{f\left(\frac{\pi}{12}\right) + f\left(\frac{\pi}{4}\right) + f\left(\frac{5\pi}{12}\right)\right\}\right.$$
$$\left. + 2\left\{f\left(\frac{\pi}{6}\right) + f\left(\frac{\pi}{3}\right)\right\} + f\left(\frac{\pi}{2}\right)\right]$$

$$= \frac{\pi}{36}[0 + 4\{0.067759 + 0.55536 + 1.2644\} + 2\{0.26180 + 0.90690\} + 1.5708]$$

$$= 0.99992$$

The exact value of the integral is 1, so Simpson's rule has achieved a remarkably accurate result.

Example 7.11

Apply (a) the trapezium rule and (b) Simpson's rule, each with four intervals, to obtain approximate values of $\int_0^1 e^{-\frac{1}{2}x}\,dx$. Which method provides the more accurate approximation?

$$f(x) = e^{-\frac{1}{2}x} \qquad a = 0 \qquad b = 1 \qquad h = 0.25$$

and we have

x	0	0.25	0.5	0.75	1
$f(x)$	1	0.88250	0.77880	0.68729	0.60653

(a) For the trapezium rule:

$$\int_0^1 e^{-\frac{1}{2}x}\,dx \approx \frac{h}{2}[f(0) + 2\{f(0.25) + f(0.5) + f(0.75)\} + f(1)]$$

$$= 0.125[1 + 2\{0.88250 + 0.77880 + 0.68729\} + 0.60653]$$

$$= 0.78796$$

(b) For Simpson's rule:

$$\int_0^1 e^{-\frac{1}{2}x}\,dx \approx \frac{h}{3}[f(0) + 4\{f(0.25) + f(0.75)\} + 2\{f(0.5)\} + f(1)]$$

$$= \frac{0.25}{3}[1 + 4\{0.88250 + 0.68729\} + 2\{0.77880\} + 0.60653]$$

$$= 0.78694$$

The exact value of the integral is 0.78693868 (to eight decimal digits), so Simpson's rule is the more accurate of the two methods. The exact value of the integral written to five decimal digits is 0.78694. Hence, Simpson's rule has once again returned an impressively accurate approximation.

Algorithm 7.2 Simpson's rule

To approximate the integral $\int_a^b f(x)\,dx$

input: end points a, b;
number of intervals m (where m is an even number)

$h := \dfrac{b-a}{m}$

$ye := f(a) + f(b)$
$y1 := 0$ (for 'odds')
$y2 := 0$ (for 'evens')
for $i = 1, 2, ..., m-1$
$x := a + ih$
if i is odd then
$y1 := y1 + f(x)$
otherwise
$y2 := y2 + f(x)$
endif
endloop

$I := \dfrac{h}{3}(ye + 4 \times y1 + 2 \times y2)$

output: I $\left(\approx \int_a^b f(x)\,dx\right)$

The trapezium rule and Simpson's rule are the methods of numerical integration most commonly in use and as can be seen from the examples these methods do provide remarkably accurate approximations. There are, however, a great many other ways of attempting numerical integration. Some of these techniques display great elegance and ingenuity and the reader is recommended to look further into this topic†.

7.8 The error term in numerical integration

To obtain an expression for the error term in numerical integration, we need to calculate

$$\int_a^b f(x)\,dx - \left\{ \text{approximation to} \int_a^b f(x)\,dx \right\} \qquad \ldots (1)$$

This may be achieved most easily by expressing both terms of Equation (1) as Taylor polynomials and regarding Equation (1) as the difference between two polynomials.

First consider $\int_a^b f(x)\,dx$.

† See Conte and de Boor 1981 *Elementary numerical analysis* 3rd edn. McGraw-Hill and Burden and Faires (1985) *Numerical analysis* 3rd edn. Prindle, Weber and Schmidt, for more details.

From Chapter 5 the Taylor polynomial of $f(x)$ centred at $x = a$ is given by:

$$f(x) = f(a) + (x-a)f'(a) + \frac{(x-a)^2}{2!}f''(a) + \frac{(x-a)^3}{3!}f'''(a) + \frac{(x-a)^4}{4!}f^{(4)}(a) + \ldots$$

Hence

$$\int_a^b f(x)\,dx = \int_a^b f(a)\,dx + \int_a^b (x-a)f'(a)\,dx + \int_a^b \frac{(x-a)^2}{2!}f''(a)\,dx$$
$$+ \int_a^b \frac{(x-a)^3}{3!}f'''(a)\,dx + \int_a^b \frac{(x-a)^4}{4!}f^{(4)}(a)\,dx + \ldots$$
$$= f(a)\Big[x\Big]_a^b + f'(a)\left[\frac{(x-a)^2}{2!}\right]_a^b + f''(a)\left[\frac{(x-a)^3}{3!}\right]_a^b$$
$$+ f'''(a)\left[\frac{(x-a)^4}{4!}\right]_a^b + f^{(4)}(a)\left[\frac{(x-a)^5}{5!}\right]_a^b + \ldots$$
$$= f(a)(b-a) + f'(a)\frac{(b-a)^2}{2!} + f''(a)\frac{(b-a)^3}{3!}$$
$$+ f'''(a)\frac{(b-a)^4}{4!} + f^{(4)}(a)\frac{(b-a)^5}{5!} + \ldots \qquad \ldots (2)$$

To go further, we need to specify which method of numerical integration is being considered.

The trapezium rule

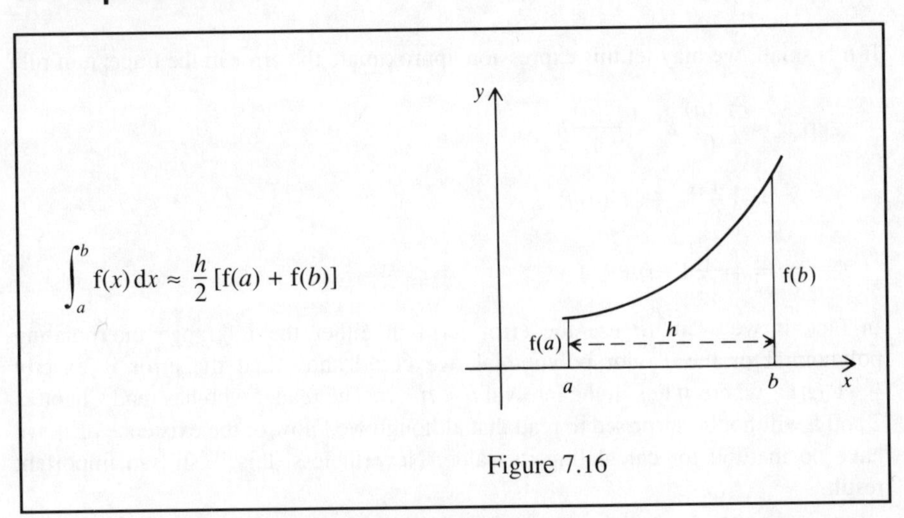

Figure 7.16

In this case, $h = b - a$ so Equation (2) becomes

$$\int_a^b f(x)\,dx = f(a)h + \frac{f'(a)}{2!}h^2 + \frac{f''(a)}{3!}h^3 + \frac{f'''(a)}{4!}h^4 + \frac{f^{(4)}(a)}{5!}h^5 + \ldots$$

and $\int_a^b f(x)\,dx$ has been written as a Taylor polynomial in h.

Now consider the trapezium rule approximation $\frac{h}{2}[f(a) + f(b)]$

To express $\frac{h}{2}[f(a) + f(b)]$ as a Taylor polynomial in h, observe that $b = a + h$ and that

$$\begin{aligned} f(b) &= f(a + h) \\ &= f(a) + hf'(a) + \frac{h^2}{2!}f''(a) + \frac{h^3}{3!}f'''(a) + \ldots \end{aligned}$$

Hence

$$\begin{aligned} \frac{h}{2}[f(a) + f(b)] &= \frac{h}{2}\left[f(a) + f(a) + hf'(a) + \frac{h^2}{2!}f''(a) + \frac{h^3}{3!}f'''(a) + \ldots\right] \\ &= hf(a) + \frac{h^2}{2}f'(a) + \frac{h^3}{4}f''(a) + \frac{h^4}{12}f'''(a) + \ldots \\ &= f(a)h + \frac{f'(a)}{2}h^2 + \frac{f''(a)}{4}h^3 + \frac{f'''(a)}{12}h^4 + \ldots \end{aligned}$$

Now $\int_a^b f(x)\,dx - \frac{h}{2}[f(a) + f(b)]$ may be written:

$$\left[f(a)h + \frac{f'(a)}{2!}h^2 + \frac{f''(a)}{3!}h^3 + \frac{f'''(a)}{4!}h^4 + \ldots\right]$$

$$-\left[f(a)h + \frac{f'(a)}{2}h^2 + \frac{f''(a)}{4}h^3 + \frac{f'''(a)}{12}h^4 + \ldots\right]$$

$$= \frac{f''(a)}{3!}h^3 - \frac{f''(a)}{4}h^3 + \ldots$$

If h is small, we may let this expression approximate the error in the trapezium rule

$$\begin{aligned} \text{error} &\approx \frac{f''(a)}{6}h^3 - \frac{f''(a)}{4}h^3 \\ &= \left(\frac{1}{6} - \frac{1}{4}\right)f''(a)h^3 \\ &= -\frac{1}{12}f''(a)h^3 \end{aligned}$$

In fact, if we were to use the error term in either the Lagrange interpolating polynomial or the Taylor polynomial, we could show that the error is exactly $-\frac{1}{12}f''(\eta)h^3$ where η lies in the interval $a < \eta < b$. The reader who has read Chapters 2 and 5 will not be surprised to read that although we know of the existence of η, we have no method for calculating its value. Nevertheless, this is still an important result.

Trapezium rule error

The error in the trapezium rule

$$\int_a^b \text{f}(x)\,\text{d}x \approx \frac{h}{2}[\text{f}(a) + \text{f}(b)]$$

is $\quad -\frac{1}{12}\text{f}''(\eta)h^3 \qquad a < \eta < b$

Observe that the error term involves the factor $\text{f}''(\eta)$. If $\text{f}(x)$ is a polynomial of degree one

$$\text{f}(x) = a_0 + a_1 x$$

then $\text{f}''(x) = 0$ for all values of x and, in this case, the error term would be identically zero. This simple argument confirms the statement in Section 7.6 that the trapezium rule is exact when $\text{f}(x)$ is a polynomial of degree one.

In the composite trapezium rule, the trapezium rule is applied to each of the n intervals $[x_r, x_{r+1}]$

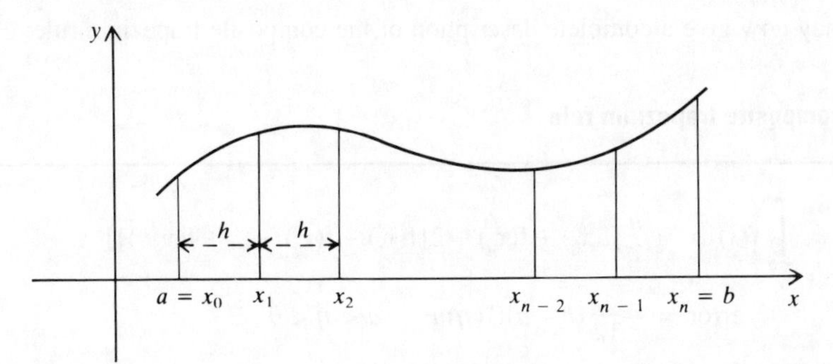

Figure 7.17

and $\displaystyle\int_{x_r}^{x_{r+1}} \text{f}(x)\,\text{d}x \approx \frac{h}{2}[\text{f}(x_r) + \text{f}(x_{r+1})]$ with error $-\frac{1}{12}\text{f}''(\eta_r)h^3$ where $x_r < \eta_r < x_{r+1}$.

The error in the composite trapezium rule is, then, the sum of the individual errors in each interval and

$$\text{error} = -\frac{1}{12}\text{f}''(\eta_0)h^3 - \frac{1}{12}\text{f}''(\eta_1)h^3 - \dots - \frac{1}{12}\text{f}''(\eta_{n-1})h^3$$

$$= -\frac{1}{12}[\text{f}''(\eta_0) + \text{f}''(\eta_1) + \dots + \text{f}''(\eta_{n-1})]h^3$$

This expression may be simplified by appealing to the intermediate value theorem. It may be shown that, if $z_0, z_1, z_2, \dots, z_{n-1}$ are n numbers, then the average of these n numbers lies between the smallest number and the largest.

So, if z_s is the smallest of the numbers and z_l the largest,

$$z_s \leqslant \frac{z_0 + z_1 + \dots + z_{n-1}}{n} \leqslant z_l$$

If now we take $z_i = f''(\eta_i)$, we have

$$f''(\eta_s) \leqslant \frac{f''(\eta_0) + f''(\eta_1) + \dots + f''(\eta_{n-1})}{n} \leqslant f''(\eta_l)$$

If $f''(x)$ is continuous in the interval $[a, b]$ then the conditions of the intermediate value theorem are satisfied and we have:

There exists a number η where $a < \eta < b$ such that

$$\frac{f''(\eta_0) + f''(\eta_1) + \dots + f''(\eta_{n-1})}{n} = f''(\eta)$$

Hence $\quad f''(\eta_0) + f''(\eta_1) + \dots + f''(\eta_{n-1}) = nf''(\eta)$

and the error term becomes

$$\text{error} = -\frac{1}{12}nf''(\eta)h^3$$

Since $nh = b - a$, the error term may also be written:

$$\text{error} = -\frac{1}{12}(b-a)f''(\eta)h^2$$

We may now give a complete description of the composite trapezium rule:

The composite trapezium rule

$$\int_a^b f(x)\,dx \approx \frac{h}{2}[f(x_0) + f(x_n) + 2\{f(x_1) + f(x_2) + \dots + f(x_{n-1})\}]$$

$$\text{error} = -\frac{1}{12}(b-a)f''(\eta)h^2 \qquad a < \eta < b$$

Example 7.12

Use the compound trapezium rule with four intervals to calculate approximately the value of $\int_1^2 e^{-\frac{1}{2}x}\,dx$.

Calculate the maximum value of the error in your answer.

With four intervals, $h = 0.25$ and we have:

x	1	1.25	1.5	1.75	2
$f(x)$	0.606 530 66	0.535 261 43	0.472 366 55	0.416 862 02	0.367 879 44

$$\int_1^2 e^{-\frac{1}{2}x}\,dx \approx \frac{h}{2}[f(1) + f(2) + 2\{f(1.25) + f(1.5) + f(1.75)\}]$$

$$= \frac{0.25}{2}[0.60653066 + 0.36787944 + 2\{0.53526143 + 0.47236655 + 0.41686202\}]$$

$$= 0.47792376$$

The error in the approximation is:

$$\text{error} = -\frac{1}{12}(b-a)f''(\eta)h^2 \qquad 1 < \eta < 2$$

$$= -\frac{1}{12}(1)f''(\eta)(0.25)^2$$

To obtain the maximum value of the error, we must determine the maximum value of $f''(\eta) \qquad a < \eta < b$.

Now $\quad f(x) = e^{-\frac{1}{2}x}$

$$f'(x) = -\frac{1}{2}e^{-\frac{1}{2}x}$$

$$f''(x) = \frac{1}{4}e^{-\frac{1}{2}x}$$

Since $\frac{1}{4}e^{-\frac{1}{2}x}$ is a decreasing function of x

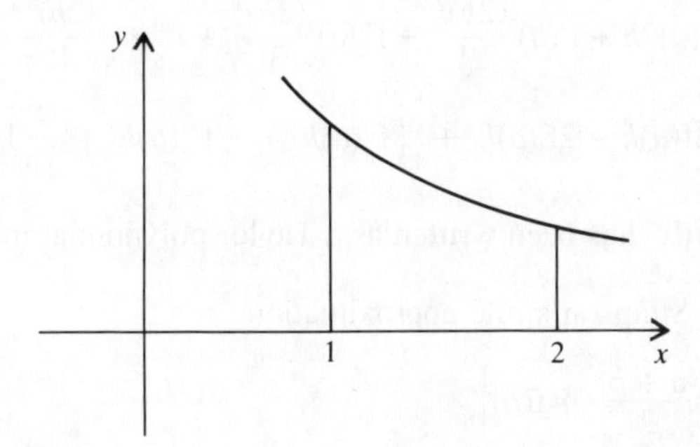

Figure 7.18

$$|f''(\eta)| < |f''(1)|$$

and $\quad |f''(\eta)| < 0.15163267$

Hence $\quad |\text{error}| < \frac{1}{12}(0.25)^2 \times 0.15163267$

$$= 7.8975 \times 10^{-4} \qquad \text{to five decimal digits}$$

In fact, the exact value of the integral, to eight decimal digits, is given by

$$\int_1^2 e^{-\frac{1}{2}x}\,dx = \left[-2e^{-\frac{1}{2}x}\right]_1^2$$

$$= -2e^{-1} + 2e^{-\frac{1}{2}}$$

$$= 0.47730244$$

The error in the trapezium rule approximation is -6.2132×10^{-4}; within the maximum predicted by the theory.

Simpson's rule

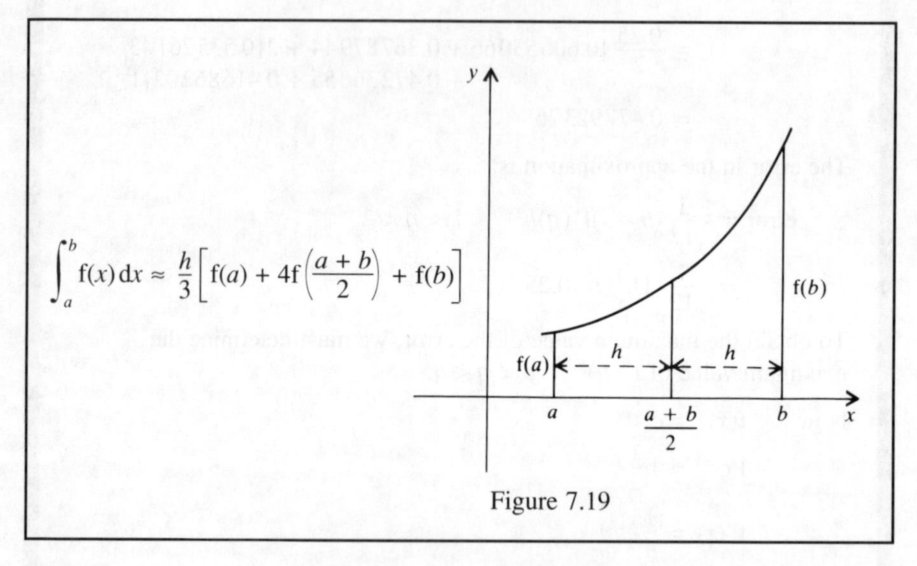

Figure 7.19

In this case $b - a = 2h$ and from Equation (2) we have

$$\int_a^b f(x)\,dx = f(a)2h + f'(a)\frac{(2h)^2}{2!} + f''(a)\frac{(2h)^3}{3!} + f'''(a)\frac{(2h)^4}{4!} + f^{(4)}(a)\frac{(2h)^5}{5!} + \ldots$$

$$= 2f(a)h + 2f'(a)h^2 + \frac{4}{3}f''(a)h^3 + \frac{2}{3}f'''(a)h^4 + \frac{4}{15}f^4(a)h^5 + \ldots$$

and, again, $\int_a^b f(x)\,dx$ has been written as a Taylor polynomial in h.

Consider now the Simpson's rule approximation

$$\frac{h}{3}\left[f(a) + 4f\left(\frac{a+b}{2}\right) + f(b)\right]$$

To write this expression as a Taylor polynomial in h observe that

$$\frac{a+b}{2} = a + h \qquad \text{and} \qquad b = a + 2h$$

Hence
$$f\left(\frac{a+b}{2}\right) = f(a+h)$$

$$= f(a) + hf'(a) + \frac{h^2}{2!}f''(a) + \frac{h^3}{3!}f'''(a) + \frac{h^4}{4!}f^{(4)}(a) + \ldots$$

$$= f(a) + f'(a)h + \frac{f''(a)}{2}h^2 + \frac{f'''(a)}{6}h^3 + \frac{f^{(4)}(a)}{24}h^4 + \ldots$$

and
$$f(b) = f(a + 2h)$$

$$= f(a) + 2hf'(a) + \frac{(2h)^2}{2!}f''(a) + \frac{(2h)^3}{3!}f'''(a) + \frac{(2h)^4}{4!}f^{(4)}(a) + \ldots$$

$$= f(a) + 2f'(a)h + 2f''(a)h^2 + \frac{4}{3}f'''(a)h^3 + \frac{2}{3}f^{(4)}(a)h^4 + \ldots$$

Putting all this together:

$$\frac{h}{3}\left[\mathrm{f}(a) + 4\mathrm{f}\left(\frac{a+b}{2}\right) + \mathrm{f}(b)\right]$$

$$= \frac{h}{3}\left[\mathrm{f}(a) + 4\{\mathrm{f}(a) + \mathrm{f}'(a)h + \frac{\mathrm{f}''(a)}{2}h^2 + \frac{\mathrm{f}'''(a)}{6}h^3 + \frac{\mathrm{f}^{(4)}(a)}{24}h^4 + \ldots\right.$$

$$\left. + \mathrm{f}(a) + 2\mathrm{f}'(a)h + 2\mathrm{f}''(a)h^2 + \frac{4}{3}\mathrm{f}'''(a)h^3 + \frac{2}{3}\mathrm{f}^{(4)}(a)h^4 + \ldots\right.$$

$$= \frac{h}{3}\left[6\mathrm{f}(a) + 6\mathrm{f}'(a)h + 4\mathrm{f}''(a)h^2 + 2\mathrm{f}'''(a)h^3 + \frac{5}{6}\mathrm{f}^{(4)}(a)h^4 + \ldots\right]$$

$$= 2\mathrm{f}(a)h + 2\mathrm{f}'(a)h^2 + \frac{4}{3}\mathrm{f}''(a)h^3 + \frac{2}{3}\mathrm{f}'''(a)h^4 + \frac{5}{18}\mathrm{f}^{(4)}(a)h^5 + \ldots$$

and we have the required Taylor polynomial.

Now $\displaystyle\int_a^b \mathrm{f}(x)\,\mathrm{d}x - \frac{h}{3}\left[\mathrm{f}(a) + 4\mathrm{f}\left(\frac{a+b}{2}\right) + \mathrm{f}(b)\right]$

may be written

$$\left[2\mathrm{f}(a)h + 2\mathrm{f}'(a)h^2 + \frac{4}{3}\mathrm{f}''(a)h^3 + \frac{2}{3}\mathrm{f}'''(a)h^4 + \frac{4}{15}\mathrm{f}^{(4)}(a)h^5 + \ldots\right]$$

$$-\left[2\mathrm{f}(a)h + 2\mathrm{f}'(a)h^2 + \frac{4}{3}\mathrm{f}''(a)h^3 + \frac{2}{3}\mathrm{f}'''(a)h^4 + \frac{5}{18}\mathrm{f}^{(4)}(a)h^5 + \ldots\right]$$

$$= \frac{4}{15}\mathrm{f}^{(4)}(a)h^5 - \frac{5}{18}\mathrm{f}^{(4)}(a)h^5 + \ldots$$

Again, if h is small, we may let this expression approximate the error in Simpson's rule.

$$\text{error} \approx \frac{4}{15}\mathrm{f}^{(4)}(a)h^5 - \frac{5}{18}\mathrm{f}^{(4)}(a)h^5$$

$$= \left(\frac{4}{15} - \frac{5}{18}\right)\mathrm{f}^{(4)}(a)h^5$$

$$= -\frac{1}{90}\mathrm{f}^{(4)}(a)h^5$$

As with the trapezium rule, if we were to be more precise in our use of Taylor polynomials and use the error term associated with these polynomials, we could show that the error in Simpson's rule is exactly $-\frac{1}{90}\mathrm{f}^{(4)}(\eta)h^5$ where $a < \eta < b$. As usual, the value of η cannot, in general be determined, but nevertheless, this error term is of considerable value.

Simpson's rule error

The error term in Simpson's rule

$$\int_a^b f(x)\,dx \approx \frac{h}{3}\left[f(a) + 4f\left(\frac{a+b}{2}\right) + f(b)\right]$$

is $\quad -\frac{1}{90}f^{(4)}(\eta)h^5 \quad$ where $\quad a < \eta < b$

Observe two points.

(a) The error in using Simpson's rule is proportional to h^5 while the error in the trapezium rule is proportional to h^3. Thus, we would expect Simpson's rule to be the more accurate of the two methods.

(b) If $f(x)$ is a polynomial of degree three or less, then the fourth derivative of $f(x)$ will be identically zero. That is: $f^{(4)}(x) \equiv 0$. In this case, the error term in Simpson's rule will also be zero and we have proved the important result:

If $f(x)$ is a polynomial of degree less than or equal to three, then Simpson's rule for approximating $\int_a^b f(x)\,dx$ is exact.

In the composite Simpson's rule, Simpson's rule is applied across each adjacent pair of intervals in the division of $[a, b]$ (Figure 7.20).

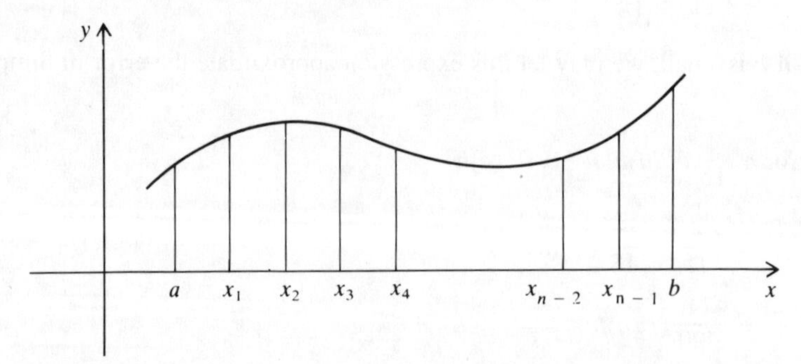

Figure 7.20

The error in the composite Simpson's rule will be the sum of the errors that arise in each single application of the rule. That is:

$$\text{error} = -\frac{1}{90}f^{(4)}(\eta_0)h^5 - \frac{1}{90}f^{(4)}(\eta_2)h^5 - \frac{1}{90}f^{(4)}(\eta_4)h^5 - \ldots - \frac{1}{90}f^{(4)}(\eta_{n-2})h^5$$

$$a < \eta_0 < x_2 \quad x_2 < \eta_2 < x_4 \quad x_4 < \eta_4 < x_6 \quad \ldots \quad x_{n-2} < \eta_{n-2} < b$$

$$= -\frac{1}{90}[f^{(4)}(\eta_0) + f^{(4)}(\eta_2) + f^{(4)}(\eta_4) + \ldots + f^{(4)}(\eta_{n-2})]h^5$$

Since n is an even number, there are $\frac{n}{2}$ error terms. To simplify this expression for the error, we repeat the argument given for the compound trapezium rule and use the intermediate value theorem to write:

there exists a number η where $a < \eta < b$ such that

$$\frac{f^{(4)}(\eta_0) + f^{(4)}(\eta_2) + f^{(4)}(\eta_4) + \dots + f^{(4)}(\eta_{n-2})}{\frac{n}{2}} = f^{(4)}(\eta)$$

This gives $f^{(4)}(\eta_0) + f^{(4)}(\eta_2) + f^{(4)}(\eta_4) + \dots + f^{(4)}(\eta_{n-2}) = \frac{n}{2} f^{(4)}(\eta)$

and hence:

$$\text{error} = -\frac{1}{90}\,\frac{n}{2}\, f^{(4)}(\eta)h^5$$

Since $nh = b - a$, the error term may be written

$$\text{error} = -\frac{1}{90}\,\frac{nh}{2}\, f^{(4)}(\eta)h^4$$

$$= -\frac{1}{180}(b-a)f^{(4)}(\eta)h^4$$

and we can give a complete description of Simpson's rule:

Simpson's rule

$$\int_a^b f(x)\,dx \approx \frac{h}{3}\,[f(a) + f(b) + 4\{f(x_1) + f(x_3) + \dots + f(x_{n-1})\} + 2\{f(x_2) + f(x_4) + \dots + f(x_{n-2})\}]$$

$$\text{error} = -\frac{1}{180}\,(b-a)f^{(4)}(\eta)h^4$$

Example 7.13

Use Simpson's rule with four intervals to calculate approximately the value of $\int_1^2 e^{-\frac{1}{2}x}\,dx$. Calculate the maximum value of the error in your approximation.

$$\int_1^2 e^{-\frac{1}{2}x}\,dx \approx \frac{h}{3}\,[f(1) + f(2) + 4\{f(1.25) + f(1.75)\} + 2\{f(1.5)\}]$$

Using the table in Example 7.12 we have

$$\int_1^2 e^{-\frac{1}{2}x}\,dx \approx \frac{0.25}{3}\,[0.60653066 + 0.36787944 + 4\{0.53526143 + 0.41686202\} + 2\{0.47236655\}]$$

$$= 0.47730308$$

The error in the approximation is:

$$\text{error} = -\frac{1}{180}(b-a)f^{(4)}(\eta)h^4$$

$$= -\frac{1}{180}(1)f^{(4)}(\eta)(0.25)^4$$

To obtain the maximum value of $f^{(4)}(\eta)$ consider

$$f(x) = e^{-\frac{1}{2}x}$$

$$f'(x) = -\frac{1}{2}e^{-\frac{1}{2}x}$$

$$f''(x) = \frac{1}{4}e^{-\frac{1}{2}x}$$

$$f'''(x) = -\frac{1}{8}e^{-\frac{1}{2}x}$$

$$f^{(4)}(x) = \frac{1}{16}e^{-\frac{1}{2}x}$$

Since $\frac{1}{16}e^{-\frac{1}{2}x}$ is a decreasing function, the maximum value in the interval [1, 2] occurs when $x = 1$. Hence

$$\max\left\{\frac{1}{16}e^{-\frac{1}{2}x} : 1 \leqslant x \leqslant 2\right\} = \frac{1}{16}e^{-\frac{1}{2}}$$

$$= 3.7908166 \times 10^{-2}$$

and max (error) $= -\frac{1}{180}(1) \times (3.7908166 \times 10^{-2})(0.25)^4$

$$= -8.2266 \times 10^{-7}$$

The exact value of the integral, as shown in Example 7.12, is 0.47730244. The error in the Simpson's rule approximation is -6.4×10^{-7} and this is less than the maximum error predicted by the theory.

Observe the very great improvement in accuracy achieved by Simpson's rule in this example over the trapezium rule approximation illustrated in Example 7.12. Ways of improving the accuracy even further are described in the Tasks.

Exercise 7

Section 7.1

1 For the data in the table below, use the forward difference formula to calculate approximately the value of $\frac{dy}{dx}$

(a) when $x = 2$ (b) when $x = 4$.

x	0	1	2	3	4	5
y	0	2	2.75	3.4	3.2	2.5

2 For the table given below, use the forward difference formula to calculate approximately the value of $\frac{dy}{dx}$ when $x = 3, 4, 5$ and 6.

x	3	4	5	6	7
y	5.8	4.5	3.2	3.3	4.7

Why can't the forward difference formula be used to estimate $\frac{dy}{dx}$ when $x = 7$? What could be done if it was required to estimate this quantity?

3 The pressure of a certain volume of gas was recorded at 2 second intervals and the results are set out below.

time (t)	0	2	4	6	8
pressure (p)	50	53	58	67	83

Use the forward difference formula to calculate approximately the value of $\frac{dp}{dt}$ at $t = 0, t = 2, t = 4, t = 6$. Plot your calculated points on a graph and sketch the graph of $\frac{dp}{dt}$ against t. Use your graph to estimate the value of $\frac{dp}{dt}$
(a) when $t = 3$ (b) when $t = 6.6$ (c) when $t = 8$.

4 Use the central difference formula and the values recorded in the table below to calculate approximately the value of $\frac{dy}{dx}$ when $x = 5$, $x = 6$, $x = 7$, $x = 8$ and $x = 9$.

x	3	4	5	6	7	8	9	10
y	4	6.5	7.8	9.2	8.7	7.6	7.0	6.5

5 If $f(x) = \sqrt{x}$ copy and complete the table shown below.

x	1	1.5	2	2.5	3	3.5
y	1			1.5811		

Use the central difference formula to calculate approximately the values of (a) $f'(1.5)$ (b) $f'(2)$ (c) $f'(2.5)$ (d) $f'(3)$. Use calculus to evaluate these quantities correct to five decimal digits and hence calculate the errors in your answers.

6 Prepare a table of values for the function $f(x) = \ln(x^2)$ taking $x = 2, 2.25, 2.5, 2.75, 3, 3.25, 3.5, 3.75, 4$.
Use (a) the forward difference formula (b) the central difference formula to calculate approximations to $\frac{dy}{dx}$ when $x = 2.25, 2.75, 3.25, 3.5, 3.75$.

Calculate the exact values of $\frac{dy}{dx}$ at these values of x and hence calculate the error in (a) and (b) above.

7 On suitable axes plot the data points

x	2	3	4	5	6
$y = f(x)$	1.49	4.21	4.80	4.42	4.01

(a) Use (i) the forward difference formula (ii) the central difference formula to estimate the value of $\frac{dy}{dx}$ when $x = 3$. Illustrate on your graph the calculation you have just performed. Provide also a geometrical interpretation of the value of $\frac{dy}{dx}$ when $x = 3$. Comment on your results.

On separate graphs, repeat the exercise for the points
(b) $x = 4$
(c) $x = 5$.

Section 7.3

8 Use the central difference formula and the table below to calculate approximately the values of $\frac{dy}{dx}$ and $\frac{d^2y}{dx^2}$ when $x = 10$, $x = 10.5$, $x = 11$ and $x = 11.5$.

x	9	9.5	10	10.5	11	11.5	12
y	16	15.3	14.1	12.8	12.3	13.4	14.6

9 Construct a table of values for the function $y = \sqrt{x^2 + 1}$ taking $x = 1, 2, 3, 4, 5$ and 6.
Use the central difference formula to calculate approximately the values of (a) $f''(2)$ (b) $f''(3)$ (c) $f''(4)$ (d) $f''(5)$.

Use calculus to determine $\frac{d^2y}{dx^2}$. Calculate the value of $\frac{d^2y}{dx^2}$ when $x = 2$, $x = 3$, $x = 4$ and $x = 5$. Hence evaluate the error in your approximations.

10 The volume of a certain mass of gas measured at one second intervals of time is given in the table.

t (s)	2	3	4	5	6	7
v (m^3)	25	26.2	27.1	26.9	26.1	25.8

Using the central difference formula, calculate approximately the values of (a) $\frac{dv}{dt}$ and (b) $\frac{d^2v}{dt^2}$ when $t = 3$, $t = 4$, $t = 5$ and $t = 6$. By interpolating polynomials of degree two to the data, calculate a second approximation to $\frac{dv}{dt}$ and $\frac{d^2v}{dt^2}$ when $t = 3$ and when $t = 6$. Comment.

11 (a) Show that the error in using the expression $\frac{f(a+h) - f(a)}{h}$ to calculate an approximation to $f'(a)$ is

$$-\frac{h}{2}f''(\eta) \qquad a < \eta < a + h$$

(b) For the data given below

x	−2	−1	0	1	2	3
f(x)	−7	−2	−3	−4	1	18

Use the forward difference formula to estimate
(i) $f'(-1)$ (ii) $f'(1)$ (iii) $f'(2)$

In fact, the data is generated by the function $f(x) = x^3 - 2x - 3$.

(c) Calculate the error in your answers to (i), (ii) and (iii) above.

(d) Calculate the maximum theoretical error in your answers.

12 Repeat Question 11 for the central difference formula. (In (a), you are required to write down an expression for the error; you should then prove that your expression is the correct one.)

Section 7.6

13 Use the trapezium rule to calculate approximately the value of $\int_1^2 f(x)\,dx$ where f(x) is as given below.

(a)

x	1	2
f(x)	3	4

(b)

x	1	2
f(x)	2	−1

(c)

x	1	2
f(x)	4.67	5.62

14 Use the trapezium rule to calculate approximately the value of $\int_a^b f(x)\,dx$ when f(x) and a and b are as given below.

(a)

x	0	1
f(x)	2	2.5

$a = 0$, $b = 1$

(b)

x	3	5
f(x)	2	3

$a = 3$, $b = 5$

(c)

x	4	4.5
f(x)	3	1.5

$a = 4$, $b = 4.5$

(d)

x	5	8
f(x)	1	−2

$a = 5$, $b = 8$

(e)

x	6	10
f(x)	−3	4

$a = 6$, $b = 10$

15 Use the trapezium rule with a single interval to calculate approximately the values of

(a) $\int_1^3 e^x\,dx$ (b) $\int_0^{0.5} 2x\,dx$ (c) $\int_4^5 \ln x\,dx$ (d) $\int_0^4 (2 - \frac{1}{2}x)\,dx$

(e) Without evaluating the integrals exactly, say which of the approximations is exact.

16 Taking four intervals and using the trapezium rule, evaluate approximately the value of $\int_1^3 f(x)\,dx$ where f(x) is as given below.

x	1	1.5	2	2.5	3
f(x)	4	5.5	4.9	4.2	3.9

17 Using the data given below, and the trapezium rule, calculate, as accurately as you can, the values of $\int_a^b f(x)\,dx$.

(a)

x	2	2.1	2.2	2.3	2.4	2.5
f(x)	3	3.6	3.9	3.7	3.4	2.9

$a = 2$, $b = 2.5$

(b)

x	5	5.5	6	6.5	7
f(x)	4	3.1	2.8	3.2	3.9

$a = 5$, $b = 7$

(c)

x	0	0.2	0.4	0.6	0.8	1
f(x)	5	5.1	4.7	4.1	3.9	4

$a = 0$, $b = 1$

(d)

x	100	101	102
f(x)	0.1	0.05	0.08

$a = 100$, $b = 102$

(e) x	0.5	0.6	0.7	0.8	0.9	1	$a = 0.5, b = 1$
f(x)	100	99.7	99.1	98.2	97.1	96.1	

18 Taking six intervals, use the trapezium rule to calculate approximately the value of $\int_1^3 e^{-\frac{1}{2}x^2}\,dx$.

19 Using the trapezium rule and eight intervals, calculate approximately the value of $\int_2^5 \frac{x}{1+\sqrt{x}}\,dx$.

20 Five values of the function f(x) are given in the table below. The value of f(2.6) is unknown. Using any technique that seems to you to be sensible, estimate the value of f(2.6). Then use the trapezium rule to calculate as accurately as possible the value of $\int_2^3 f(x)\,dx$.

x	2	2.2	2.4	2.6	2.8	3.0
f(x)	57	61	63		65.2	65.4

21 (a) Using the trapezium rule with just one interval, calculate approximately the value of $\int_0^{0.5} \sqrt{1+x}\,dx$.

Calculate the greatest theoretical value of the error in your approximation.

(b) Perform the integration exactly and calculate the error in your approximation. Comment.

22 Evaluate (approximately) the integral $\int_4^{4.25} e^{-2x}\,dx$ using the trapezium rule with one interval. Find the greatest theoretical error in your approximation. Evaluate the integral exactly and comment on your results.

23 Use the trapezium rule with four intervals to calculate approximately the value of $\int_2^4 \frac{1}{(1+x)^4}\,dx$.

Calculate the maximum value of the error in your approximation.
Check this result by evaluating the integral exactly.

24 Calculate approximately the value of $\int_0^1 10e^{3x}\,dx$ using the trapezium rule and

(a) one interval (b) two intervals (c) four intervals.

In each case, calculate the greatest theoretical error in your approximation. Perform the integration exactly and comment on your results.

25 The trapezium rule is to be used to calculate approximately the value of $\int_0^1 \sin 2x\,dx$ (x is measured in radians). By using the error term for the compound trapezium rule, calculate how many intervals would have to be chosen if the error was to be less than 0.01.

Perform the integration exactly and check your answer.

Section 7.7

26 Use Simpson's rule to calculate approximately the value of $\int_2^4 f(x)\,dx$ when f(x) is as given below.

(a)

x	2	3	4
f(x)	6	4.5	2

(b)

x	2	3	4
f(x)	1.25	3.02	2.57

(c)

x	2	3	4
f(x)	−2	−1.3	−0.9

27 Use Simpson's rule to calculate approximately the value of $\int_a^b f(x)\,dx$ when f(x), a and b are as given below.

(a) $a = 0$, $b = 2$

x	0	1	2
f(x)	3	4.1	3.8

(b) $a = 1$, $b = 2$

x	1	1.5	2
f(x)	5.2	4.8	3.6

(c) $a = 3$, $b = 5$

x	3	4	5
f(x)	−4.2	−5.1	−5.7

(d) $a = 10$, $b = 11$

x	10	10.5	11
f(x)	−0.5	−0.1	0

28 Using a single application of Simpson's rule calculate approximately the values of

(a) $\int_1^2 6\sqrt{x+1}\,dx$ (b) $\int_{-2}^{-1} 2e^{-x^2}\,dx$ (c) $\int_0^{0.5} \frac{5}{x^3+1}\,dx$

(d) $\int_2^3 (\tfrac{1}{2}x^3 + 4)\,dx$ (e) $\int_{-1}^{-0.5} 4\ln(x+10)\,dx$ (f) $\int_0^1 (\sin x + 2\cos x)\,dx$

Are any of these approximations exact?

29 Use the composite Simpson's rule to calculate approximately the values of

(a) $\int_1^7 \frac{3}{(1+\sqrt{x})}\,dx$ using six intervals

(b) $\int_0^4 0.5e^{-\frac{1}{2}x}\,dx$ using four intervals

(c) $\int_{-5}^5 \frac{1}{4}x^2\,dx$ using two intervals

(d) $\int_1^2 \frac{1}{3+\ln x}\,dx$ using four intervals

(e) $\int_0^{\frac{\pi}{4}} \sin x\,dx$ using six intervals (working in radians)

(f) $\int_1^2 \left(3x^3 + \frac{1}{x}\right) dx$ using four intervals

(g) $\int_{-2}^2 \frac{3x}{(4+x)^2} dx$ using eight intervals

Are any of these approximations exact?

30 Calculate approximately the value of $\int_a^b f(x)\,dx$ where $f(x)$, a and b are as given below.

(a)

x	1	1.5	2	2.5	3
$f(x)$	6	5.2	4.9	4.8	5.1

$a = 1$, $b = 3$, four intervals

(b)

x	4	5	6	7	8	9	10
$f(x)$	0.1	0.4	0.5	0.55	0.49	0.45	0.43

$a = 4$, $b = 10$, six intervals

(c)

x	0	0.25	0.5	0.75	1
$f(x)$	101	103.1	102.8	102.1	101.8

$a = 0$, $b = 1$, four intervals

(d)

x	2	2.2	2.4	2.6	2.8	3.0	3.2
$f(x)$	−4.2	−3.9	−3.7	−3.6	−3.8	−4.0	−4.1

$a = 2$, $b = 3.2$, six intervals

(e)

x	−3	−2.9	−2.8	−2.7	−2.6
$f(x)$	10	8.8	7.9	8.2	8.7

$a = -3$, $b = -2.6$, four intervals

(f)

x	5	5.25	5.5	5.75	6	6.25	6.5
$f(x)$	0	−0.3512	−0.4537	−0.3714	−0.2413	−0.1328	−0.1170

$a = 5$, $b = 6.5$, six intervals

31 Calculate approximately the value of $\int_1^5 4e^{-x}\,dx$ using Simpson's rule with four intervals. Calculate the integral exactly and calculate the error in your Simpson's rule approximation.

32 Calculate the value of $\int_4^{10} 4\ln(x+1)\,\mathrm{d}x$ given by an application of Simpson's rule using six intervals. Given that the exact value of the integral, to six decimal digits (rounded) is 49.3186, calculate the error in your approximation.

33 Using Simpson's rule, calculate approximately the value of $\int_0^1 \frac{3\sin 2x}{x+1}\,\mathrm{d}x$ using (a) four (b) six (c) eight intervals. What is the value of this integral correct to four decimal digits (rounded)?

34 Calculate the value $\int_1^3 \frac{3}{(x+1)}\,\mathrm{d}x$ using (a) two intervals (b) four intervals (c) six intervals. Using your answers is it possible to deduce the value of the integral correct to four decimal digits (rounded)? Calculate the exact value of the integral and check your answer.

35 Find how many intervals are necessary if Simpson's rule is to be used to calculate $\int_2^4 4(3x^2 + 2x - 1)^{\frac{1}{2}}\,\mathrm{d}x$ correct to six decimal digits (using rounding). Do the calculation.

36 Use (a) the trapezium rule and (b) Simpson's rule, each with four intervals, to calculate approximately the value of $\int_5^7 10e^{-\frac{1}{2}x}\,\mathrm{d}x$. Which is the more accurate approximation?

37 Use (a) the trapezium rule and (b) Simpson's rule, each with six intervals, to calculate approximately the value of $\int_{\frac{\pi}{4}}^{\frac{\pi}{2}} \sin x\,\mathrm{d}x$ (x measured in radians). Which is the more accurate approximation?

Section 7.8

38 Use the trapezium rule with two intervals to calculate approximately the value of $\frac{1}{2}\int_5^7 \frac{1}{x+4}\,\mathrm{d}x$. Write down an expression for the error in your approximation. Find the greatest possible value of the error. Calculate the exact value of the integral. Does the exact value lie within the predicted range?

39 Using Simpson's rule with $h = 0.5$, calculate approximately the value of $\int_{-2}^{-1} \sqrt{x+3}\,\mathrm{d}x$. Write down an expression for the error term. Find the greatest value that the error term can achieve. Evaluate the integral exactly. Comment.

40 It is required to calculate the value $\int_1^3 3e^{-2x}\,\mathrm{d}x$ so that the error is less than 0.000 05. How many intervals must be taken if (a) the trapezium rule and (b) Simpson's rule is used?

41 The integral $\int_{-1}^{1} \frac{1}{\sqrt{x+3}} dx$ is to be evaluated so that the error is less than 0.0001.

How many intervals must be used if (a) the trapezium rule and (b) Simpson's rule is used? If this number of intervals is used in each case, what would be the error in the approximations?

Miscellaneous

42 The following is a table of values for the first Bessel function $J_1(x)$. Calculate tables for $J_1'(x)$ and $J_1''(x)$.

x	$J_1(x)$
0	0.0000
0.1	0.0499
0.2	0.0995
0.3	0.1483
0.4	0.1960
0.5	0.2423

43 Illustrate graphically the approximations

(a) $\frac{dy}{dx} \approx \frac{f(a+h) - f(a)}{h}$ when $x = a$

(b) $\frac{dy}{dx} \approx \frac{f(a+h) - f(a-h)}{2h}$ when $x = a$

and comment briefly on the accuracy of the approximations.

The data below gives the temperature of a certain mass of gas at 0.2 second intervals.

x (s)	9.6	9.8	10	10.2	10.4	10.6
y (°C)	51	51.6	52.4	53.8	55.1	55.0

Estimate as accurately as you can, the rate of increase of temperature with respect to time at $x = 9.8$, $x = 10.2$, $x = 10.4$. Estimate also the value of $\frac{d^2y}{dx^2}$ at $x = 10$, and at $x = 10.2$.

44 Copy and complete the table of values given below for the function $f(x) = \frac{1}{\sqrt{x}}$.

x	1	1.5	2	2.5	3
f(x)	1				

(a) Use your table to estimate, as accurately as you can, the value of f'(2).

(b) Write down the error term for the approximation method given in (a).

(c) Calculate the greatest value that can be taken by the error term for $1 \leqslant x \leqslant 3$.

Does the error in your answer to (a) lie within the bounds predicted by your answer to (c)?

45 Prove that the error in using

$$\frac{f(a+h)-f(a)}{h}$$

to approximate f'(a) is $-\frac{h}{2}f''(\eta) \qquad a<\eta<a+h$

For the data given below, use the above expression to calculate approximately the value of $\frac{dy}{dx}$ when $x = 2$.

x	1	1.5	2	2.5
y	−1.5	2	0.8	0.1

Interpolate a polynomial of degree two to three points of the data and use your polynomial to obtain a second estimate of the value of $\frac{dy}{dx}$ when $x = 2$. Which of the two estimates is likely to be the more accurate?

46 Show that the finite difference approximation

$$y''(x) \approx \frac{1}{h^2}[y(x+h) - 2y(x) + y(x-h)]$$

is exact when y(x) is a cubic polynomial in x.

The function V(x) satisfies the differentiation equation V" = xV and the conditions V(0) = 1, V(0.4) = 3. Using the above approximation and a step length of $h = 0.1$, estimate the value of V(0.2). (MEI)

47 Use
(a) the trapezium rule
(b) Simpson's rule

to estimate the value of $\int_3^5 f(x)\,dx$ when f(x) is as given in the table below.

x	3	3.5	4	4.5	5
f(x)	14	11.3	12.6	12.9	13.8

Write down expressions for the error terms for both methods and say which is likely to produce the more accurate answer.

48 Experimental values of a continuous function f(x) were found to be given by

x	0	0.2	0.4	0.7	0.8	1.0	1.2
f(x)	1.285	1.114	0.944	0.706	0.634	0.500	0.384

Use linear interpolation to estimate
(a) f(0.6)
(b) a value of x for which $f(x) = 1$.

Estimate $\int_0^{1.2} f(x)\,dx$ using Simpson's rule with seven ordinates. (MEI)

49 Find, using Simpson's rule with five ordinates, an approximate value for the area of the finite region bounded by the curve $y = \sin x$ and the x-axis between the values $x = 0$ and $x = \pi$.
Find the percentage error in your result by comparing it with the exact value obtained by integration. (MEI)

50 From the following table of values of a function $f(x)$, calculate
(a) f(0.48) by linear interpolation
(b) $\int_{0.1}^{0.5} f(x)\,dx$ by Simpson's rule
giving your results to three places of decimals.

x	0.1	0.2	0.3	0.4	0.5
$f(x)$	1.2840	1.3499	1.4191	1.4918	1.5683

If it is known that the errors in the tabulated values of $f(x)$ are exactly kx at each point, where k is a constant, find expressions, in the terms of k, for the errors in the calculated values of f(0.48) and $\int_{0.1}^{0.5} f(x)\,dx$ resulting from these errors in tabulation. (MEI)

51 Given that it is required to use Simpson's rule to evaluate $\int_0^6 f(x)\,dx$, where $f(x) = \sqrt{(400 + x^2)} - 20$, show that it is preferable, when using tables to obtain values of $f(x)$, to use $f(x)$ in the equivalent form

$$f(x) = \frac{x^2}{\sqrt{(400 + x^2)} + 20}$$

Use Simpson's rule with three ordinates to evaluate the integral, giving the result to four significant figures.
An alternative method uses the approximation

$$f(x) \approx \frac{x^2}{40}$$

and direct integration. Show that the two methods give results differing by less than 2%. (MEI)

52 Draw up a table of values of e^{-x^2} for values of x from 0 to 2 in steps of 0.2.
(a) Using Simpson's rule, with ten equal intervals, find an approximate value for $\int_0^2 e^{-x^2}\,dx$, giving your answer to four decimal places, taking care to show all your relevant working.

(b) It can be shown that, if $R \geqslant 2$,

$$\int_R^{\infty} e^{-x^2}\,dx \approx e^{-R^2}\left[\frac{1}{2R} - \frac{1}{4R^3}\right]$$

Using this formula, obtain an estimate for the area under the graph of e^{-x^2} from $x = 2$ to ∞, giving your answer to four decimal places.

(c) It is known that $\int_0^{\infty} e^{-x^2}\,dx = \frac{1}{2}\sqrt{\pi}$.

Using the results in (a) and (b), estimate the value of π, giving your answer to two decimal places. (MEI)

53 The value of $\int_1^3 \frac{1}{x}\,dx$ is estimated, using the trapezium rule with n intervals. Show that an estimate for the absolute error bound due to the use of the trapezium rule is $\frac{2}{3n}$. If in addition each ordinate is evaluated correct to $\pm\,\varepsilon$, find an absolute error bound for the result.

Find a value of n which will ensure that the result is accurate to two places of decimals if $\varepsilon = 0.0005$. (MEI)

54 (a) Show that Simpson's rule with three ordinates gives an exact result for the integral from $-h$ to $+h$ of a cubic polynomial.
Show further that the error in using Simpson's rule with three ordinates for the integral of $f(x) = Ax^4$ from $-h$ to $+h$ is

$$\left|\frac{4}{15}Ah^5\right| = \left|\frac{h^5}{90}f^{(4)}(0)\right|$$

(b) Show that if $f(x) = e^{-\frac{1}{2}x^2}$ then $f^{(4)}(x) = e^{-\frac{1}{2}x^2}\,[(x^2 - 3)^2 - 6]$ and deduce that $\left|f^{(4)}(x)\right| \leqslant 3$ in the interval $0 \leqslant x \leqslant 1$.

Assuming that in this case the error in applying Simpson's rule with three ordinates over a small interval $2h$ is less than $\frac{1}{90}h^5M$, where M is the greatest value of $\left|f^{(4)}(x)\right|$ in the interval, use Simpson's rule to calculate

$$\int_0^1 e^{-\frac{1}{2}x^2}\,dx$$

in two steps (using five ordinates) and give an upper bound for the error. (MEI)

55 Derive the trapezium rule for finding the approximate integral of $y = f(x)$ for $a \leqslant x \leqslant b$ using n strips with a fixed step interval of h.
By considering the integral over a single interval $x_r \leqslant x \leqslant x_{r+1}$ show that

$$\int_{x_r}^{x_{r+1}} f(x)\,dx = hf(x_r) + \tfrac{1}{2}h^2f'(x_r) + \tfrac{1}{6}h^3f''(x_r) + \ldots$$

Deduce that the modulus of the truncation error which occurs by approximating to this integral by using the trapezium rule with a single strip is approximately

$$\tfrac{1}{12}h^3\left|\mathrm{f}''(x_r)\right|$$

Show that an approximate upper bound to the modulus of the truncation error over $a \leqslant x \leqslant b$ is given by

$$\tfrac{1}{12}(b-a)h^2M$$

where M is the maximum value of $\left|\mathrm{f}''(x)\right|$ for $a \leqslant x \leqslant b$. (MEI)

Chapter 8

Numerical solution of differential equations

Introduction

Differential equations permeate the whole of mathematics; they occur with great regularity in mechanics, probability and statistics as well as in many branches of pure mathematics. They also occur frequently in physics and economics. In fact if continuous variation, in any form, is being considered, it is highly likely that the variation will be described by means of a differential equation. Here are just a few examples.

In mechanics, the differential equation

$$m\frac{\mathrm{d}v}{\mathrm{d}t} = mg - kv^2$$

describes the motion of a body falling freely under gravity through a medium which offers resistance proportional to the square of the velocity.

In a study of conflicting species, Volterra published the following two differential equations

$$\frac{\mathrm{d}x}{\mathrm{d}t} = a(x - xy) \qquad \frac{\mathrm{d}y}{\mathrm{d}t} = -c(y - xy)$$

where x and y represent the population sizes of the two conflicting species.

In probability theory, the differential equation

$$\frac{\mathrm{d}p_r}{\mathrm{d}t} = (r + 1)(n - r)p_{r+1} - r(n - r + 1)p_r$$

is used to study the progress of an epidemic. $p_r(t)$ is the probability that in a population of size n, r people are still uninfected at time t.

In physics, we have the differential equation

$$\frac{\mathrm{d}^2y}{\mathrm{d}x^2} + y = a + by^2$$

the solution of which was used by Einstein in his investigation of the orbital motion of planets under the assumptions of general relativity.

A differential equation usually associated with physics is Van der Pol's equation

$$\frac{d^2y}{dt^2} - \varepsilon(1 - y^2)\frac{dy}{dt} + ay = 0$$

But this equation has also been used to model the working of the heart.

In economics, a differential equation which provides a simple model of economic growth is

$$\frac{dx}{dt} = sf(x) - g(x)$$

where x = ratio of supply of capital to supply of labour, f(x) is the productivity function, g(x) represents the growth in the supply of labour and s is a constant.

In fact, it is difficult to over-emphasise the importance of the role played by differential equations in mathematical investigations.

It is a remarkable fact that although it is often relatively simple to write down a differential equation, it is frequently very difficult indeed to solve that differential equation. Indeed, serious problems can arise in finding solutions to four of the six equations written above.

So what are we to do when we encounter a differential equation that we cannot solve or, what is almost as bad, a differential equation whose solution is so awkward and cumbersome as to the almost unusable? One answer is to calculate a numerical solution. This means that instead of producing an algebraic expression for the solution (an analytical solution) we calculate (perhaps approximately) the numerical values taken by the solution. For a solution, we finish up, not with an algebraic expression, but with the numerical values taken by that expression. This is known as a numerical solution to the differential equation.

Before we look at how a numerical solution can be calculated, some basic facts about differential equations will be set out.

8.1 Differential equations

Definition

A differential equation is an equation containing one or more derivatives.

Hence $\frac{dy}{dx} = 6\sin xy$

$$\frac{d^2y}{dx^2} - 3\frac{dy}{dx} + x^2y = \sin x$$

$$\left(\frac{d^2y}{dx^2}\right)^2 + \left(\frac{dy}{dx}\right)^3 = x^4$$

are examples of differential equations.

If a differential equation contains only first derivatives, it is called a first order differential equation. For example:

$$\frac{dy}{dx} + 3y = e^x \qquad \text{and} \qquad 5 - t^2 \frac{dy}{dt} + y^2 = 0$$

If a differential equation contains second derivatives but no higher order derivatives, it is called a second order differential equation. For example:

$$4\frac{d^2y}{dx^2} - 12y^2 = 0 \qquad \text{and} \qquad x\frac{d^2x}{dt^2} + 4\frac{dx}{dt} - 5x = e^{-t}$$

In general, if a differential equation contains an *n*th order derivative but no derivative of higher order, it is called an *n*th order differential equation.

This book is concerned mainly with first order differential equations but the chapter will conclude with a brief discussion of second order differential equations and it will be seen that many of the techniques described will readily extend to higher order differential equations.

Definition

> An algebraic expression is called a solution (an analytic solution) if the expression satisfies the differential equation.

Example A

The expression $y = x\sin x$ is a solution of the differential equation

$$\frac{dy}{dx} = x\cos x + \sin x$$

Proof

Let $\quad y = x\sin x$

Then $\quad \dfrac{dy}{dx} = x\cos x + \sin x$

and it may be seen immediately that $y = x\sin x$ satisfies the differential equation

$$\frac{dy}{dx} = x\cos x + \sin x.$$

Example B

The expression $y = 3e^{2x} + e^x$ is a solution of the differential equation

$$\frac{d^2y}{dx^2} - 3\frac{dy}{dx} + 2y = 0$$

Proof

Let $y = 3e^{2x} + e^x$

Then $\dfrac{dy}{dx} = 6e^{2x} + e^x$ and $\dfrac{d^2y}{dx^2} = 12e^{2x} + e^x$

Hence
$$\frac{d^2y}{dx^2} - 3\frac{dy}{dx} + 2y = 12e^{2x} + e^x - 3(6e^{2x} + e^x) + 2(3e^{2x} + e^x) = 0$$

and $y = 3e^{2x} + e^x$ satisfies the differential equation

$$\frac{d^2y}{dx^2} - 3\frac{dy}{dx} + 2y = 0$$

It may be shown that if there is one solution to a differential equation, then there are infinitely many solutions. Consider Example B.

If $y = 5e^{2x} - 4e^x$

then $\dfrac{dy}{dx} = 10e^{2x} - 4e^x$ and $\dfrac{d^2y}{dx^2} = 20e^{2x} - 4e^x$

Hence
$$\frac{d^2y}{dx^2} - 3\frac{dy}{dx} + 2y = 20e^{2x} - 4e^x - 3(10e^{2x} - 4e^x) + 2(5e^{2x} - 4e^x) = 0$$

and $y = 5e^{2x} - 4e^x$ is also a solution of the differential equation

$$\frac{d^2y}{dx^2} - 3\frac{dy}{dx} + 2y = 0$$

To write all solutions simultaneously, the idea of arbitrary constants is introduced. It may be shown that every solution of the equation

$$\frac{d^2y}{dx^2} - 3\frac{dy}{dx} + 2y = 0$$

may be written in the form

$$y = Ae^{2x} + Be^x$$

where A and B are (generally unknown and hence arbitrary) constants. This expression is known as the general solution of the differential equation. This deserves a definition.

Definition

> An algebraic expression is said to be the general solution of a differential equation if:
> (a) the expression is a solution of the differential equation
> (b) the expression contains one or more arbitrary constants.
> The number of arbitrary constants present in the expression must equal the order of the differential equation.

In Example A, $y = x\sin x + c$ is a solution of the differential equation $\frac{dy}{dx} = x\cos x + \sin x$. Further, $x\sin x + c$ contains one arbitrary constant and the order of the differential equation is one. Hence, $y = x\sin x + c$ is the general solution of the differential equation $\frac{dy}{dx} = x\cos x + \sin x$.

When the arbitrary constant(s) in the general solution are assigned numerical values, the resulting solution is called a **particular solution**. So, in Example A above, if c was given the value $c = 3$, we should have the particular solution $y = x\sin x + 3$.

The usual way of deriving a particular solution is to state a condition (or conditions) which must be satisfied by the solution $y(x)$. Then, when the solution is made to satisfy these conditions, particular values are assigned to the arbitrary constants. So in Example A, if the condition was that the solution must pass through the point $x = 0$, $y = 3$ we should have:

$$3 = 0\sin 0 + c$$

which forces $c = 3$ and gives the particular solution

$$y = x\sin x + 3$$

In the case of a first order differential equation, only one arbitrary constant occurs in the general solution so, to obtain a particular solution, only one condition on the solution is required. This usually takes the form of defining the value of the solution $y(x)$ at a particular value of x. Hence we might have: when $x = 1$, the value of the solution is 3, or in symbols, $y(1) = 3$.

For a second order differential equation two arbitrary constants occur in the general solution, so to obtain a particular solution, two conditions on the solution are required. One of these defines the value of the solution at a given value of x. As above, we might be given that $y(1) = 3$. The second condition, in this book, will describe the behaviour of the solution at this value of x. Usually, this takes the form of defining the value of the derivative of the solution. We might be given for example, that $\frac{dy}{dx} = -0.5$ when $x = 1$, or in a more concise notation $y'(1) = -0.5$. Hence we would be given the two conditions $y(1) = 3$ and $y'(1) = -0.5$.

It is important to realise that the information given about the solution is absolutely vital to finding a numerical solution to a differential equation. It is only at the point at which the information is given that we know with certainty what the solution is, and all methods for finding a numerical solution begin at this point. To emphasise the importance of this, the given conditions will always be described at the same time that we state the differential equation to be solved. In this way, the conditions become part of the statement of the problem.

Throughout the chapter, the techniques will be illustrated on differential equations that we can solve. At first sight this may appear to be a waste of time; if we can obtain an algebraic solution to a certain differential equation, why bother to calculate a numerical solution? The reason, of course, is that knowing the exact solution allows us to calculate the errors in the numerical solution and so form an impression of the accuracy of the method. But first, some notation must be established.

Notation

Let $[a, b]$ represent an interval on the x-axis and let $y(x)$ denote a function defined on this interval. We divide the interval $[a, b]$ into a number of equal subintervals, each having length h. Denote the points of division by $x_1, x_2, \ldots$ and for convenience, write $x_0 = a$. This is shown in Figure 8.1.

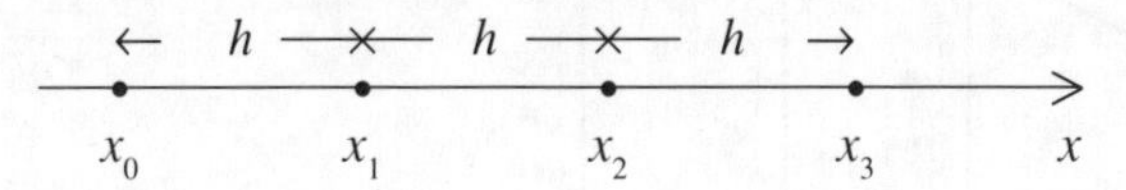

Figure 8.1

h will be called **the step size**. It is clear that:

$$x_1 = x_0 + h$$
$$x_2 = x_0 + 2h$$
$$x_3 = x_0 + 3h$$

When the function $y(x)$ is evaluated at x_n we denote the value of the function by y_n. So $y_n = y(x_n)$.

We shall let y_n denote the exact value of the solution of a differential equation when $x = x_n$. Hence we have, for an exact solution $y(x)$, the diagram in Figure 8.2.

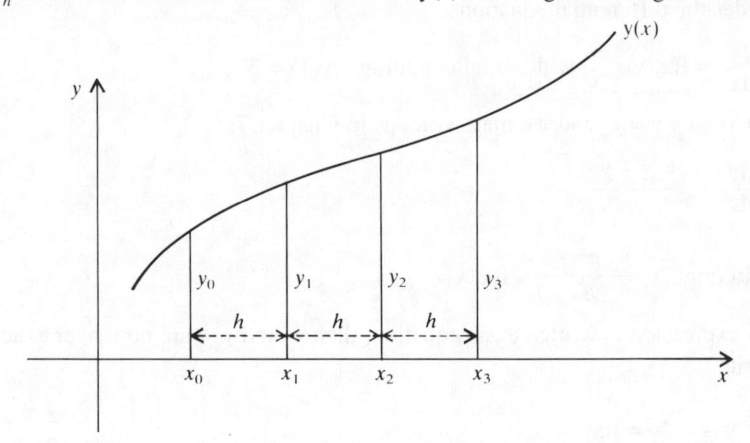

Figure 8.2

Normally, however, we shall not be able to calculate the exact value of the solution. The best that we can hope for is that we shall be able to calculate good approximations to the values of $y_1, y_2, \ldots, y_n$ and that we shall have some idea of the error involved in the approximation. We shall let w_n denote the calculated approximation to y_n. Hence we may write

$$y_n = w_n + \text{error}$$

and we have the diagram shown in Figure 8.3.

Observe that $w_0 = y_0$ and that the error in the approximation w_0 is zero.

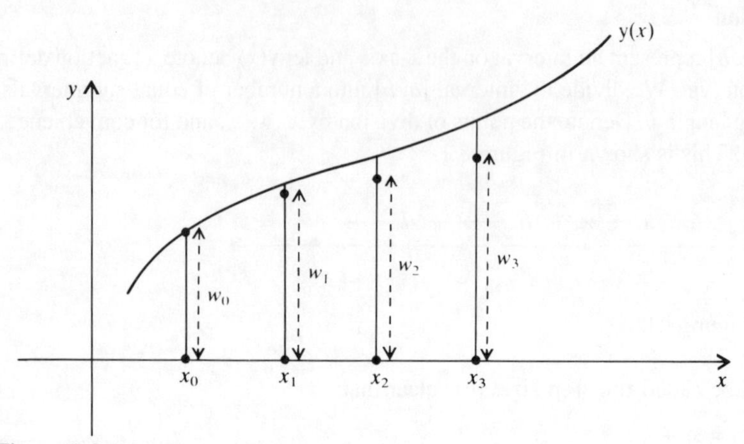

Figure 8.3

The remainder of the chapter will consider different ways of calculating the approximations w_1, w_2, ... and the errors involved in these approximations.

8.2 Euler's method

Consider the differential equation

$$\frac{dy}{dx} = f(x, y) \qquad \text{with initial condition} \quad y(1) = 3$$

When $x = x_n$, $y = y_n$ and we may write, as in Chapter 7:

$$\frac{dy}{dx} \approx \frac{y_{n+1} - y_n}{h}$$

Substituting $\dfrac{y_{n+1} - y_n}{h} \approx f(x, y)$

If this expression is written as an equation, then y_n and y_{n+1} are no longer exact and we write

$$\frac{w_{n+1} - w_n}{h} = f(x_n, w_n)$$

This expression is known as the **difference equation** associated with the differential equation $\dfrac{dy}{dx} = f(x, y)$. We may write this difference equation in the form:

$$w_{n+1} - w_n = hf(x_n, w_n)$$

or

$$\boxed{w_{n+1} = w_n + hf(x_n, w_n)}$$

The idea now is to let n take successively the values $n = 0$, $n = 1$, $n = 2$, ... and use the initial condition to calculate successively w_1, w_2, w_3, ... Recall that the initial condition tells us that when $x_0 = 1$, $y_0 = w_0 = 3$.

From the difference equation:

$n = 0$ gives $w_1 = w_0 + hf(x_0, w_0) = 3 + hf(1, 3)$
$n = 1$ gives $w_2 = w_1 + hf(x_1, w_1)$
$n = 2$ gives $w_3 = w_2 + hf(x_2, w_2)$

and so on.

This method of calculating the approximations $w_1, w_2, w_3, \ldots$ to the solution $y_1, y_2, y_3, \ldots$ is known as Euler's method. An example will help to make the method clear.

Example 8.1

Use Euler's method to find approximate values of the solution of the differential equation

$$\frac{dy}{dx} = y - x + 5 \quad \text{with initial condition} \quad y(2) = 1$$

at the points where $x = 2.1$, 2.2, and 2.3.

The differential equation to be solved is $\frac{dy}{dx} = y - x + 5$ so in this case, $f(x, y) = y - x + 5$.
The difference equation associated with Euler's method is

$$w_{n+1} = w_n + hf(x_n, w_n)$$

which becomes $$w_{n+1} = w_n + h(w_n - x_n + 5)$$

From the initial conditions, $x_0 = 2$ and $y_0 = w_0 = 1$. Since the solution is required at $x_1 = 2.1$, $x_2 = 2.2$, $x_3 = 2.3$, we take $h = 0.1$.

If $n = 0$ $w_1 = w_0 + (0.1)(w_0 - x_0 + 5) = 1 + (0.1)(1 - 2 + 5) = 1.4$
If $n = 1$ $w_2 = w_1 + (0.1)(w_1 - x_1 + 5) = 1.4 + (0.1)(1.4 - 2.1 + 5) = 1.83$
If $n = 2$ $w_3 = w_2 + (0.1)(w_2 - x_2 + 5) = 1.83 + (0.1)(1.83 - 2.2 + 5) = 2.293$

Hence the approximate values of the solution are:

$x = 2$	$x = 2.1$	$x = 2.2$	$x = 2.3$
$y = 1$	$y \approx w_1 = 1.4$	$y \approx w_2 = 1.83$	$y \approx w_3 = 2.293$

Since the differential equation is known to have the solution $y = x - 4 + 3e^{x-2}$ we can compare the computed solution with the exact solution.

n	x_n	y_n	w_n	| error |
0	2	1	1	0
1	2.1	1.415512754	1.4	1.5512754×10^{-2}
2	2.2	1.864208274	1.83	3.4208274×10^{-2}
3	2.3	2.349576423	2.293	5.6576423×10^{-2}

As might have been expected, the approximate solution becomes less accurate as the points move away from the initial value of the solution. What is happening is that the errors incurred by the use of the difference equation are compounding and growing in magnitude. The situation may be improved by reducing the step size h. The table below gives the approximate solution when a step size $h = 0.05$ is used.

n	x_n	y_n	w_n	| error |
0	2	1	1	0
1	2.05	1.20381329	1.2	3.813×10^{-3}
2	2.1	1.41551275	1.4075	8.013×10^{-3}
3	2.15	1.63550273	1.622875	1.263×10^{-2}
4	2.2	1.864208274	1.84651875	1.769×10^{-2}
5	2.25	2.10207625	2.07884469	2.323×10^{-2}
6	2.3	2.34957642	2.32028692	2.929×10^{-2}

As can be seen, the improvement is considerable, but observe that this improvement in accuracy has been achieved at the cost of doubling the number of calculations. We shall return to such considerations later in the chapter, but now we will look at an algorithm for Euler's method and a second worked example.

Algorithm 8.1 Euler's method

To obtain an approximation solution of the differential equation $\frac{dy}{dx} = f(x, y)$ with initial condition $y(a) = y_0$.

input: initial value of $x = x_0$;
 initial value of $y = y_0$;
 step size h

$w_0 := y_0$

for $n = 0, 1, ...,$ until satisfied

$w_{n+1} := w_n + hf(x_n, w_n)$

$x_{n+1} := x_0 + (n + 1)h$

endloop

output: $w_0, w_1, w_2, ...$

Example 8.2

Use Euler's method to calculate approximately the solution of the differential equation $\frac{dy}{dx} = \sqrt{y}\, e^x$ with initial condition $y(1) = 2$ at the points $x = 1.1, 1.2, 1.3$ and 1.4.

Since in this problem, $f(x, y) = \sqrt{y}\, e^x$, the difference equation for Euler's method is:

$$w_{n+1} = w_n + h\sqrt{w_n}\, e^{x_n}$$

We have $x_0 = 1, x_1 = 1.1, x_2 = 1.2, x_3 = 1.3, x_4 = 1.4$

Hence $h = 0.1$ and $w_0 = y(1) = 2$

Taking $n = 0, 1, 2$ and 3 in turn gives

$$w_1 = w_0 + h\sqrt{w_0}e^{x_0} = 2 + (0.1)\sqrt{2}\,e^1 = 2.3844$$
$$w_2 = w_1 + h\sqrt{w_1}e^{x_1} = 2.8483$$
$$w_3 = w_2 + h\sqrt{w_2}e^{x_2} = 3.4086$$

and $w_4 = 4.0861$

where all the answers are written to five decimal digit accuracy (the calculations, however, were performed to eight digit accuracy).

The exact solution is $y = \dfrac{(e^x + 2\sqrt{2} - e)^2}{4}$ and in the table below we compare the approximate and the exact solutions of the differential equation.

n	x_n	y_n	w_n	$\lvert$ error $\rvert$
0	1	2	2	0
1	1.1	2.4247	2.3844	4.03×10^{-2}
2	1.2	2.9417	2.8483	9.34×10^{-2}
3	1.3	3.5710	3.4086	1.624×10^{-1}

If, however, we take h = 0.05 we obtain the following:

n	x_n	y_n	w_n	$\lvert$ error $\rvert$
0	1	2	2	0
1	1.05	2.2020	2.1922	9.80×10^{-3}
2	1.1	2.4247	2.4038	2.09×10^{-2}
3	1.15	2.6705	2.6366	3.39×10^{-2}
4	1.2	2.9417	2.8931	4.86×10^{-2}
5	1.25	3.2409	3.1754	6.55×10^{-2}
6	1.3	3.5710	3.4864	8.46×10^{-2}

and it may be seen that greater accuracy has again been achieved at the expence of increased calculation.

Geometrical interpretation of Euler's method

Consider the differential equation $\dfrac{dy}{dx} = f(x, y)$ with initial condition $y(x_0) = y_0$ (Figure 8.4).

The initial condition ensures that the solution $y = y(x)$ passes through the point $P(x_0, y_0)$. The problem is to find the value of the solution when $x = x_1$. That is, we want to find $BR = y_0 + k$. Since y_0 is known we want, in fact, to find k.

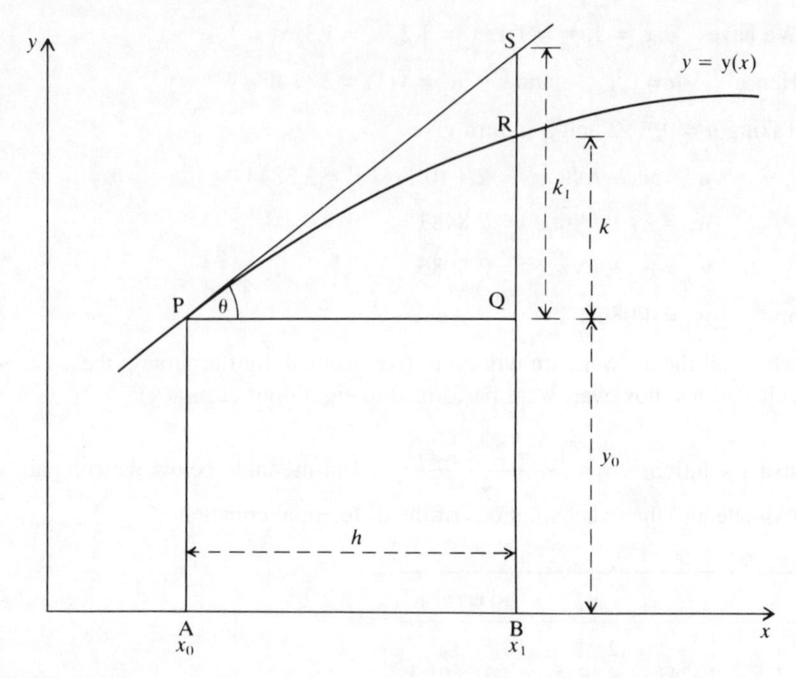

Figure 8.4

The argument behind Euler's method is this: let PS be the tangent to the curve at P. Then, if the step size h is not too large, QS = k_1 should provide a reasonable approximation to k. That is, $k_1 \approx k$. The approximate solution at $x = x_1$ is then given by

$$w_1 = y_0 + k_1 \quad \ldots (1)$$

The length k_1 may easily be calculated. From elementary trigonometry, $\dfrac{\text{QS}}{\text{PQ}} = \tan\theta$. But QS = k_1; PQ = h and this equation gives $k_1 = h\tan\theta$. But $\tan\theta$ is the gradient of the tangent at P.

Hence $\quad \tan\theta = \dfrac{dy}{dx}$ evaluated at $x = x_0$

$$= f(x_0, y_0) \qquad \text{from the differential equation}$$

$$\therefore\ k_1 = hf(x_0, y_0)$$

Writing this expression for k_1 in Equation (1) and writing $w_0 = y_0$ gives

$$w_1 = w_0 + hf(x_0, w_0)$$

which is the first stage of Euler's method.

We shall return to this geometric interpretation when we begin to study more sophisticated methods later in the chapter.

It is hoped that the reader now feels prepared to carry out a calculation using Euler's method. The 'mechanics' of the procedure have been established and we now begin to think about the errors.

8.3 The error in Euler's method

We notice something very interesting when comparing the error and the step size h. In both examples, when the step size was halved, the error at each step was approximately halved, also. This leads us to suspect that the error is of the same order as the step size h; and so it turns out to be.

Recall that in Chapter 7 we established that the error in representing $\frac{dy}{dx}$ by $\frac{y_{n+1} - y_n}{h}$ was $\frac{-h}{2}y''(\eta)$ where $x_n < \eta < x_{n+1}$.

We showed, in fact, that

$$\frac{dy}{dx} = \frac{y_{n+1} - y_n}{h} - \frac{h}{2}y''(\eta)$$

So, when considering the differential equation $\frac{dy}{dx} = f(x, y)$, substitution yields

$$\frac{y_{n+1} - y_n}{h} - \frac{h}{2}y''(\eta) = f(x_n, y_n)$$

and we may write

$$\frac{y_{n+1} - y_n}{h} - f(x_n, y_n) = \frac{h}{2}y''(\eta)$$

The quantity $\frac{y_{n+1} - y_n}{h} - f(x_n, y_n)$ is called the **local discretisation error**. It provides a measure of the error that arises when $\frac{dy}{dx}$ is replaced by $\frac{y_{n+1} - y_n}{h}$ and hence it measures the error in using a difference equation to calculate a solution of the differential equation. The word 'local' is used because the error is specified only in the interval $[x_n, x_{n+1}]$.

Local discretisation error

In Euler's method:

$$\text{Local discretisation error} = \frac{y_{n+1} - y_n}{h} - f(x_n, y_n)$$

$$= \frac{h}{2}y''(\eta) \qquad x_n < \eta < x_{n+1}$$

The difficulty with this expression for the local discretisation error, as noted in previous chapters, is that we have no method for evaluating $y''(\eta)$. We do not know the value of η and even if we did, we do not know the form of $y''(x)$. The discretisation error is, then, of limited practical use. Its great value, however, is that it enables us to compare different techniques for solving differential equations numerically. We shall see more of this later in the chapter.

But now, consider the following example:

Example 8.3

A solution to the differential equation $\frac{dy}{dx} = -50y$ is required when $x = 1.1$, $x = 1.2$, $x = 1.3$ and $x = 1.4$. Use Euler's method to calculate approximations to the solution when it is given that $y = 1$ when $x = 1$.

The difference equation associated with Euler's method is

$$w_{n+1} = w_n + hf(x_n, w_n)$$

Since $f(x, y) = -50y$ this becomes

$$w_{n+1} = w_n + h(-50w_n)$$
$$= (1 - 50h)w_n$$

Taking $h = 0.1$ and $w_0 = 1$

if $n = 0$ $w_1 = (1 - 50(0.1))w_0 = (-4) \times 1 = -4$
if $n = 1$ $w_2 = (-4)w_1 = 16$
if $n = 2$ $w_3 = (-4)w_2 = -64$
if $n = 3$ $w_4 = (-4)w_3 = 256$

The solution to the differential equation $\frac{dy}{dx} = -50y$ is $y = e^{50(1-x)}$ and the approximate and exact solutions are printed in the table below.

n	x_n	y_n	w_n	| error |
0	1	1	1	0
1	1.1	6.738×10^{-3}	−4	4.007
2	1.2	4.540×10^{-5}	16	15.99995
3	1.3	3.059×10^{-7}	−64	64.00
4	1.4	2.061×10^{-9}	256	256.0

Clearly, the calculated solution is hopelessly inaccurate. One way out of this difficulty is to take a smaller step size and perform the calculation again. With $h = 0.05$ we have the table shown below:

n	x_n	y_n	w_n	| error |
0	1	1	1	0
2	1.1	6.738×10^{-3}	2.25	2.243
4	1.2	4.540×10^{-5}	5.0625	5.062
6	1.3	3.059×10^{-7}	11.391	11.391
8	1.4	2.061×10^{-9}	25.629	25.629

Again this is hopelessly inaccurate. With $h = 0.01$ we have:

n	x_n	y_n	w_n	| error |
0	1	1	1	0
1	1.01	0.606 530 667	0.5	0.106 530 666
2	1.02	0.367 879 449	0.25	0.117 879 449
3	1.03	0.223 130 168	0.125	9.81302×10^{-2}
4	1.04	0.135 335 289	0.0625	7.28353×10^{-2}
5	1.05	8.20850×10^{-2}	0.031 25	5.08350×10^{-2}
6	1.06	4.97871×10^{-2}	0.015 625	3.41621×10^{-2}
7	1.07	3.01974×10^{-2}	0.007 8125	2.23849×10^{-2}
8	1.08	1.83156×10^{-2}	0.003 906 25	1.44094×10^{-2}
9	1.09	1.11090×10^{-2}	0.001 953 125	9.15587×10^{-3}
10	1.1	6.73795×10^{-3}	9.76563×10^{-4}	5.76139×10^{-3}
...				
20	1.2	4.53999×10^{-5}	9.53674×10^{-7}	4.44463×10^{-5}
...				
30	1.3	3.05902×10^{-7}	9.31323×10^{-10}	3.04971×10^{-7}
...				
40	1.4	2.06115×10^{-9}	9.09495×10^{-13}	2.06025×10^{-9}

and this is much more acceptable.

An analysis of why the method failed when $h = 0.1$ and $h = 0.05$ but was successful (if time consuming) when $h = 0.01$ is beyond the scope of this book. We shall say only that for any method of finding a numerical solution of a differential equation, a differential equation can be found for which that method will fail. A great deal of current research is aimed at trying to find a technique that will warn the numerical analyst if the method that they are proposing to use is likely to fail for a particular differential equation. But while some progress has been made, the problem remains generally unsolved. Perhaps the best advice that can be given is: for any method, repeat the calculations using several different values of h. If the solutions appear to be broadly similar, everything is probably right, but if not ...

8.4 An alternative approximation to $\frac{dy}{dx}$

In Chapter 7, a second method of approximating $\frac{dy}{dx}$ was presented. This consisted of writing:

$$\text{when} \quad x = x_n \quad \frac{dy}{dx} = \frac{y_{n+1} - y_{n-1}}{2h} - \frac{h^2}{3!} y'''(\eta)$$

$$\text{where} \quad x_{n-1} < \eta < x_{n+1}$$

Since the error in this approximation is proportional to h^2, the reader may ask whether substituting $\frac{y_{n+1} - y_{n-1}}{2h}$ for $\frac{dy}{dx}$ in the differential equation will provide a more accurate solution than is provided by Euler's method.

If the differential equation is written as $\frac{dy}{dx} = f(x, y)$, we may write

$$\frac{y_{n+1} - y_{n-1}}{2h} - \frac{h^2}{3!}y'''(\eta) = f(x_n, y_n)$$

giving $$\frac{y_{n+1} - y_{n-1}}{2h} - f(x_n, y_n) = \frac{h^2}{3!}y'''(\eta)$$

and the local discretisation error in this method is proportional to h^2. Since we may then expect the local discretisation error in this new method to be smaller than in Euler's method, we have grounds for believing that this technique will obtain greater accuracy than Euler's method.

However, to take an example:

when $\frac{y_{n+1} - y_{n-1}}{2h}$ is substituted for $\frac{dy}{dx}$ in the differential equation $\frac{dy}{dx} = 2y$ we get the difference equation

$$\frac{w_{n+1} - w_{n-1}}{2h} = 2w_n \qquad \ldots (2)$$

giving

$$w_{n+1} - 4hw_n - w_{n-1} = 0$$

when $n = 1$ $\quad w_2 - 4hw_1 - w_0 = 0$

when $n = 2$ $\quad w_3 - 4hw_2 - w_1 = 0$

when $n = k - 1$ $\quad w_k - 4hw_{k-1} - w_{k-2} = 0$

The problem now is that, to solve these equations, we need to know the values of two of the w_i. In particular, we should like to know the values of w_0 and w_1; then, w_2 could be calculated from the first equation. Using the known values of w_1 and w_2, w_3 could then be calculated from the second equation and so on. Unfortunately, since $\frac{dy}{dx} = 2y$ is a first order differential equation, only one of the w_i values, w_0 is given. Thus we have a problem.

One way out of the difficulty is to use Taylor's theorem to calculate w_1. This may be achieved by writing:

$$y(x_0 + h) \approx y(x_0) + hy'(x_0) + \frac{h^2}{2!}y''(x_0) + \ldots$$

$$= y(x_0) + hf(x_0, y_0) + \frac{h^2}{2!}y''(x_0) + \ldots$$

and calculating as many terms of the right-hand side as are required for the efficient estimation of $y(x_0 + h)$. The value that we obtain for $y(x_0 + h)$ is then, of course, w_1. Then we may proceed as suggested in the paragraph above.

Example 8.4

Use the approximation:

when $x = x_n$ $\quad \frac{dy}{dx} \approx \frac{y_{n+1} - y_{n-1}}{2h}$

to obtain numerically the solution of the differential equation

$$\frac{dy}{dx} = y - x + 5 \qquad y(2) = 1$$

at the points $x = 2.1$, $x = 2.2$, $x = 2.3$.

Substitution gives the difference equation

$$\frac{w_{n+1} - w_{n-1}}{2h} = w_n - x_n + 5$$

Hence $\quad w_{n+1} - 2h(w_n - x_n + 5) - w_{n-1} = 0 \qquad \ldots (3)$

To obtain the value of w_1, use Taylor's theorem to write:

$$y(2 + h) = y(2) + hy'(2) + \frac{h^2}{2!}y''(2) + \frac{h^3}{3!}y'''(2) + \ldots \qquad \ldots (4)$$

where $\quad y'(x) = y - x + 5 \quad$ (from the differential equation)

$y''(x) = y' - 1$

$y'''(x) = y''$

This gives $\quad y(2) = 1 \quad$ (the initial condition)

$y'(2) = 1 - 2 + 5 = 4$

$y''(2) = 4 - 1 = 3$

$y'''(2) = 3$

and, with $\quad h = 0.1$

$$y(2 + 0.1) \approx 1 + 0.1 \times 4 + \frac{(0.1)^2}{2} \times 3 + \frac{(0.1)^3}{6} \times 3$$

$$= 1.4155$$

Hence $\quad w_1 = 1.4155$

From the difference Equation (3)

when $\quad n = 1 \quad w_2 - 2 \times (0.1)(w_1 - x_1 + 5) - w_0 = 0$

$w_2 = 0.2(1.4155 - 2.1 + 5) + 1 = 1.8631$

when $\quad n = 2 \quad w_3 - 2 \times (0.1)(w_2 - x_2 + 5) - w_1 = 0$

$w_3 = 0.2(1.8631 - 2.2 + 5) + 1.4155 = 2.348\,12$

Hence we have the table

n	x_n	y_n	w_n	| error |
0	2	1	1	0
1	2.1	1.415 512 8	1.4155	1.28×10^{-5}
2	2.2	1.864 208 3	1.8631	1.1083×10^{-3}
3	2.3	2.349 576 4	2.348 12	1.4564×10^{-3}

A comparison of these results with the table of results of Example 8.1 shows the improvement in accuracy achieved by this method.

8.5 Taylor methods

The method used in Example 8.4 to calculate w_1 could also have been used to calculate w_2 and w_3. In terms of x_0, y_0 and h, Taylor's theorem gives

$$y(x_0 + h) = y(x_0) + hy'(x_0) + \frac{h^2}{2!}y''(x_0) + \frac{h^3}{3!}y'''(x_0) + \ldots$$

But
$$\begin{aligned} y'(x_0) &= y_0 - x_0 + 5 \\ y''(x_0) &= y'(x_0) - 1 = y_0 - x_0 + 5 - 1 \\ y'''(x_0) &= y''(x_0) \quad = y_0 - x_0 + 4 \end{aligned}$$

Hence, in terms of x_0, w_0, w_1 and h, we may write

$$\begin{aligned} w_1 &= w_0 + h(w_0 - x_0 + 5) + \frac{h^2}{2}(w_0 - x_0 + 4) + \frac{h^3}{6}(w_0 - x_0 + 4) \\ &= w_0 + h(w_0 - x_0 + 4 + 1) + \frac{h^2}{2}(w_0 - x_0 + 4) + \frac{h^3}{6}(w_0 - x_0 + 4) \\ &= w_0 + h + \left(h + \frac{h^2}{2} + \frac{h^3}{6}\right)(w_0 - x_0 + 4) \end{aligned}$$

Similarly we have

$$w_2 = w_1 + h + \left(h + \frac{h^2}{2} + \frac{h^3}{6}\right)(w_1 - x_1 + 4)$$

$$w_3 = w_2 + h + \left(h + \frac{h^2}{2} + \frac{h^3}{6}\right)(w_2 - x_2 + 4)$$

and in general

$$w_{n+1} = w_n + h + \left(h + \frac{h^2}{2} + \frac{h^3}{6}\right)(w_n - x_n + 4)$$

This represents a difference equation for the differential equation

$$\frac{dy}{dx} = y - x + 5 \qquad y(2) = 1$$

and the method by which the difference equation was obtained is called the Taylor method of order three. The method is said to be 'of order three' because terms in h^3 (but not terms in h^4) have been included in the calculations. As will be seen in the next section, this means that the local discretisation error is proportional to h^3.

Using the difference equation to calculate w_2 and w_3 and taking $w_1 = 1.4155$ (from Example 8.4), $h = 0.1$, $x_1 = 2.1$ and $x_2 = 2.2$,

$$\begin{aligned} w_2 &= 1.4155 + 0.1 + \left(0.1 + \frac{(0.1)^2}{2} + \frac{(0.1)^3}{6}\right)(1.4155 - 2.1 + 4) \\ &= 1.864\,180\,1 \end{aligned}$$

$$\begin{aligned} w_3 &= 1.864\,180\,1 + 0.1 + \left(0.1 + \frac{(0.1)^2}{2} + \frac{(0.1)^3}{6}\right)(1.864\,180\,1 - 2.2 + 4) \\ &= 2.349\,529\,7 \end{aligned}$$

A glance at the results of Example 8.4 shows that a further improvement in accuracy has been achieved. This was to have been expected. Although in Example 8.4, w_1 was calculated to third order accuracy, by using $\dfrac{y_{n+1} - y_{n-1}}{2h}$ to approximate $\dfrac{dy}{dx}$, w_2 and w_3 were calculated to only second order accuracy.

Formal description of the Taylor method

To obtain a numerical solution to the differential equation

$$\frac{dy}{dx} = f(x, y) \qquad y(x_0) = y_0$$

Write a Taylor polynomial for the solution $y(x)$ centred at $x = x_n$ (we choose a polynomial of degree three):

$$y(x_n + h) = y(x_n) + hy'(x_n) + \frac{h^2}{2!}y''(x_n) + \frac{h^3}{3!}y'''(x_n) + \frac{h^4}{4!}y^{(4)}(\eta)$$

where $\quad x_n < \eta < x_n + h$

Since $\quad y'(x_n) = f(x_n, y_n) \quad$ (from the differential equation)

$$y''(x_n) = f'(x_n, y_n)$$
$$y'''(x_n) = f''(x_n, y_n) \quad \text{etc.}$$

we have

$$y(x_n + h) = y(x_n) + hf(x_n, y_n) + \frac{h^2}{2}f'(x_n, y_n) + \frac{h^3}{6}f''(x_n, y_n) + \frac{h^4}{24}y^{(4)}(\eta) \quad \ldots (5)$$

From Equation (5):

writing $\quad w_{n+1} \approx y(x_n + h) \qquad w_n \approx y(x_n)$

we get the difference equation associated with the Taylor method of order three:

$$\boxed{w_{n+1} = w_n + hf(x_n, w_n) + \frac{h^2}{2}f'(x_n, w_n) + \frac{h^3}{6}f''(x_n, w_n)}$$

Also from Equation (5),

writing $\quad y_{n+1} = y(x_n + h) \qquad y_n = y(x_n)$

we have

$$y_{n+1} = y_n + hf(x_n, y_n) + \frac{h^2}{2}f'(x_n, y_n) + \frac{h^3}{6}f''(x_n, y_n) + \frac{h^4}{24}y^{(4)}(\eta)$$

giving

$$\frac{y_{n+1} - y_n}{h} - \left[f(x_n, y_n) + \frac{h}{2}f'(x_n, y_n) + \frac{h^2}{6}f''(x_n, y_n)\right]$$
$$= \frac{h^3}{24}y^{(4)}(\eta)$$

By comparison with Euler's method in Section 8.3, we see that the local discretisation error is $\frac{h^3}{24}y^{(4)}(\eta)$. Hence the justification for saying that this method 'is of order three'.

Observe that for a Taylor method of order one, we would have

$$y(x_n + h) = y(x_n) + hy'(x_n) + \frac{h^2}{2!}y''(\eta) \qquad x_n < \eta < x_n + h$$

or, using the above notation

$$y(x_n + h) = y(x_n) + hf(x_n, y_n) + \frac{h^2}{2!}y''(\eta)$$

The difference equation for this Taylor method is

$$w_{n+1} = w_n + hf(x_n, w_n)$$

and the discretisation error is

$$\frac{y_{n+1} - y_n}{h} - f(x_n, y_n) = \frac{h}{2}y''(\eta)$$

This is, of course, Euler's method and we have the important result:

Euler's method is the Taylor method of order one.

Example 8.5

Use the Taylor method of order two to obtain an approximate solution of the differential equation

$$\frac{dy}{dx} = x^2 + y \qquad y(3) = 1$$

at the points given by $x = 3.2$, $x = 3.4$, $x = 3.6$.

Write down an expression for the discretisation error and suggest how the accuracy of your answers could be improved.

Clearly $x_0 = 3$ and $y_0 = w_0 = 1$.
For the Taylor method of order two, write

$$y(x_n + h) = y(x_n) + hy'(x_n) + \frac{h^2}{2!}y''(x_n) + \frac{h^3}{3!}y'''(\eta) \qquad \ldots (6)$$

where $\quad x_n < \eta < x_n + h$

$$y'(x) = x^2 + y \qquad \text{(from the differential equation)}$$
$$y''(x) = 2x + y'$$
$$= 2x + x^2 + y$$

Substituting these expressions in Equation (6), we obtain the difference equation associated with the Taylor method of order two:

$$w_{n+1} = w_n + h(x_n^2 + w_n) + \frac{h^2}{2}(2x_n + x_n^2 + w_n)$$

Take $h = 0.2$

$n = 0$ $\quad w_1 = w_0 + 0.2(x_0^2 + w_0) + \frac{(0.2)^2}{2}(2x_0 + x_0^2 + w_0)$

$$= 1 + 0.2(3^2 + 1) + \frac{(0.2)^2}{2}(2 \times 3 + 3^2 + 1)$$

$$= 3.32$$

$n = 1$ $\quad w_2 = w_1 + 0.2(x_1^2 + w_1) + \frac{(0.2)^2}{2}(2x_1 + x_1^2 + w_1)$

$$= 3.32 + 0.2(3.2^2 + 3.32) + \frac{(0.2)^2}{2}(2 \times 3.2 + 3.2^2 + 3.32)$$

$$= 6.4312$$

$n = 2$ $\quad w_3 = 6.4312 + 0.2(3.4^2 + 6.4312)$

$$+ \frac{(0.2)^2}{2}(2 \times 3.4 + 3.4^2 + 6.4312)$$

$$= 10.525\,264$$

Hence we obtain approximately

x	3	3.2	3.4	3.6
y	1	3.32	6.4312	10.525 264

The discretisation error for this method is (from Equation (6))

$$\frac{y_{n+1} - y_n}{h} - \left[y'(x_n) + \frac{h}{2}y''(x_n)\right] = \frac{h^2}{3!}y'''(\eta) \qquad x_n < \eta < x_n + h$$

The accuracy could be improved by:

(a) reducing the value of h and taking more steps to obtain the solution at the required points,

(b) using a Taylor method of order higher than two.

Example 8.6

The solution of the differential equation $\frac{dy}{dx} = xy \quad y(1) = 5$ is required at the points given by $x = 1.1$, $x = 1.2$, $x = 1.3$. Use the Taylor method of order four to calculate approximately the value of the solution at these points.

The analytical solution is $y = 5e^{\frac{1}{2}(x^2-1)}$. Calculate the error in your answers.

$x_0 = 1$ and $y_0 = w_0 = 5$

We shall take $h = 0.1$.

For the Taylor method of order four consider:

$$y(x_n + h) \approx y(x_n) + hy'(x_n) + \frac{h^2}{2!}y''(x_n) + \frac{h^3}{3!}y'''(x_n) + \frac{h^4}{4!}y^{(4)}(x_n)$$

But
$$y'(x) = xy \quad \text{(from the differential equation)}$$
$$y''(x) = xy' + y \quad \text{(using the product rule)}$$
$$y'''(x) = xy'' + y' + y' = xy'' + 2y'$$
$$y^{(4)}(x) = xy''' + y'' + 2y'' = xy''' + 3y''$$

In terms of x_n, w_n:

$$y'(x_n) = x_n w_n$$
$$y''(x_n) = x_n(x_n w_n) + w_n = (x_n^2 + 1)w_n$$
$$y'''(x_n) = x_n(x_n^2 + 1)w_n + 2x_n w_n = (x_n^3 + 3x_n)w_n$$
$$y^{(4)}(x_n) = x_n(x_n^3 + 3x_n)w_n + 3(x_n^2 + 1)w_n = (x_n^4 + 6x_n^2 + 3)w_n$$

The difference equation is:

$$w_{n+1} = w_n + h(x_n w_n) + \frac{h^2}{2}(x_n^2 + 1)w_n + \frac{h^3}{6}(x_n^3 + 3x_n)w_n + \frac{h^4}{24}(x_n^4 + 6x_n^2 + 3)w_n$$

$n = 0$
$$w_1 = w_0 + (0.1)(x_0 w_0) + \frac{(0.1)^2}{2}(x_0^2 + 1)w_0 + \frac{(0.1)^3}{6}(x_0^3 + 3x_0)w_0 + \frac{(0.1)^4}{24}(x_0^4 + 6x_0^2 + 3)w_0$$

$$= 5 + (0.1)(1 \times 5) + \frac{(0.1)^2}{2}(1^2 + 1)5 + \frac{(0.1)^3}{6}(1^3 + 3 \times 1)5 + \frac{(0.1)^4}{24}(1^4 + 6 \times 1^2 + 3)5$$

$$= 5.553\,541\,7$$

$$n = 1 \quad w_2 = 5.5535417 + (0.1)(1.1 \times 5.5535417)$$
$$+ \frac{(0.1)^2}{2}(1.1^2 + 1) \times 5.5535417$$
$$+ \frac{(0.1)^3}{6}(1.1^3 + 3 \times 1.1)5.5535417$$
$$+ \frac{(0.1)^4}{24}(1.1^4 + 6 \times 1.1^2 + 3) \times 5.5535417$$
$$= 6.2303556$$

$$n = 2 \quad w_3 = 6.2303556 + (0.1) \times (1.2 \times 6.2303556)$$
$$+ \frac{(0.1)^2}{2}(1.2^2 + 1) \times 6.2303556$$
$$+ \frac{(0.1)^3}{6}(1.2^3 + 3 \times 1.2) \times 6.2303556$$
$$+ \frac{(0.1)^4}{24}(1.2^4 + 6 \times 1.2^2 + 3) \times 6.2303556$$
$$= 7.0598972$$

Hence we have:

n	x	y	w	| error |
0	1	5	5	0
1	1.1	5.5535531	5.5535417	1.14×10^{-5}
2	1.2	6.2303837	6.2303556	2.81×10^{-5}
3	1.3	7.0599496	7.0598972	5.25×10^{-5}

Advantages and disadvantages of the Taylor methods

The great appeal of the Taylor methods is that they provide methods of orders one, two, three, four, ... for obtaining the numerical solution of a differential equation. Hence we have a sequence of techniques which can be used to achieve (in principle) an arbitrarily high degree of accuracy in the solution. As we saw in Example 8.6, impressive accuracy can be achieved by the use of these methods.

The serious disadvantage to the Taylor methods is that if the function $f(x, y)$ is at all complicated, the derivatives $y''(x)$, $y'''(x)$, $y^{(4)}(x)$, ... become cumbersome and difficult to evaluate. For this reason, in spite of the above remarks, the Taylor methods are rarely used in practice.

What is needed is a second sequence of techniques which have the accuracy of the Taylor methods but which do not require the evaluation of $y''(x)$, $y'''(x)$, ...

Such a collection of techniques is provided by the Runge-Kutta methods. These are studied in the next section.

8.6 Runge-Kutta methods

To solve the differential equation $\dfrac{dy}{dx} = f(x, y) \quad y(x_0) = y_0$:

The aim is to produce a sequence of methods that have the accuracy of the Taylor methods, but which do not require the differentiation of $f(x, y)$. We approach the problem geometrically, so consider again the diagram illustrating Euler's method (Figure 8.4, page 292).

An estimate is needed for the value of y when $x = x_1$. Since, at this point, $y_1 = BR = y_0 + k$ and y_0 is known, the problem reduces to finding a good approximation to the value of k. In Euler's method, this approximation was

$$\begin{aligned} k_1 &= h \times \left[\frac{dy}{dx} \text{ evaluated at } x_0\right] \\ &= hf(x_0, y_0) \end{aligned}$$

It might be argued, however, that a better approximation would be achieved if we had used the value of $\dfrac{dy}{dx}$ at some intermediate point on the curve PR and not at the initial point. Unfortunately, we do not know any of the intermediate points. But we could write:

$$k_2 = hf(x_0 + ah, y_0 + bk_1)$$

where a and b are (as yet) unknown numbers. The aim is to assign values to a and b so as to produce a more accurate approximation to k. In fact, we go a little further and take, as our approximation to k a weighted average of k_1 and k_2. That is, we write

$$k \approx Ak_1 + Bk_2$$

where A and B are also (as yet) unknown numbers. Since $w_0 = y_0$, this gives the difference equation

$$w_1 = w_0 + Ak_1 + Bk_2$$

At this point, geometric intuition runs out and we are left with the need for some messy and fairly advanced algebra, which is confined to Appendix 3. The reader should be aware, however, that the idea is to choose A, B, a, b so that the difference equation agrees with a Taylor method of as high an order as possible. In this case, we shall consider methods of order two, so we shall choose A, B, a, b to match the Taylor method of order two. It is shown in Appendix 3 that A, B, a, b must then satisfy

$$A + B = 1 \qquad \text{and} \qquad Ba = Bb = \tfrac{1}{2}$$

If we choose $A = \frac{1}{2}$, $B = \frac{1}{2}$, $a = 1$, $b = 1$ then these equations are satisfied and the difference equation

$$\begin{aligned} w_1 &= w_0 + \tfrac{1}{2}[k_1 + k_2] \\ &= w_0 + \tfrac{1}{2}[h\mathrm{f}(x_0, w_0) + h\mathrm{f}(x_0 + h, w_0 + k_1)] \end{aligned}$$

has discretisation error proportional to h^2. Thus we have established a method of order two which does not require differentiation of $\mathrm{f}(x, y)$. Part of the aim has been achieved.

In general terms: when w_n has been calculated,

set $\quad k_1 = h\mathrm{f}(x_n, w_n)$

and $\quad k_2 = h\mathrm{f}(x_n + h, w_n + k_1)$

$$\boxed{w_{n+1} = w_n + \tfrac{1}{2}(k_1 + k_2)}$$

Or, when written out in full,

$$w_{n+1} = w_n + \frac{h}{2}[\mathrm{f}(x_n, w_n) + \mathrm{f}(x_n + h, w_n + h\mathrm{f}(x_n, w_n))]$$

Although this difference equation looks fearsome, the calculations are easy to perform, as is shown in the algorithm that follows. This method is often called the simple Runge-Kutta method, or the modified Euler method.

Algorithm 8.2 The simple Runge-Kutta method (modified Euler method)

To obtain approximately the solution of the differential equation $\frac{\mathrm{d}y}{\mathrm{d}x} = \mathrm{f}(x, y)$ with initial condition $y(a) = y_0$.

input: initial value of $x = x_0$;
initial value of $y = y_0$;
step size h

$w_0 := y_0$

for $n = 0, 1, \ldots,$ until satisfied

$k_1 := h\mathrm{f}(x_n, w_n)$

$k_2 := h\mathrm{f}(x_n + h, w_n + k_1)$

$w_{n+1} := w_n + \tfrac{1}{2}(k_1 + k_2)$

$x_{n+1} := x_0 + (n + 1)h$

endloop

output: $w_0, w_1, w_2, \ldots$

The following examples illustrate the Runge-Kutta methods. To show the improvement in accuracy achieved by these methods, we shall compare the solutions with those achieved by Euler's method.

Example 8.7

Use the simple Runge-Kutta method to obtain an approximate solution to the differential equation $\frac{dy}{dx} = y - x + 5$ with initial condition $y(2) = 1$ at the points $x = 2.1, 2.2, 2.3$.

$x_0 = 2 \quad y_0 = 1$ and step size $h = 0.1$

Set $w_0 = y_0 = 1$

Since $f(x, y) = y - x + 5$

$$k_1 = hf(x_0, w_0) = 0.1(1 - 2 + 5) = 0.4$$

Now $x_0 + h = 2 + 0.1 = 2.1$

and $w_0 + k_1 = 1 + 0.4 = 1.4$

$$\therefore \quad k_2 = hf(x_0 + h, w_0 + k_1) = 0.1(1.4 - 2.1 + 5) = 0.43$$

and

$$w_1 = w_0 + \tfrac{1}{2}(k_1 + k_2) = 1 + 0.5(0.4 + 0.43) = 1.415$$

Now take

$$k_1 = hf(x_1, w_1) = 0.1(1.415 - 2.1 + 5) = 0.4315$$

$$x_1 + h = 2.1 + 0.1 = 2.2$$

and $w_1 + k_1 = 1.415 + 0.4315 = 1.8465$

$$\therefore \quad k_2 = hf(x_1 + h, w_1 + k_1) = 0.1(1.8465 - 2.2 + 5) = 0.46465$$

and

$$w_2 = w_1 + \tfrac{1}{2}(k_1 + k_2) = 1.415 + 0.5(0.4315 + 0.46465) = 1.863075$$

Now take

$$k_1 = hf(x_2, w_2) = 0.1(1.863075 - 2.2 + 5) = 0.4663075$$

$$k_2 = hf(x_2 + h, w_2 + k_1) = 0.1(2.3293825 - 2.3 + 5) = 0.50293825$$

$$w_3 = w_2 + \tfrac{1}{2}(k_1 + k_2) = 2.347697875$$

The different methods are compared in the table below.

			← Runge-Kutta →		← Euler →	
n	x_n	y_n	w_n	| error |	w_n	| error |
0	2	1	1	0	1	0
1	2.1	1.41551275	1.415	5.12753×10^{-4}	1.4	1.55128×10^{-2}
2	2.2	1.86420827	1.863075	1.13327×10^{-3}	1.83	3.42083×10^{-2}
3	2.3	2.34957642	2.34769788	1.87854×10^{-3}	2.293	5.65764×10^{-2}

If the step size is reduced, the results are even more dramatic. If $h = 0.05$, we have:

			← Runge-Kutta →		← Euler →	
n	x_n	y_n	w_n	| error |	w_n	| error |
0	2	1	1	0	1	0
1	2.05	1.20381329	1.20375	6.32894×10^{-5}	1.2	3.81329×10^{-3}
2	2.1	1.41551275	1.41537969	1.33065×10^{-4}	1.4075	8.01275×10^{-3}
3	2.15	1.63550273	1.6352929	2.09830×10^{-4}	1.622875	1.26277×10^{-2}
4	2.2	1.86420827	1.86391416	2.94114×10^{-4}	1.84651875	1.76895×10^{-2}
5	2.25	2.10207625	2.10168976	3.86488×10^{-4}	2.07884469	2.32316×10^{-2}
6	2.3	2.34957642	2.34908886	4.87559×10^{-4}	2.32028692	2.92895×10^{-2}

and it is clear that the simple Runge-Kutta method shows a great improvement over Euler's method. This was to be expected, of course: the simple Runge-Kutta method has discretisation error of order two, while Euler's method is of order one. The claim made earlier that the discretisation error was useful for comparing the accuracy of different methods has been seen to work in practice.

Other values may be taken by A, B, a and b. If we take $A = \frac{1}{4}$, $B = \frac{3}{4}$, $a = b = \frac{2}{3}$, we obtain Heun's method. In this case,

$$k_1 = hf(x_n, w_n)$$

$$k_2 = hf(x_n + \tfrac{2}{3}h, w_n + \tfrac{2}{3}k_1)$$

and the difference equation becomes

$$\begin{aligned} w_{n+1} &= w_n + \tfrac{1}{4}k_1 + \tfrac{3}{4}k_2 \\ &= w_n + \frac{h}{4}[f(x_n, w_n) + 3f(x_n + \tfrac{2}{3}h, w_n + \tfrac{2}{3}k_1)] \end{aligned}$$

Algorithm 8.3 Heun's method

To obtain approximately the solution of the differential equation $\frac{dy}{dx} = f(x, y)$ with initial condition $y(a) = y_0$.

input: initial value of $x = x_0$;
initial value of $y = y_0$;
step size h

$w_0 := y_0$
for $n = 0, 1, \ldots$, until satisfied
$k_1 := hf(x_n, w_n)$
$k_2 := hf(x_n + \frac{2}{3}h, w_n + \frac{2}{3}k_1)$
$w_{n+1} := w_n + \frac{1}{4}(k_1 + 3k_2)$
$x_{n+1} := x_0 + (n + 1)h$
endloop
output: $w_0, w_1, w_2, \ldots$

Example 8.8

Use Heun's method to obtain an approximate solution to the differential equation $\frac{dy}{dx} = -y + 2e^x \quad y(0) = 2$ at the points $x = 0.05, 0.1, 0.15$ and 0.2. Given that the exact solution is $y = e^{-x} + e^x$ calculate the error in the approximation.

$x_0 = 0 \quad y_0 = 2 \quad$ step size $h = 0.05$

Set $w_0 = 2$

$$\begin{aligned}
k_1 &= hf(x_0, w_0) = 0.05(-2 + 2e^0) &&= 0 \\
k_2 &= hf(x_0 + \tfrac{2}{3}h, w_0 + \tfrac{2}{3}k_1) \\
&= 0.05(-2 + 2e^{0.03333333}) &&= 0.003\,389\,51 \\
w_1 &= w_0 + \tfrac{1}{4}(k_1 + 3k_2) &&= 2.002\,542\,13 \\
k_1 &= hf(x_1, w_1) \\
&= 0.05(-2.002\,542\,13 + 2e^{0.05}) &&= 0.005\,000\,00 \\
k_2 &= hf(x_1 + \tfrac{2}{3}h, w_1 + \tfrac{2}{3}k_1) \\
&= 0.05(-2.005\,875\,46 + 2e^{0.08333333}) &&= 0.008\,396\,63 \\
w_2 &= w_1 + \tfrac{1}{4}(k_1 + 3k_2) &&= 2.010\,089\,61
\end{aligned}$$

Similarly, we have for the next two steps

$$\begin{aligned}
k_1 &= 0.010\,012\,6 \\
k_2 &= 0.013\,424\,8 \\
w_3 &= 2.022\,661\,4
\end{aligned}$$

and

$$\begin{aligned}
k_1 &= 0.015\,050\,4 \\
k_2 &= 0.018\,486\,7 \\
w_4 &= 2.040\,289\,0
\end{aligned}$$

A table giving the approximate solution, the exact solution and the error, is given below.

n	x_n	y_n	w_n	| error |
0	0	2	2	0
1	0.05	2.002 500 52	2.002 542 13	$4.161\,24 \times 10^{-5}$
2	0.1	2.010 008 34	2.010 089 61	$8.127\,19 \times 10^{-5}$
3	0.15	2.022 542 22	2.022 661 4	$1.191\,77 \times 10^{-4}$
4	0.2	2.040 133 51	2.040 289 03	$1.555\,18 \times 10^{-4}$

and it is clear that Heun's method also produces a remarkably accurate solution.

To achieve completely the aim described at the beginning of this section, we must produce Runge-Kutta methods of orders three, four, ... In fact, Runge-Kutta methods of order three do exist, but are rarely used in practice.

The method most commonly used is the Runge-Kutta method of order four and this is demonstrated next. Methods of order higher than four do exist, but because of their complexity and because the method of order four is so accurate, these higher order methods are rarely used.

The Runge-Kutta method of order four

The difference equation is constructed as follows:

If w_n has been calculated, set

$$k_1 = hf(x_n, w_n)$$
$$k_2 = hf(x_n + \tfrac{1}{2}h, w_n + \tfrac{1}{2}k_1)$$
$$k_3 = hf(x_n + \tfrac{1}{2}h, w_n + \tfrac{1}{2}k_2)$$
$$k_4 = hf(x_n + h, w_n + k_3)$$

Then set

$$w_{n+1} = w_n + \tfrac{1}{6}(k_1 + 2k_2 + 2k_3 + k_4)$$

Observe that when using this method, it is necessary to calculate k_1 before attempting to calculate k_2; to calculate k_2 before attempting to calculate k_3 and so on.

Algorithm 8.4 The Runge-Kutta method of order four

To obtain approximately the solution of the differential equation $\frac{dy}{dx} = f(x, y)$ with initial condition $y(a) = y_0$.

input: initial value of $x = x_0$;
initial value of $y = y_0$;
step size h

$w_0 := y_0$

for $n = 0, 1, ...,$ until satisfied

$k_1 := hf(x_n, w_n)$

$k_2 := hf(x_n + \tfrac{1}{2}h, w_n + \tfrac{1}{2}k_1)$

$k_3 := hf(x_n + \tfrac{1}{2}h, w_n + \tfrac{1}{2}k_2)$

$k_4 := hf(x_n + h, w_n + k_3)$

$w_{n+1} := w_n + \tfrac{1}{6}(k_1 + 2k_2 + 2k_3 + k_4)$

$x_{n+1} := x_0 + (n + 1)h$

endloop

output: $w_0, w_1, w_2, ...$

Example 8.9

Use the Runge-Kutta method of order four to obtain an approximate solution to the differential equation

$$\frac{dy}{dx} = y - x + 5 \qquad y(2) = 1$$

at the points $x = 2.1$, 2.2 and 2.3. Given that the exact solution is $y = x - 4 + 3e^{x-2}$ calculate the error in the approximation and comment on the accuracy.

$$x_0 = 2 \qquad y_0 = 1 \qquad \text{and} \qquad \text{take } h = 0.1$$

$$f(x, y) = y - x + 5 \qquad \text{and} \qquad w_0 = 1$$

$$k_1 = hf(x_0, w_0) = h(w_0 - x_0 + 5) = 0.1(1 - 2 + 5) = 0.4$$

$$k_2 = hf(x_0 + \tfrac{1}{2}h, w_0 + \tfrac{1}{2}k_1) = 0.1(1.2 - 2.05 + 5) = 0.415$$

$$k_3 = hf(x_0 + \tfrac{1}{2}h, w_0 + \tfrac{1}{2}k_2) = 0.1(1.2075 - 2.05 + 5) = 0.41575$$

$$k_4 = hf(x_0 + h, w_0 + k_3) = 0.1(1.41575 - 2.1 + 5) = 0.431575$$

$$w_1 = w_0 + \tfrac{1}{6}(k_1 + 2k_2 + 2k_3 + k_4)$$

$$= 1 + \tfrac{1}{6}(0.4 + 2 \times 0.415 + 2 \times 0.41575 + 0.431575) = 1.4155125$$

w_2 and w_3 may be calculated similarly. Since this method would normally be implemented on a computer, we shall not provide further details of the calculation but jump straight to the usual table illustrating the results.

n	x_n	y_n	w_n	| error |
0	2	1	1	0
1	2.1	1.41551275	1.4155125	2.52854×10^{-7}
2	2.2	1.86420827	1.86420771	5.59259×10^{-7}
3	2.3	2.34957642	2.34957549	9.25735×10^{-7}

It is clear that the Runge-Kutta method of order four is very accurate indeed as would be expected from a technique having discretisation error proportional to h^4.

To complete this section, the solutions obtained from the Runge-Kutta method of order four and Euler's method are shown side by side

			← Runge-Kutta →		← Euler →	
n	x_n	y_n	w_n	| error |	w_n	| error |
0	2	1	1	0	1	0
1	2.1	1.41551275	1.4155125	2.52854×10^{-7}	1.4	1.55128×10^{-2}
2	2.2	1.86420827	1.86420771	5.59259×10^{-7}	1.83	3.42083×10^{-2}
3	2.3	2.34957642	2.34957549	9.25735×10^{-7}	2.293	5.65764×10^{-2}

It is hoped that this table will illustrate the remarkable progress that has been made.

8.7 Second order differential equations

This section shows how the techniques for obtaining approximate solutions to first order differential equations may be applied to second order differential equations. Bearing in mind the enormous improvement in accuracy that was provided by these techniques, we might hope that their application to second order differential equations might produce a similar improvement in accuracy.

The central idea is to replace a second order differential equation by two first order differential equations and then apply the first order techniques to these first order equations. This reduction is achieved by means of two substitutions:

set $u = y$

and $v = \dfrac{dy}{dx}$

Differentiation then gives

$$\frac{du}{dx} = \frac{dy}{dx} = v$$

$$\frac{dv}{dx} = \frac{d^2y}{dx^2}$$

The technique is best illustrated by means of an example.

Example 8.10

Reduce the differential equation

$$\frac{d^2y}{dx^2} + \frac{dy}{dx} - 6y = \sin x$$

when $x = 1$ $y = 3$ and $\dfrac{dy}{dx} = -1$

to two first order differential equations.

Let $u = y$

and $v = \dfrac{dy}{dx}$

Differentiating gives

$$\frac{du}{dx} = \frac{dy}{dx} = v$$

$$\frac{dv}{dx} = \frac{d^2y}{dx^2}$$

But $\dfrac{d^2y}{dx^2} = -\dfrac{dy}{dx} + 6y + \sin x$

so we may write:

$$\frac{du}{dx} = v$$

$$\frac{dv}{dx} = -\frac{dy}{dx} + 6y + \sin x$$

Or, writing $\dfrac{dy}{dx} = v$ and $y = u$

$$\frac{du}{dx} = v$$

$$\frac{dv}{dx} = -v + 6u + \sin x$$

Note how easily the initial conditions are amalgamated into this scheme. When $x = 1$, $y = 3$ and $\dfrac{dy}{dx} = -1$. Hence $u(1) = 3$ and $v(1) = -1$. Thus the second order differential equation

$$\frac{d^2y}{dx^2} + \frac{dy}{dx} - 6y = \sin x \qquad y(1) = 3,\ y'(1) = -1$$

may be replaced by the two first order differential equations with the given initial conditions:

$$\frac{du}{dx} = v \qquad u(1) = 3$$

$$\frac{dv}{dx} = -v + 6u + \sin x \qquad v(1) = -1$$

These equations may now be solved using any of the first order techniques described in Sections 8.2 and 8.6. Euler's method and the simple Runge-Kutta method will be used to illustrate those ideas. But first, some notation.

Notation

To obtain an approximate solution to the system of differential equations

$$\frac{du}{dx} = v$$

$$\frac{dv}{dx} = f(x, u, v)$$

with initial conditions $u(a) = u_0$, $v(a) = v_0$.

(In the above example $\quad f(x, u, v) = -v + 6u + \sin x \qquad a = 1,\ u_0 = 3,\ v_0 = -1$)

Let $x_0 = a$, $x_1 = a + h$, $x_2 = a + 2h$, $x_3 = a + 3h$... be the points at which the value of the solution is required.

Let $u_0, u_1, u_2, \ldots$ represent the exact values of u (and hence the exact values of the solution y) at the points $x_0, x_1, x_2, \ldots$ Similarly, let $v_0, v_1, v_2, \ldots$ represent the exact values of v (and hence of $\dfrac{dy}{dx}$) at $x_0, x_1, x_2, \ldots$

Let $w_0, w_1, w_2, \ldots$ represent the approximate values of u at $x_0, x_1, x_2, \ldots$ and let $z_0, z_1, z_2, \ldots$ represent the approximate values of v at these points.

Algorithm 8.5 Euler's method

To obtain the approximate solution of the differential equation $\frac{d^2y}{dx^2} = f\left(x, y, \frac{dy}{dx}\right)$ with initial conditions $y(a) = u_0$, $y'(a) = v_0$.

input: initial value of $x = x_0$;
initial value of $y = u_0$;
initial value of $\frac{dy}{dx} = v_0$;
step size h

$w_0 := u_0$
$z_0 := v_0$
for $n = 0, 1, \ldots$, until satisfied

$$w_{n+1} := w_n + hz_n \qquad \left(\text{for } \frac{du}{dx} = v\right)$$

$$z_{n+1} := z_n + hf(x_n, w_n, z_n) \qquad \left(\text{for } \frac{dv}{dx} = f(x, u, v)\right)$$

$$x_{n+1} := x_0 + (n+1)h$$

endloop
output: the approximate values of the solution $w_0, w_1, w_2, \ldots$;
the approximate values of $\frac{dy}{dx}$, $z_0, z_1, z_2, \ldots$

Example 8.11

Reduce the differential equation

$$\frac{d^2y}{dx^2} + \frac{dy}{dx} - 6y = \sin x \qquad y(1) = 3 \qquad y'(1) = -1$$

to a system of first order differential equations. Use Euler's method to calculate approximate solutions at the points $x = 1.1, 1.2, 1.3$.

We have seen that the system of first order differential equations is:

$$\left.\begin{aligned} \frac{du}{dx} &= v \\ \frac{dv}{dx} &= -v + 6u + \sin x \end{aligned}\right\} \qquad \begin{aligned} u(1) &= 3 \\ v(1) &= -1 \end{aligned}$$

Set $u_0 = 3$, $v_0 = -1$ and take $h = 0.1$.

$$f(x, u, v) = -v + 6u + \sin x \qquad w_0 = 3 \qquad \text{and} \qquad z_0 = -1$$

for $n = 0$ $\quad w_1 = w_0 + hz_0 = 3 + (0.1) \times (-1) \qquad = 2.9$

$$z_1 = z_0 + h(-z_0 + 6w_0 + \sin x_0)$$
$$= -1 + 0.1[-(-1) + 6 \times 3 + \sin 1] \qquad = 0.984\,147\,10$$

for $n = 1$ $w_2 = w_1 + hz_1 = 2.9 + 0.1 \times 0.98414710 = 2.99841471$

$z_2 = z_1 + h(-z_1 + 6w_1 + \sin x_1) = 2.71485312$

for $n = 2$ $w_3 = w_2 + hz_2 = 3.26990002$

$z_3 = z_2 + h(-z_2 + 6w_2 + \sin x_2) = 4.33562055$

The approximate solution, the exact solution and the error are given in the table below.

n	x_n	u_n	w_n	| error |
0	1	3	3	0
1	1.1	2.99556534	2.9	9.55653×10^{-2}
2	1.2	3.17196887	2.99841471	0.173554158
3	1.3	3.52251367	3.26990002	0.252613651

Below is an algorithm for the Simple Runge-Kutta method. The justification for the algorithm will follow the worked example.

Algorithm 8.6 The simple Runge-Kutta method

To obtain approximately the solution of the differential equation

$$\frac{d^2y}{dx^2} = f\left(x, y, \frac{dy}{dx}\right) \text{ with initial conditions } y(a) = u_0,\ y'(a) = v_0.$$

input: initial value of $x = x_0$;
initial value of $y = u_0$;
initial value of $\frac{dy}{dx} = v_0$;
step size h

$w_0 := u_0$
$z_0 := v_0$
for $n = 0, 1, \ldots$, until satisfied
$k_1 := hz_n$
$l_1 := hf(x_n, w_n, z_n)$
$k_2 := h(z_n + l_1)$
$l_2 := hf(x_n + h, w_n + k_1, z_n + l_1)$
$w_{n+1} := w_n + \frac{1}{2}(k_1 + k_2)$
$z_{n+1} := z_n + \frac{1}{2}(l_1 + l_2)$
$x_{n+1} := x_0 + (n + 1)h$
endloop
output: the approximate values of the solution $w_0, w_1, w_2, \ldots$;
the approximate values of $\frac{dy}{dx}$, $z_0, z_1, z_2, \ldots$

Example 8.12

Use the simple Runge-Kutta method to calculate approximate solutions to the differential equation

$$\frac{d^2y}{dx^2} + \frac{dy}{dx} - 6y = \sin x \qquad y(1) = 3 \qquad y'(1) = -1$$

at the points $x = 1.1$, $x = 1.2$, $x = 1.3$.

As before:

$$\left.\begin{aligned} \frac{du}{dx} &= v \\ \frac{dv}{dx} &= -v + 6u + \sin x \end{aligned}\right\} \qquad \begin{aligned} u(1) &= 3 \\ v(1) &= -1 \end{aligned}$$

$w_0 = 3 \qquad z_0 = -1 \qquad$ and we take $\qquad h = 0.1$

$f(x, u, v) = -v + 6u + \sin x$

for $n = 0$

$k_1 = hz_0 = 0.1 \times (-1) = -0.1$

$l_1 = h(-z_0 + 6w_0 + \sin x_0)$
$= 0.1(-(-1) + 6 \times 3 + \sin 1) = 1.98414710$

$k_2 = h(z_0 + l_1)$
$= 0.1(-1 + 1.98414710) = 0.09841471$

$l_2 = h(-(z_0 + l_1) + 6(w_0 + k_1) + \sin(x_0 + h)) = 1.73070603$

hence

$w_1 = w_0 + \frac{1}{2}(k_1 + k_2) = 2.99920736$

$z_1 = z_0 + \frac{1}{2}(l_1 + l_2) = 0.85742656$

for $n = 1$

$k_1 = hz_1 = 0.08574266$

$l_1 = h(-z_1 + 6w_1 + \sin x_1) = 1.80290249$

$k_2 = h(z_1 + l_1) = 0.26603291$

$l_2 = h(-(z_1 + l_1) + 6(w_1 + k_1) + \sin x_2) = 1.67814101$

$w_2 = w_1 + \frac{1}{2}(k_1 + k_2) = 3.17509514$

$z_2 = z_1 + \frac{1}{2}(l_1 + l_2) = 2.59794831$

for $n = 2$

$k_1 = hz_2 = 0.25979483$

$l_1 = h(-z_2 + 6w_2 + \sin x_2) = 1.73846616$

$k_2 = h(z_2 + l_1) = 0.43364145$

$l_2 = h(-(z_2 + l_1) + 6(w_2 + k_1) + \sin x_3) = 1.72364835$

$w_3 = w_2 + \frac{1}{2}(k_1 + k_2) = 3.52181328$

$z_3 = z_2 + \frac{1}{2}(l_1 + l_2) = 4.32900557$

A table illustrates the results

n	x_n	u_n	w_n	| error |
0	1	3	3	0
1	1.1	2.995 565 34	2.999 207 36	$3.642 01 \times 10^{-3}$
2	1.2	3.171 968 87	3.175 095 14	$3.126 27 \times 10^{-3}$
3	1.3	3.522 513 67	3.521 813 28	$7.003 97 \times 10^{-3}$

The justification of the Runge-Kutta algorithm involves a marvellous piece of simplification. All the variables are written as components of two dimensional vectors and from this it is shown that every second (or higher) order differential equation may be written as a first order vector differential equation. To see this, suppose that the second order differential equation under consideration is written in the form

$$\frac{d^2y}{dx^2} = f\left(x, y, \frac{dy}{dx}\right)$$

In the example above

$$\frac{d^2y}{dx^2} + \frac{dy}{dx} - 6y = \sin x$$

so to write this equation in the required form,

write $$\frac{d^2y}{dx^2} = -\frac{dy}{dx} + 6y + \sin x = f\left(x, y, \frac{dy}{dx}\right)$$

We have the substitutions

$$u = y$$

$$v = \frac{dy}{dx}$$

leading to

$$u' = \frac{dy}{dx} = v$$

$$v' = \frac{d^2y}{dx^2} = f\left(x, y, \frac{dy}{dx}\right)$$

$$= f(x, u, v)$$

Now, write these equations in the more general form

$$u' = f_1(x, u, v)$$

$$v' = f_2(x, u, v)$$

where, in the above equations,

$$f_1(x, u, v) = v$$

and $$f_2(x, u, v) = f(x, u, v)$$

To express these quantities vectorially, write

$$\mathbf{u} = \begin{pmatrix} u \\ v \end{pmatrix} \quad \text{and} \quad \mathbf{f}(x, u, v) = \begin{pmatrix} f_1(x, u, v) \\ f_2(x, u, v) \end{pmatrix}$$

Then $\mathbf{u}' = \begin{pmatrix} u' \\ v' \end{pmatrix}$

$$= \begin{pmatrix} f_1(x, u, v) \\ f_2(x, u, v) \end{pmatrix}$$

or $\mathbf{u}' = \mathbf{f}(x, u, v)$

In fact, we go one step further and write

$$\mathbf{u}' = \mathbf{f}(x, \mathbf{u})$$

If the initial conditions are $y(a) = u_0$, $y'(a) = v_0$, then $u(a) = u_0$ and $v(a) = v_0$ and the initial conditions also may be written in vector form as

$$\mathbf{u}(a) = \begin{pmatrix} u_0 \\ v_0 \end{pmatrix} = \mathbf{u}_0$$

Putting all this together, we have the following rather impressive result.

> Given the second order differential equation
>
> $$\frac{d^2y}{dx^2} = f\left(x, y, \frac{dy}{dx}\right)$$
>
> with initial conditions $y(a) = u_0$ $y'(a) = v_0$
>
> write $u = y$ $v = \dfrac{dy}{dx}$ and $\mathbf{u} = \begin{pmatrix} u \\ v \end{pmatrix}$
>
> Then the given differential equation may be written as the first order vector differential equation
>
> $$\mathbf{u}' = \mathbf{f}(x, \mathbf{u})$$
>
> with initial condition $\mathbf{u}(a) = \mathbf{u}_0$

Observe the very great similarity between this differential equation and the first order differential equation

$$y' = f(x, y) \qquad y(a) = y_0$$

which we considered earlier (section 8.2).

This suggests that if in each first order technique, y is replaced by the vector $\mathbf{u}$ and $f(x, y)$ is replaced by the vector $\mathbf{f}(x, \mathbf{u})$ the technique will follow through to be applicable to a second (or higher) order differential equation. And so it turns out to be.

This technique will be applied to Euler's method and to the simple Runge-Kutta method. In the process, we will see the justification for the algorithms for these methods given above. First, however, more notation and some algebra.

Notation

Write the approximations w_n, z_n in vector form as $\mathbf{w}_n = \begin{pmatrix} w_n \\ z_n \end{pmatrix}$

then we have

$$\mathbf{f}(x_n, \mathbf{w}_n) = \begin{pmatrix} \mathrm{f}_1(x_n, \mathbf{w}_n) \\ \mathrm{f}_2(x_n, \mathbf{w}_n) \end{pmatrix}$$

$$= \begin{pmatrix} \mathrm{f}_1(x_n, w_n, z_n) \\ \mathrm{f}_2(x_n, w_n, z_n) \end{pmatrix}$$

But $\quad \mathrm{f}_1(x_n, w_n, z_n) = z_n$

and $\quad \mathrm{f}_2(x_n, w_n, z_n) = \mathrm{f}(x_n, w_n, z_n)$

so $\quad \mathbf{f}(x_n, \mathbf{w}_n) = \begin{pmatrix} z_n \\ \mathrm{f}(x_n, w_n, z_n) \end{pmatrix}$

Using this equation, we consider:

Euler's method

First order		**Second order**	
$y' = \mathrm{f}(x, y)$	$y(a) = y_0$	$\mathbf{u}' = \mathbf{f}(x, \mathbf{u})$	$\mathbf{u}(a) = \mathbf{u}_0$
$w_{n+1} = w_n + h\mathrm{f}(x_n, w_n)$		$\mathbf{w}_{n+1} = \mathbf{w}_n + h\mathbf{f}(x_n, \mathbf{w}_n)$	

The difference equation $\mathbf{w}_{n+1} = \mathbf{w}_n + h\mathbf{f}(x_n, \mathbf{w}_n)$ may be written

$$\begin{pmatrix} w_{n+1} \\ z_{n+1} \end{pmatrix} = \begin{pmatrix} w_n \\ z_n \end{pmatrix} + h \begin{pmatrix} z_n \\ \mathrm{f}(x_n, w_n, z_n) \end{pmatrix}$$

or

$$w_{n+1} = w_n + hz_n$$
$$z_{n+1} = z_n + h\mathrm{f}(x_n, w_n, z_n)$$

which are precisely the difference equations established earlier.

The simple Runge-Kutta method

First order		**Second order**	
$y' = \mathrm{f}(x, y)$	$y(a) = y_0$	$\mathbf{u}' = \mathbf{f}(x, \mathbf{u})$	$\mathbf{u}(a) = \mathbf{u}_0$
$k_1 = h\mathrm{f}(x_n, w_n)$		$\mathbf{k}_1 = h\mathbf{f}(x_n, \mathbf{w}_n)$	
$k_2 = h\mathrm{f}(x_n + h, w_n + k_1)$		$\mathbf{k}_2 = h\mathbf{f}(x_n + h, \mathbf{w}_n + \mathbf{k}_1)$	
$w_{n+1} = w_n + \frac{1}{2}(k_1 + k_2)$		$\mathbf{w}_{n+1} = \mathbf{w}_n + \frac{1}{2}(\mathbf{k}_1 + \mathbf{k}_2)$	

Observe that, in the second order equations, $\mathbf{k}_1$ and $\mathbf{k}_2$ are vectors.

Write $\quad \mathbf{k}_1 = \begin{pmatrix} k_1 \\ l_1 \end{pmatrix} \quad$ and $\quad \mathbf{k}_2 = \begin{pmatrix} k_2 \\ l_2 \end{pmatrix}$

Since $\quad \mathbf{k}_1 = \mathrm{h}\mathbf{f}(x_n, \mathbf{w}_n)$,

we have $\quad \begin{pmatrix} k_1 \\ l_1 \end{pmatrix} = h \begin{pmatrix} z_n \\ \mathrm{f}(x_n, w_n, z_n) \end{pmatrix}$

giving $\quad k_1 = hz_n$

and $\quad l_1 = h\mathrm{f}(x_n, w_n, z_n)$

Observe now that since $\mathbf{w}_n + \mathbf{k}_1 = \begin{pmatrix} w_n \\ z_n \end{pmatrix} + \begin{pmatrix} k_1 \\ l_1 \end{pmatrix} = \begin{pmatrix} w_n + k_1 \\ z_n + l_1 \end{pmatrix}$

we have $$\mathbf{f}(x_n + h, \mathbf{w}_n + \mathbf{k}_1) = \begin{pmatrix} f_1(x_n + h, \mathbf{w}_n + \mathbf{k}_1) \\ f_2(x_n + h, \mathbf{w}_n + \mathbf{k}_1) \end{pmatrix}$$

$$= \begin{pmatrix} f_1(x_n + h, w_n + k_1, z_n + l_1) \\ f_2(x_n + h, w_n + k_1, z_n + l_1) \end{pmatrix}$$

$$= \begin{pmatrix} z_n + l_1 \\ f(x_n + h, w_n + k_1, z_n + l_1) \end{pmatrix}$$

Since $\mathbf{k}_2 = h\mathbf{f}(x_n + h, \mathbf{w}_n + \mathbf{k}_1)$

we have $$\begin{pmatrix} k_2 \\ l_2 \end{pmatrix} = h \begin{pmatrix} z_n + l_1 \\ f(x_n + h, w_n + k_1, z_n + l_1) \end{pmatrix}$$

giving $k_2 = h(z_n + l_1)$

and $l_2 = hf(x_n + h, w_n + k_1, z_n + l_1)$

Now $\mathbf{w}_{n+1} = \mathbf{w}_n + \frac{1}{2}(\mathbf{k}_1 + \mathbf{k}_2)$ becomes

$$\begin{pmatrix} w_{n+1} \\ z_{n+1} \end{pmatrix} = \begin{pmatrix} w_n \\ z_n \end{pmatrix} + \frac{1}{2}\left[\begin{pmatrix} k_1 \\ l_1 \end{pmatrix} + \begin{pmatrix} k_2 \\ l_2 \end{pmatrix}\right]$$

or $$w_{n+1} = w_n + \tfrac{1}{2}(k_1 + k_2)$$
$$z_{n+1} = z_n + \tfrac{1}{2}(l_1 + l_2)$$

as given in Algorithm 8.6.

This technique may be extended to the Runge-Kutta method of order four and an algorithm is provided for this method. The algebraic details are rather tedious and are omitted.

Algorithm 8.7 The Runge-Kutta method of order four

To obtain approximately the solution of the differential equation

$\frac{d^2y}{dx^2} = f\left(x, y, \frac{dy}{dx}\right)$ with initial conditions $y(a) = u_0$ $y'(a) = v_0$.

input: initial value of $x = x_0$;
initial value of $y = u_0$;
initial value of $\frac{dy}{dx} = v_0$;
step size h

$w_0 := u_0$
$z_0 := v_0$

for $n = 0, 1, ...,$ until satisfied

$k_1 := hz_n$

$l_1 := hf(x_n, w_n, z_n)$

$k_2 := h(z_n + \frac{1}{2}l_1)$

$l_2 := hf(x_n + \frac{1}{2}h, w_n + \frac{1}{2}k_1, z_n + \frac{1}{2}l_1)$

$k_3 := h(z_n + \frac{1}{2}l_2)$

$l_3 := hf(x_n + \frac{1}{2}h, w_n + \frac{1}{2}k_2, z_n + \frac{1}{2}l_2)$

$k_4 := h(z_n + l_3)$

$l_4 := hf(x_n + h, w_n + k_3, z_n + l_3)$

$w_{n+1} := w_n + \frac{1}{6}(k_1 + 2k_2 + 2k_3 + k_4)$

$z_{n+1} := z_n + \frac{1}{6}(l_1 + 2l_2 + 2l_3 + l_4)$

$x_{n+1} := x_0 + (n + 1)h$

endloop

output: the approximate values of the solution $w_0, w_1, w_2, ...;$

the approximate values of $\frac{dy}{dx}$ $z_0, z_1, z_2, ...$

The Runge-Kutta method of order four is illustrated on the, by now standard, differential equation

$$\frac{d^2y}{dx^2} + \frac{dy}{dx} - 6y + \sin x \qquad y(1) = 3 \qquad y'(1) = -1$$

Because the method is really only practical when a computer is available, only the results of the calculation are shown.

n	x_n	u_n	w_n	| error |
0	1	3	3	0
1	1.1	2.995 565 34	2.995 588 5	$2.315 45 \times 10^{-5}$
2	1.2	3.171 968 87	3.171 998 69	$2.982 09 \times 10^{-5}$
3	1.3	3.522 513 67	3.522 538 58	$2.490 73 \times 10^{-5}$

It is clear that this represents a considerable improvement in accuracy over the Runge-Kutta method of order two and a remarkable improvement over Euler's method. This demonstrates, it is hoped conclusively, that, by exercising a measure of mathematical intelligence, quite dramatic improvements in accuracy can be made.

One of the author's objectives in writing this book was to establish this very point. It is hoped that the twin aims of numerical analysis which are to provide methods of calculation that

(a) achieve the desired level of accuracy, and

(b) are economic in the time and effort required for their application

have been understood and appreciated and that the case for the study of numerical analysis has been made.

There is more, much more, that could have been written and it is the author's hope that this book may have given the reader the urge to explore further in this exciting and rapidly expanding subject.

Exercise 8

Introduction

1 Describe, in general terms
(a) a solution
(b) the general solution
(c) a particular solution
of a differential equation.

2 Show that $y = x^2 + 5x$ is a solution of the differential equation $x\dfrac{dy}{dx} - 2y + 5x = 0$. Show that $y = x^2 + 5x + k$, where k is a constant, is the general solution of this differential equation. If it is known that $y = 3$ when $x = 1$, find a particular solution for this differential equation.

3 Show that $y = x\tan x + c$ is the general solution of the differential equation $\dfrac{dy}{dx} - x\sec^2 x = \tan x$.
If $y = 0$ when $x = 0$, find a particular solution.

4 Describe the difference between an analytical solution of a differential equation and a numerical solution.

Section 8.2

5 Write down the difference equation associated with Euler's method for the differential equation

$$\frac{dy}{dx} = 3x - 2y \qquad y(1) = 1$$

Use the difference equation to calculate approximately the value of the solution of the differential equation when $x = 1.1$, $x = 1.2$, $x = 1.3$ and $x = 1.4$.

6 Use Euler's method to obtain a numerical solution of the differential equation

$$\frac{dy}{dx} = 4x^2 + 3\ln x \qquad y = 10 \text{ when } x = 2$$

at the points $x = 2.1, 2.2, 2.3, 2.4, 2.5$.
Plot the values you obtain on x- y-axes. Join the points and so obtain the graph of your (approximate) solution.

7 (a) Use Euler's method to calculate approximately the solution of the differential equation

$$\frac{dy}{dx} = y + e^x \qquad y(0) = 0.5$$

when $x = 0.1$, $x = 0.2$, $x = 0.3$, $x = 0.4$. Take step size $h = 0.1$.

(b) Now take step size $h = 0.05$ and calculate again the approximate solution at $x = 0.1, 0.2, 0.3, 0.4$.
The analytical solution of the differential equation is $y = (x + \frac{1}{2})e^x$. Comment on the accuracy of your two sets of solutions.

8 Use Euler's method to calculate approximately the solution of the differential equation

$$\frac{dy}{dx} = x^2 + 3y \qquad y(0) = 2$$

when $x = 0.1$, $x = 0.2$, $x = 0.3$, $x = 0.4$. Use:
(a) $h = 0.1$
(b) $h = 0.05$
The analytical solution is $y = -\frac{1}{3}x^2 - \frac{2}{9}x - \frac{2}{27} + \frac{56}{27}e^{3x}$.
Calculate the errors in your numerical solution and comment on your results.

9 The differential equation

$$x\frac{dy}{dx} + y = 10 \qquad y = 8 \text{ when } x = 5$$

has the solution $y = 10 - \dfrac{10}{x}$. Plot the graph $y = 10 - \dfrac{10}{x}$ for $5 \leqslant x \leqslant 5.5$.

(a) Calculate approximations to the solution of the differential equation at $x = 5$, 5.1, 5.2, 5.3, 5.4, 5.5 using Euler's method with step size $h = 0.1$. Plot these points on your graph.
(b) Now calculate approximations to the solution using step size $h = 0.05$. Plot these points on your graph also. Comment on your results for (a) and (b).

10 Use Euler's method to calculate approximately the value of the solution of the differential equation

$$\frac{dy}{dx} = y - x + 10 \qquad y = 4.5 \qquad \text{when} \qquad x = 1$$

at the points where $x = 1.5$, $x = 2$, $x = 2.5$, $x = 3$, $x = 3.5$. Experiment with several step sizes.
The analytical solution is $y = -9 + x + 12.5e^{x-1}$. Calculate the errors in your approximations and comment on your solution.

11 It is required to obtain the numerical solution of the differential equation

$$\frac{dy}{dx} = 2x^2y \qquad y(3) = 1$$

at the points $x = 3.2$, $x = 3.4$, $x = 3.6$, $x = 3.8$, $x = 4$.
Investigate the solutions obtained using Euler's method with several different step sizes. It is known that the solution of the differential equation is $y = e^{\frac{2}{3}x^3 - 18}$.
Calculate the error in your answers and comment.

12 Obtain the approximate solution of the differential equation

$$\frac{dy}{dx} = xy \qquad y(5) = 2$$

at the points $x = 5.1$, $x = 5.2$, $x = 5.3$, $x = 5.4$ and $x = 5.5$ using Euler's method.

It is required to know the value of y when
(a) $x = 5.17$ and (b) $x = 5.32$. Suggest how these values may be calculated.

13 If $\dfrac{dy}{dt} = \dfrac{1}{t^2} - \dfrac{y}{t} - y^2 \quad y(1) = -1$

use Euler's method (a) with $h = 0.1$ (b) with $h = 0.05$ to approximate the solution $y(t)$ for $1 \leqslant t \leqslant 2$. Use your answers and polynomial interpolation to calculate approximately the values of (a) $y(1.051)$ (b) $y(1.555)$ (c) $y(1.9978)$. Compare your answers with the exact values of the solution given by $y = -\dfrac{1}{t}$.

14 Use Euler's method and a step size of 0.1 to obtain an approximate numerical solution of the differential equation

$$\frac{dy}{dx} = -2y \qquad y(1) = 1$$

at the points $x = 1.1, 1.2, 1.3$. The exact solution of the differential equation is $y = e^{2(1-x)}$. Calculate the error in your answers.
Write down an expression for the discretisation error for this method and calculate the greatest value that the discretisation error can achieve in the interval $1 \leqslant x \leqslant 1.1$. Without using the exact solution, write down an interval within which $y(1.1)$ must lie. Check that your interval does contain the value $y(1.1)$.

Section 8.4

15 Use the approximation

$$\frac{dy}{dx} \approx \frac{y_{n+1} - y_{n-1}}{2h}$$

to obtain a difference equation for the differential equation

$$\frac{dy}{dx} = \frac{4e^x}{y} \qquad y = 1 \quad \text{when} \quad x = 1$$

It is given that $w_0 = 1$. Why is this information insufficient to obtain (approximately) the solution when $x = 1.1, 1.2, 1.3, 1.4$ and 1.5?
Use a Taylor polynomial of degree one to calculate approximately the value $y(1.1)$. Taking this to be the value of w_1 calculate approximately the values of (a) $y(1.2)$ (b) $y(1.3)$ (c) $y(1.4)$ and (d) $y(1.5)$. The analytical solution is given by $y^2 = 8(e^x - e) + 1$. How could the accuracy be improved?

16 It is required to use a numerical technique to solve the differential equation

$$\frac{dy}{dx} = y - 2x \qquad y = 2 \quad \text{when} \quad x = 1$$

at $x = 1.2$, $x = 1.4$, $x = 1.6$. Show that the approximation

$$\frac{dy}{dx} \approx \frac{y_{n+1} - y_{n-1}}{2h}$$

leads to the difference equation

$$w_{n+1} - 0.4w_n + 0.8x_n - w_{n-1} = 0$$

If $w_0 = 2$, $h = 0.2$, use Taylor polynomials (a) of degree one (b) of degree two (c) of degree three to calculate w_1.
For each value of w_1 use the difference equation to calculate w_2, w_3.
If the analytical solution of the differential equation is $y = 2x + 2 - 3e^{x-1}$ compare the accuracy of your approximations in (a), (b) and (c) above.

Section 8.5

17 Use the Taylor method of order two to obtain approximately the solution of the differential equation

$$\frac{dy}{dx} = x^2 + y^2 \qquad y(2) = 2$$

at the points $x = 2.1, 2.2, 2.3$.

18 Complete the table, using the Taylor method of order two

x	1	1.4	1.5
y	0		

where the relationship between x and y is given by

$$\frac{dy}{dx} = x + \sin y \qquad y(1) = 0$$

19 Obtain numerically the solution of the differential equation

$$\frac{dy}{dx} = 3x + 4y \qquad y(0) = 0$$

at $x = 0.2, 0.4, 0.6, 0.8, 1$ using the Taylor method of order three.
The analytical solution for this differential equation is

$$y = -\tfrac{3}{4}x - \tfrac{3}{16} + \tfrac{3}{16}e^{4x}$$

Calculate the errors in your solution and comment.

20 It is required to obtain a numerical solution for the differential equation

$$\frac{dy}{dx} = \frac{x}{y} \qquad y(3) = 1$$

when $x = 3.2$ and when $x = 3.4$. Use the Taylor method of order two with varying step sizes to calculate approximately the values of y(3.2) and y(3.4).
The analytical solution is given by $y^2 = x^2 - 8$. Calculate the error in your answers and comment.

21 Use the Taylor method (a) of order one (b) of order two (c) of order three to calculate approximately the solution of the differential equation

$$5\frac{dy}{dx} = 3x^2 - 2y \qquad y(5) = 1$$

at the points given by $x = 5.1$, $x = 5.2$, $x = 5.3$.

22 Use (a) Euler's method (b) the Taylor method of order two to obtain numerically the solution of the differential equation

$$\frac{dy}{dx} = x(1 - y) \qquad y(-1) = 2$$

for the values of x, $x = -0.9$, $x = -0.8$, $x = -0.7$, $x = -0.6$ and $x = -0.5$.

The exact solution is $y = 1 + e^{\frac{1}{2}(1-x^2)}$. Calculate the error in your answers and comment on your results.

23 A mathematician wishes to know the solution of the differential equation

$$x\frac{dy}{dx} = (1 + x)y \qquad y(1) = 1$$

when $x = 1.4$. Use:
(a) Euler's method with step size 0.1
(b) The Taylor method of order two with step size 0.2
(c) The Taylor method of order four with step size 0.4
to calculate the value of y(1.4). The analytical solution is $y = xe^{x-1}$. Comment on your solutions.

24 Consider the differential equation

$$\frac{dy}{dx} = \sqrt{y} \qquad y = 9 \quad \text{when} \quad x = 5$$

Experiment with different step sizes and obtain the solution to this differential equation when $x = 5.1$, $x = 5.2$, $x = 5.3$, $x = 5.4$ and $x = 5.5$ using
(a) Euler's method
(b) the approximation $\dfrac{dy}{dx} \approx \dfrac{y_{n+1} - y_{n-1}}{2h}$ and estimating y(0.5) as accurately as possible.
(c) The Taylor method of order two

The analytical solution is $2\sqrt{y} = x + 1$. Compare the accuracy of your methods.

Section 8.6

25 Using the simple Runge-Kutta method of order two, obtain at the points indicated, approximate solutions to the following differential equations

(a) $\dfrac{dy}{dx} = 2xy \qquad y(1) = 3 \qquad \text{at} \qquad x = 1.2, 1.4, 1.6, 1.8, 2$

(b) $\dfrac{dy}{dx} = 4x^2 + 2xy \qquad y(0) = 4 \qquad \text{at} \qquad x = 0.1, 0.2, 0.3, 0.4, 0.5$

(c) $\dfrac{dy}{dx} = \dfrac{5}{\sqrt{x^2 + y^2}} \qquad y(5) = 0.35 \qquad \text{at} \qquad x = 5.2, 5.4, 5.6, 5.8, 6$

(d) $\dfrac{dy}{dx} = 3x^2e^{-y} + 2 \qquad y(-3) = -1 \qquad \text{at} \qquad x = -2.9, -2.8, -2.7, -2.6$

26 Repeat Questions 7, 8, 9, 10, 11, 13 using the Runge-Kutta method or order two instead of Euler's method.

27 Use the simple Runge-Kutta method of order two and several different step sizes to obtain the solution of the differential equation

$$\frac{dy}{dx} = xy^2 \qquad y(1) = 2$$

at the points given by $x = 1, 1.1, 1.2, 1.3, 1.4$.

The analytical solution is $y = \dfrac{2}{2 - x^2}$. Calculate the error in your answers.

28 Use (a) Euler's method (b) the Runge-Kutta method of order two to obtain the approximate solution of the differential equation

$$\frac{dy}{dx} = x + 2y - 6 \qquad y(0) = 3$$

when $x = 0.2, 0.4, 0.6, 0.8, 1$. Experiment with different step sizes. The analytical solution is $y = \frac{1}{4}e^{2x} - \frac{1}{2}x + \frac{11}{4}$. Comment on the accuracy of your solutions.

29 Use Heun's method to calculate approximately the solution of the differential equation

$$\frac{dy}{dx} = \sqrt{x}\ln(y + 1) + 0.5 \qquad y(1) = 1$$

at the points given by $x = 1.2, 1.4, 1.6, 1.8, 2$.

30 Use Heun's method to obtain approximately the solution of the differential equation

$$\frac{dy}{dx} = \frac{x}{y} \qquad y(1) = 10$$

at the points where $x = 1.2, x = 1.4, x = 1.6, x = 1.8, x = 2$.

The exact solution of the differential equation is $y = \sqrt{x^2 + 99}$.

Plot the graphs of the exact solution and the calculated solution. Comment on your graphs.

31 Use Heun's method to solve approximately the differential equation

$$\frac{dy}{dx} = x^2 + 3y - 2 \qquad y(0) = 1\tfrac{16}{27}$$

when $x = 0.2, x = 0.4, x = 0.6, x = 0.8, x = 1$.

The analytical solution is $y = e^{3x} - \frac{1}{3}x^2 - \frac{2}{9}x + \frac{16}{27}$.

Draw the graphs of your approximate solution and of the analytical solution. Comment on your results.

32 Use

(a) Euler's method

(b) the Runge-Kutta method of order two

(c) the Taylor method of order three

to solve numerically the differential equation

$$\frac{dy}{dx} = (xy)^2 \qquad y(0.5) = 4$$

at the points given by $x = 0.6, 0.7, 0.8, 0.9$.

The exact solution is given by $y = \dfrac{24}{7 - 8x^3}$.

Compare your approximations with the exact values.

33 Use

(a) the simple Runge-Kutta method of order two with step size 0.1

(b) the Taylor method of order four with step size 0.2

to obtain a numerical solution for the differential equation

$$\frac{dy}{dx} = 2y - x \qquad y(0) = -2$$

at the points given by $x = 0.2$, $x = 0.4$, $x = 0.6$, $x = 0.8$, $x = 1$.
The exact solution is $y = \frac{1}{2}x + \frac{1}{4} - \frac{9}{4}e^{2x}$. Comment on the accuracy and the efficiency of the methods.

34 Use the Runge-Kutta method of order four to obtain approximately the solution of the following differential equations at the points indicated.

(a) $\dfrac{dy}{dx} + 2y = e^{-2x} \qquad y(1) = 10 \qquad \text{at} \qquad x = 1.1, 1.2, 1.3, 1.4, 1.5$

(b) $\dfrac{dy}{dx} = e^x - \dfrac{y}{x} \qquad y(4) = 1.5 \qquad \text{at} \qquad x = 4.2, 4.4, 4.6, 4.8, 5$

35 Using the Runge-Kutta method of order four obtain numerically the solution of the differential equation

$$\frac{dy}{dx} = \frac{y + x - 1}{x} \qquad y(2) = -3$$

at $x = 2.1, 2.2, 2.3, 2.4, 2.5$.
The analytical solution of this differential equation is $y = x\ln x + 1 - 2.693\,147\,181x$.
Calculate the errors in the numerical values of your solution.

36 The current i in an electric circuit of resistance R and self induction L satisfies the differential equation

$$L\frac{di}{dt} + Ri = 40\sin 100t \qquad i = 0 \qquad \text{when} \qquad t = 0$$

If $L = 0.20$ and $R = 20$, use the Runge-Kutta method of order four to calculate approximately the value of i when $t = 0.15$.

37 Use the Runge-Kutta method of order four to obtain a numerical solution of the differential equation

$$\frac{dy}{dx} = \frac{x - xy}{1 + x^2} \qquad y(5) = 3$$

at the points where $x = 5.1$, $x = 5.2$, $x = 5.3$, $x = 5.4$, $x = 5.5$. It is required to estimate the solution when $x = 5.24$. Use polynomial interpolation to estimate the value of the solution at this point.

Check your answers with the analytical solution $y = 1 + \dfrac{2\sqrt{26}}{\sqrt{1 + x^2}}$.

38 Use the Runge-Kutta method of order four to obtain an approximation to the solution of the differential equation

$$\frac{dy}{dx} = (xy)^2 \qquad y(0.5) = 4$$

at the points where $x = 0.6, 0.7, 0.8, 0.9$.
Compare your solution with those obtained in Question 32.

39 Use
(a) Euler's method
(b) the Runge-Kutta method of order two
(c) the Runge-Kutta method of order four
to obtain numerically, the solution of the differential equation

$$\frac{dy}{dt} = \frac{1}{t^4} - \frac{2y}{t} - y^2 \qquad y(1) = 1$$

at the points where $t = 1.2, 1.4, 1.6, 1.8, 2$.
Take, successively, $h = 0.2$, $h = 0.1$, $h = 0.05$.
The exact solution is $y = -\frac{1}{t^2}$. Comment on your solutions.

40 Use
(a) Euler's method
(b) a Runge-Kutta method of order two
(c) the Runge-Kutta method of order four
all with step size 0.2 to obtain a numerical solution to the differential equation

$$\frac{dy}{dx} = -100y \qquad y(1) = 1$$

at $x = 1.2$, $x = 1.4$, $x = 1.6$.
Comment on your solutions. Experiment with different step sizes. Describe your findings.

Section 8.7

41 Write each of the following second order differential equations as a system of two first order differential equations:

(a) $\frac{d^2y}{dx^2} - 4\frac{dy}{dx} + 2y = x \qquad y(3) = -1 \qquad y'(3) = 0$

(b) $\frac{d^2y}{dx^2} + 2\frac{dy}{dx} - 7y = \ln x + 1 \qquad y(-1) = 2 \qquad y'(-1) = 3$

(c) $x\frac{d^2y}{dx^2} + 3y^2 = 1 \qquad y(50) = 0.2 \qquad y'(50) = 0.4$

(d) $\sqrt{(x-1)}\frac{d^2y}{dx^2} + x\frac{dy}{dx} - 5y^2 = 0 \qquad y(10) = 1.4 \qquad y'(10) = -0.2$

42 Use Euler's method to obtain the solutions of the following differential equations at the points indicated.

(a) $\dfrac{d^2y}{dx^2} - \dfrac{dy}{dx} - 2y = -(4x + 2)$ $\quad y(0) = 2 \quad y'(0) = 9$

at $x = 0.1, 0.2, 0.3, 0.4, 0.5$.

(b) $\dfrac{d^2y}{dx^2} + 3\dfrac{dy}{dx} - 4y = 5e^x$ $\quad y(0) = 0 \quad y'(0) = 1$

at $x = 0.2, 0.4, 0.6, 0.8, 1$.

(c) $x\dfrac{d^2y}{dx^2} - \dfrac{dy}{dx} = 2x$ $\quad y(1) = 0 \quad y'(1) = 1$

at $x = 1.3, 1.6, 1.8, 2.1, 2.4$.

(d) $x\dfrac{d^2y}{dx^2} - 2\dfrac{dy}{dx} + xy = -2\sin x$ $\quad y\left(\dfrac{\pi}{4}\right) = \dfrac{\pi}{4\sqrt{2}} \quad y'\left(\dfrac{\pi}{4}\right) = \dfrac{\pi + 4}{4\sqrt{2}}$

at $x = \dfrac{\pi}{4} + 0.1,\ \dfrac{\pi}{4} + 0.2,\ \dfrac{\pi}{4} + 0.3,\ \dfrac{\pi}{4} + 0.4,\ \dfrac{\pi}{4} + 0.5$ (radians).

(e) $(x^2 + 1)\dfrac{d^2y}{dx^2} + 4x\dfrac{dy}{dx} + 2y = 0$ $\quad y(1) = 0.5 \quad y'(1) = -0.5$

at $x = 1.1, 1.2, 1.3, 1.4, 1.5$.

43 Repeat Question 42 using the simple Runge-Kutta method of order two.

44 Repeat Question 42 using the Runge-Kutta method of order four.

45 The analytical solutions of the differential equations in Questions 42 are:

(a) $y = 3e^{2x} - e^{-x} + 2x$

(b) $y = xe^x$

(c) $y = x^2 \ln x$

(d) $y = x \sin x$

(e) $y = \dfrac{1}{(x^2 + 1)}$

Calculate the errors in your approximations in Questions 42, 43 and 44 and comment (i.e. be impressed by the improvement in accuracy).

46 Use

(a) Euler's method

(b) the simple Runge-Kutta method

(c) the Runge-Kutta method of order four

to obtain the approximate solution of the differential equation

$$\frac{d^2y}{dx^2} - 2\frac{dy}{dx} + y = 2e^x \qquad y(3) = 1 \qquad y'(3) = 1$$

at the points given by $x = 3.2, 3.4, 3.6, 3.8, 4$.

The exact solution of the differential equation is

$$y = [(9 + e^{-3}) - 6x + x^2]e^x$$

Comment on the accuracy achieved by the three methods.

47 Consider the differential equation

$$\frac{d^2y}{dx^2} - \frac{dy}{dx} - 2y = 1 + x \qquad y(0) = 1 \qquad y'(0) = 0.25$$

The exact solution is $y = \frac{2}{3}e^{2x} + \frac{7}{12}e^{-x} - \frac{x}{2} - \frac{1}{4}$.

Using any numerical method or technique, obtain the values of the solution at the points given by $x = 0.1, 0.2, 0.3, 0.4, 0.5$ as accurately as you can. Calculate the errors in your solution.

48 Use
(a) the simple Runge-Kutta method
(b) the Runge-Kutta method of order four
to obtain approximations to the solution of the differential equation

$$\frac{d^2y}{dx^2} + 101\frac{dy}{dx} + 100y = 0 \qquad y(0) = 1 \qquad y'(0) = -1$$

at the points $x = 0.1, 0.2, 0.3, ..., 0.9, 1$.
Take $h = 0.1$ and observe that the numerical solution bears no resemblance to the analytical solution $y = e^{-x}$.
Repeat your calculations with $h = 0.05$ and with $h = 0.01$. Comment.

49 Consider the differential equation

$$\frac{d^2y}{dx^2} + 51\frac{dy}{dx} + 49y = 0 \qquad y(0) = 0 \qquad y'(0) = 2$$

It is required to obtain a numerical solution at the points given by $x = 0.2, 0.4, 0.6, 0.8, 1$. Using a numerical method of your choice and different values of h, calculate approximations to the solution at these points.
The analytical solution is given by $y = Ae^{-0.979\,600\,33x} - Ae^{-50.020\,399\,65x}$ where $A = 0.040\,782\,369\,53$. Comment on the accuracy of your solution.

Miscellaneous

50 Let $y = f(x)$ represent a function of x. Let $x_0 = 1$ and write $x_n = x_0 + nh$ and $y_n = f(x_n)$.
(a) Indicate why $\frac{y_{n+1} - y_n}{h}$ may be taken to be an approximation on $\frac{dy}{dx}$ when $x = x_n$. Comment on the accuracy of this approximation.
(b) Suppose now that this approximation is used to obtain approximately the numerical solution of the differential equation

$$\frac{dy}{dx} = 2x + y^2 \qquad \text{where} \qquad y(1) = 2$$

when $x = 1.05, 1.1, 1.15, 1.2$.
Find the value of the solution generated in this way and comment on the errors in your answers.

51 It is required to obtain the solution of the equation

$$\frac{dy}{dx} = \sqrt{xy^2} \qquad y(1) = 1 \qquad \text{at the point } x = 1.364.$$

Use Euler's method (with $h = 0.1$) to calculate the value of the solution at $x = 1.1, 1.2, 1.3, 1.4$ and 1.5. Then use polynomial interpolation to obtain the required answer as accurately as you can.

52 Consider the differential equation

$$\frac{dy}{dx} = 2x^2y \qquad y(2) = 1$$

(a) Use Euler's method (with different values taken for h) to obtain the solution when $x = 2.1, 2.2, 2.3$.

(b) Consider a more accurate method using Taylor polynomials.

Write $\quad y(a + h) \approx y(a) + hy'(a) + \dfrac{h^2}{2!}y''(a)$

Then, if $\quad a = 2 \quad$ and $\quad h = 0.1$

$$y(2.1) \approx y(2) + (0.1)y'(2) + \frac{(0.1)^2}{2}y''(2)$$

From the initial conditions $\quad y(2) = 1$

From the differential equation $\quad \dfrac{dy}{dx} = 2x^2y$

we get $\quad y'(2) = 2 \times 2^2 \times 1 = 8$

By differentiating the differential equation

$$\frac{d^2y}{dx^2} = 2\left(x^2\frac{dy}{dx} + 2xy\right)$$

we get $\quad y''(2) = 2(2^2 \times 8 + 2 \times 2 \times 1)$
$\qquad\qquad = 72$

Hence $\quad y(2.1) \approx 1 + 0.1 \times 8 + \dfrac{(0.1)^2}{2} \times 72 = 2.16$

In a similar way, estimate (a) $y(2.2)$ and (b) $y(2.3)$.

How could the accuracy in this method be improved?

53 Use the simple Runge-Kutta method to solve the differential equation

$$\frac{dy}{dx} = 2x^2y \qquad y(2) = 1$$

at the points $x = 2.1, 2.2, 2.3$.
Compare your answers with those obtained in Question 52. Comment on your observations.

54 The solution of the differential equation

$$\frac{dy}{dx} = x^2 - \cos y$$

for which $y = 0$ when $x = 1$ is to be found numerically. Use a Taylor series method with $h = 0.2$, and keeping terms up to and including h^3, to find the values of y for $x = 1.2$, and $x = 1.4$, giving your answers to three places of decimals.

(C)

55 Find the solution of the differential equation

$$\frac{dy}{dx} = y^2$$

which is such that $y = 1$ when $x = 0$.
Establish the approximation

$$\left(\frac{dy}{dx}\right)_{x=x_n} = \frac{1}{h}(y_{n+1} - y_n)$$

where y_n and y_{n+1} are the ordinates at $x = x_n$ and $x = x_n + h$ respectively. Hence find an approximation to the solution of the equation

$$\frac{dy}{dx} = y^2$$

by a step-by-step method using a step length $h = 0.25$ and taking $x = 0$, 0.25, 0.5, 0.75, 1 and 1.25. Compare the results with those obtained from the exact solution.
Explain why, however small the step length h is chosen, this method cannot be used to give an accurate determination of y for values of x in the above range.
(MEI)

56 Write down the first two terms of the Taylor expansion of $y(x + h)$. Hence, or otherwise, show that

$$y'_r \approx \frac{1}{h}(y_{r+1} - y_r)$$

Use this approximation to estimate y(1.1) and y(1.2) if

$$y' = x^2 - \frac{y}{x} \qquad \text{and} \qquad y(1) = 1$$

Explain why, in general, the use of the approximation

$$y'_r \approx \frac{1}{2h}(y_{r+1} - y_{r-1})$$

is to be preferred. Use this second approximation to solve

$$y' = x^2 - \frac{y}{x}$$

for $x = 1$, 1.2, 1.4, 1.6, 1.8, 2 where y(1) = 1 and the estimate of y(1.2) is taken from the first approximation.
Compare the value of y(2) with that obtained from an exact solution of the differential equation.
(MEI)

57 Use (a) Euler's method (b) the simple Runge-Kutta method to obtain the approximate value of the solution of the differential equation

$$\frac{dy}{dx} = x^3 + y \qquad y(0) = 2$$

at the points $x = 0.2$, $x = 0.4$, $x = 0.6$, $x = 0.8$, $x = 1$. (In each method, use several values of h and aim to obtain a very accurate solution.)
The exact solution is $y = 8e^x - 6 - 6x - 3x^2 - x^3$.
Calculate the errors in your approximations and comment on the accuracy.

58 The differential equation

$$\frac{dy}{dx} = x(1 + y)$$

has initial condition y(6) = 0.

Use

(a) Heun's method

(b) the Runge-Kutta method of order four

to calculate numerically the solution of the differential equation at $x = 6.1, 6.2, 6.3, 6.4, 6.5$. For each method, experiment with different step sizes.

The analytical solution is $y = e^{\frac{1}{2}(x^2-36)} - 1$. Calculate the errors in your approximations and comment on the accuracy of the methods.

59 Solve, as accurately as you can, the differential equation

$$\frac{dy}{dx} = x^3 - y^3 \qquad y(2) = 1$$

at the points $x = 2.2$, $x = 2.4$, $x = 3$.

60 (a) Using the Taylor method of order two, obtain a numerical solution of the differential equation

$$\frac{dy}{dx} = \frac{1}{x^2} - \frac{2y}{x} \qquad y(1) = 1$$

at $x = 1.1$, $x = 1.2$, $x = 1.3$, $x = 1.4$ and $x = 1.5$.

The exact solution of the differential equation is $y = \dfrac{1}{x}$.

Calculate the error in your answers.

(b) Write down an expression for the error term in your method.

(c) Use the exact solution to calculate the greatest value that can be taken by the error in approximating the value of y(1.1).

Does your error lie within the limits predicted by the theory? How could you estimate the greatest error that might occur in calculating (approximately) the value of y(1.2)?

61 Variables x and y satisfy the differential equation

$$y'' = x^2y$$

Given that y(1) = 0, y(1.2) = 1, use the approximation

$$y'' \approx \frac{1}{h^2}\{y(x + h) - 2y(x) + y(x - h)\}$$

to estimate y(1.1), with (a) $h = 0.1$, (b) $h = 0.05$, working to six decimal places throughout.

Given that the exact solution y(x) of the differential equation and an approximate solution Y(x) using a step length h are approximately related by

$$y(x) = Y(x) + Ah^2$$

where A is independent of h, estimate a better value for y(1.1) by suitably combining the two approximate solutions already obtained.

(MEI)

62 Van der Pol's equation has the form

$$y'' - \mu(1 - y^2)y' + y = 0$$

where μ is a parameter such that $0 < \mu \leqslant 1$.
Given that $\mu = 0.1$, $y(0) = 1$ and $y'(0) = 0$ use an appropriate Taylor series expansion of $y(x)$ to show that $y(0.2) \approx 0.980$.
Using the approximations

$$y'_r \approx \frac{1}{2h}(y_{r+1} - y_{r-1}) \qquad \text{and} \qquad y''_r \approx \frac{1}{h^2}(y_{r+1} - 2y_r + y_{r-1})$$

find an approximate expression for y_{r+1} in terms of y_r, y_{r-1} and h which satisfies Van der Pol's equation with $\mu = 0.1$.
With the values of μ, y(0) and y(0.2) given above, tabulate the solution of Van der Pol's equation for values of x from 0 to 1.6, inclusive, in steps of 0.2.
Without further calculation indicate briefly how the accuracy of the solution could be assessed.

(MEI)

63 Use (a) Euler's method (b) the Runge-Kutta method of order two (c) the Runge-Kutta method of order four to obtain approximate values for the solution of the differential equation

$$\frac{d^2y}{dt^2} - 4y = 5e^{3t} \qquad y(0) = -2 \qquad y'(0) = -3$$

at $t = 0.1$, $t = 0.2$, $t = 0.3$, $t = 0.35$.
Use several step sizes and aim to produce a highly accurate solution.
The analytical solution is $y = e^{3t} - 3e^{2t}$. Calculate the errors in your answers and comment.

64 Obtain, as accurately as you can, the solution of the white dwarf equation

$$xy'' + 2y' + x(y^2 - 1)^{\frac{3}{2}} = 0 \qquad y(1) = 1 \qquad y'(1) = 1$$

when $x = 1.2, 1.4, 1.6, 1.8, 2$.

65 Consider Bessel's equation

$$x^2\frac{d^2y}{dx^2} + x\frac{dy}{dx} + (x^2 - n)y = 0 \qquad y(1) = 1 \qquad y'(1) = 0.5$$

where n is not an integer.
Take $n = 1.5$ and investigate the solution in the interval $1 \leqslant x \leqslant 2$.
Take different values of n and use varying values of step size to investigate the solutions of this differential equation.

Tasks

Task A

f(x) is a function whose values are known only at $x = 2$, $x = 2.5$, $x = 3.5$ and at $x = 4$. These values are given in the table below.

x	2	2.5	3.5	4
f(x)	1.5	3.9	2.4	2.8

Write down $p_3(x)$, the polynomial of degree three that interpolates this data. Estimate the value of f(3). Comment on the accuracy of your solution.

It is required to find $\int_2^4 f(x)\,dx$.

Estimate the value of this integral:

(a) by evaluating $\int_2^4 p_3(x)\,dx$

(b) by using the trapezium rule

(c) by using Simpson's rule

Describe carefully some of the advantages and some of the disadvantages of each of the three methods.

Task B

A cinema manager uses the formula

$$p = 100 - \frac{1200}{x + 5} - 2x \qquad 10 \leqslant x \leqslant 30$$

to calculate his expected profit p (in pounds) per 100 customers on the sale of ice cream when the temperature is set at $x°$ C. During the week that he was showing 'Rocky XXX' (Rocky gives up boxing and becomes a great mathematician), the cinema manager decided to replace the above expression by a polynomial. Suggest how he might do this. Illustrate your answers and comment on the accuracy of your answers.

Task C

From mechanics, we have that the coordinates (x, y) of a projectile fired with speed $v\,\mathrm{m\,s^{-1}}$ at an angle θ radians to the horizontal satisfy the equation

$$y = x\tan\theta - \frac{g}{2(v\cos\theta)^2}x^2$$

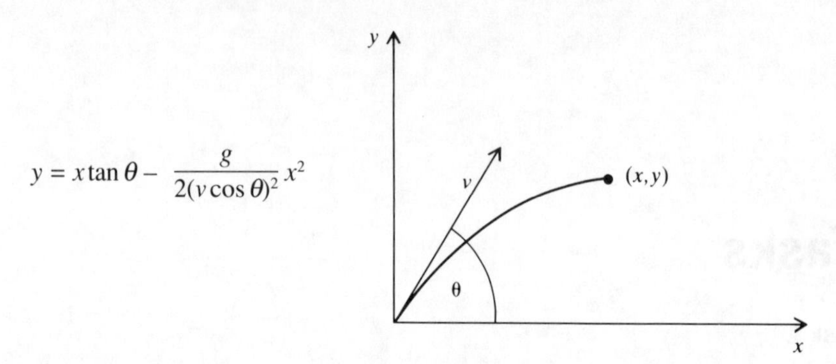

If $x = 100\,\mathrm{m}$, $v = 200\,\mathrm{m\,s^{-1}}$, $g = 9.81\,\mathrm{m\,s^{-2}}$, $\theta = \frac{\pi}{4}$ radians, calculate y. Suppose now that the values of v and θ are known only approximately. If v is subject to an error of 2% and θ is subject to an error of 0.5%, investigate the limits within which y must lie.

To see which of the three variables x, v, θ is the most sensitive:
(a) Calculate the percentage change in y due to an error of 1% in v.
(b) Calculate the percentage change in y due to an error of 1% in θ.
(c) Repeat for an error of 1% in x.
Comment on your answers.

Use the approximation

$$\mathrm{f}(u) - \mathrm{f}(u^*) \approx (u - u^*)\mathrm{f}'(u^*)$$

and consider approximately the change in y due to changes of 1% in
(d) v
(e) θ
(f) x
Compare your answers with (a), (b) and (c) and comment.

Task D

A function satisfies the differential equation

$$\frac{\mathrm{d}y}{\mathrm{d}x} = x - y \qquad \mathrm{y}(1) = 2 \qquad \ldots (1)$$

By differentiating (1) find the values of

$$\frac{\mathrm{d}^2y}{\mathrm{d}x^2}\ (= y_2),\ \frac{\mathrm{d}^3y}{\mathrm{d}x^3}\ (= y_3),\ \frac{\mathrm{d}^4y}{\mathrm{d}x^4}\ (= y_4),\ \frac{\mathrm{d}^5y}{\mathrm{d}x^5}\ (= y_5),\ \frac{\mathrm{d}^6y}{\mathrm{d}x^6}\ (= y_6) \quad \text{when} \quad x = 1$$

Writing $(1 + h)$ for x, use a Taylor polynomial for $\mathrm{y}(x)$ to calculate to four decimal digit accuracy the value of y when $x = 1.1$.

In the same way, start at the new value ($y = 1.9097$, $x = 1.1$) and use a Taylor polynomial for y near $x = 1.1$ to calculate a value for y when $x = 1.2$. Then start at this new value and use a Taylor polynomial to calculate a value for y when $x = 1.3$.

The algorithm below is designed to use this technique to calculate y(1.1), y(1.2), y(1.3), y(1.4), y(1.5) (using a Taylor polynomial of degree four). $x_0, x_1,$... denote the successive values of x; w_0, w_1, ... denote approximations to y at $x = x_0, x_1$, ... and h is the step size. Complete the algorithm.

input: x_0, w_0
$n := 1$
$h := 0.1$

● $y_1 := x - y$
$y_2 := 1 - y_1$
y :=
...
x_n := ...
$w_n := w_{n-1} +$
print x_n, w_n
if n = ... then
  stop
otherwise
  $n := n + 1$ and goto ●
endif

Use a second technique to obtain approximate numerical solutions to this differential equation at $x = 1.1$, $x = 1.2$, ..., $x = 1.5$

Compare your answers and comment fully on the method.

Task E

Use the Newton-Raphson method to obtain, correct to four decimal digits, the solutions of the equation

$$3x^3 + 4x^2 - 2x - 1 = 0$$

The Newton-Raphson method for obtaining solutions of the equation $f(x) = 0$ may be generalised by writing:

$$u_n = \frac{f(x_n)}{f'(x_n)}$$

$$x_{n+1} = x_n - \frac{u_n}{(au_n + b)}$$

for some constants a and b.

Show that if $a = 0$ and $b = 1$, the above iterative equation reduces to the Newton-Raphson method.

In relation to the equation

$$3x^3 + 4x^2 - 2x - 1 = 0$$

investigate other values of a and b and report fully on your findings.

Write $\quad g(x) = x - \dfrac{u}{au + b} \quad$ where $\quad u = \dfrac{f(x)}{f'(x)}$

and use Taylor's theorem to make a statement about the error term for this method.

Task F

Consider the system of equations

$$\begin{aligned} 3.02x + 6.03y + 1.99z &= 1 \\ 1.27x + 4.15y - 1.23z &= 1 \\ 0.987x - 4.80y + 9.34z &= 1 \end{aligned}$$

(a) Solve this system of equations, achieving as high a degree of accuracy as possible.
(b) Now solve the equations working in three digit rounded arithmetic.
(c) Compare your answers and comment.
(d) What is meant by a system of ill-conditioned equations? Give examples to illustrate your answers.
(e) Some of the coefficients are altered slightly and a new system is produced:

$$\begin{aligned} 3.01x + 6.03y + 1.99z &= 1 \\ 1.27x + 4.15y - 1.23z &= 1 \\ 0.99x - 4.80y + 9.34z &= 1 \end{aligned}$$

Solve this system as accurately as possible and comment fully on your solutions.

Task G

The table given below gives values taken by a function which depends on both x and y. So the numbers in the table are values of z where $z = f(x, y)$.

y	$x = 3$	3.5	4	4.5	5
1	10.15	12.01	13.87	15.73	17.59
1.2	11.76	13.92	16.08	18.24	20.40
1.4	13.77	16.29	18.82	21.35	23.88
1.6	16.26	19.24	22.21	25.19	28.17
1.8	19.35	22.87	26.40	29.92	33.45
2	23.17	27.36	31.56	35.75	39.95

We want to estimate the value of z when $x = 3.8$ and $y = 1.55$. Using any method that seems to you to be suitable, estimate this value of z.

A second method is the following. Choose several x values close to $x = 3.8$ (in this case, we might choose $x = 3.5$, $x = 4$, $x = 4.5$). Take the first ($x = 3.5$) and 'fix' x at this value. The y values associated with this (now fixed) value of x are:

$x = 3.5$

y	1	1.2	1.4	1.6	1.8	2
z	12.01	13.92	16.29	19.24	22.87	27.36

Use polynomial interpolation and find a polynomial that passes through some (or all) of these points. If you do not choose to interpolate all six points, which points would it be sensible to choose and why? Let $y = 1.55$ in your interpolating polynomial. The value of your polynomial at $y = 1.55$ will provide an estimate of the value of z when $x = 3.5$ and $y = 1.55$.

Take a second x value ($x = 4$) and repeat the process. This will provide an estimate of z at the point where $x = 4$ and $y = 1.55$. Take a third value of x and repeat the process.

You will now have an estimate of z (a) at (3.5, 1.55) (b) at (4, 1.55) (c) at (4.5, 1.55) How could you calculate the required estimate?

Use this method to achieve as accurate an estimate of f(3.8, 1.55) as you can.

The data is generated by the equation $z = x - y + xe^y$. Comment on your answers.

Task H

Show graphically, or otherwise, that the equation

$$2x^3 + 5x^2 - 0.01 = 0$$

has three real solutions. Find the values of these solutions, correct to three decimal digits (rounded), using the Newton-Raphson method.

If, instead, the fixed point iterative method was to be used to solve this equation, comment on the suitability of the iterative equation

$$x_{n+1} = k\left[\frac{0.01 - 5x_n^2}{x_n^2}\right]$$

where k represents an (as yet) undetermined constant.

Find all three solutions of the cubic equation

$$2x^3 + 5x^2 - 0.01 = 0$$

using the fixed point iterative method. Describe carefully your reasons for choosing the iterative equation (or equations) that you use.

Task I

The aim of this task is to compare Euler's method with the Runge-Kutta methods. To do this, consider the differential equation

$$\frac{dy}{dx} = -30y \qquad y = 1 \qquad \text{when} \qquad x = 0$$

(a) Use Euler's method with a step size of 0.2 to obtain a numerical solution to the differential equation when $x = 0.2, 0.4, 0.6, 0.8, 1$. The exact solution of the differential equation is $y = e^{-30x}$. Calculate the error in your approximations.

Experiment with different values of h and using Euler's method calculate, as accurately as you can, the value of y when $x = 1$.

Plot a graph of 'error in the value of y at $x = 1$' against h and comment on your results.

(b) Repeat part (a) using a Runge-Kutta method.

Task J

Investigate the difference equation:

$$x_{n+1} = kx_n^4 - 3kx_n^2 + (2k + 1)x_n - k$$

Task K

Evaluate the integral $\int_0^2 e^{\sqrt{x}}\,dx$ using the trapezium rule with (a) 1 interval (b) 2 intervals (c) $2^2 = 4$ intervals (d) $2^3 = 8$ intervals (e) $2^4 = 16$ intervals.

The exact solution is 5.407 528 185. Comment on the accuracy of your solutions.

Write R_{11}, R_{21}, R_{31}, R_{41} and R_{51} to represent your answers to (a) – (e) respectively.

Now form a second sequence:

$$R_{22} = \frac{4R_{21} - R_{11}}{4 - 1}$$

$$R_{32} = \frac{4R_{31} - R_{21}}{4 - 1}$$

$$R_{42} = \frac{4R_{41} - R_{31}}{4 - 1}$$

$$R_{52} = \frac{4R_{51} - R_{41}}{4 - 1}$$

Calculate the value of R_{22}, R_{32}, R_{42}, and R_{52},

Now form a third sequence:

$$R_{33} = \frac{4^2R_{32} - R_{22}}{4^2 - 1}$$

$$R_{43} = \frac{4^2R_{42} - R_{32}}{4^2 - 1}$$

What should be the expression for R_{53}?
Calculate the values of R_{33}, R_{43}, and R_{53}.

Write a third sequence:

$R_{44} = ...$
$R_{54} = ...$

and a fourth:

$R_{55} = ...$

Calculate the values of R_{44}, R_{54}, R_{55}
What can you say about R_{11}, R_{22}, R_{33}, R_{44}, and R_{55}?

An algorithm for calculating R_{11}, R_{22}, R_{33}, ... is as follows:

input: a,b,n
$h := b - a$
$R_{11} := \frac{h}{2}[\mathrm{f}(a) + \mathrm{f}(b)]$
for $i = 2, 3, ..., n$
  $s := 0$
  for $k = 1, 2, ..., 2^{i-2}$
    $s := \mathrm{f}(a + (k - 0.5)h) + s$
  endloop
  $R_{21} := \frac{1}{2}[R_{11} + hs]$
  for $j = 2, 3, ...,$ i
    $R_{2j} := \dfrac{4^{j-1}R_{2j-1} - R_{1j-1}}{4^{j-1} - 1}$
  endloop
  $h := \frac{h}{2}$
  for $j = 1, 2, ..., i$
    $R_{1j} := R_{2j}$
  endloop
endloop
output: R_{nn}

Consider a definite integral of your choice. Calculate as accurately as you can the value of this integral using:
(a) the trapezium rule
(b) Simpson's rule
(c) the technique described above (called **Romberg integration**).

Comment on these three methods of integration and on your answers.

Task L

Express $x^3 - 3x^2 + x + 2$ in the form

$$p(x - 2)^3 + q(x - 2)^2 + r(x - 2) + s$$

where p, q, r and s are constants to be found.

Given a cubic polynomial $p(x) = Ax^3 + Bx^2 + Cx + D$, the following (incomplete) algorithm is intended, by dividing $p(x)$ repeatedly by $(x - H)$, to find the constants appropriate for expressing $p(x)$ 'in terms of $(x - H)$'.

```
    input: A, B, C, D, H
    N := 0
●   N := N + 1
    B := B + HA
    if N = 3 then print A, B and stop
    C := C + HB
    if N = 2 then print C and goto ●
    D := ...
    ...
```

(a) Describe the purpose of N.
(b) Complete the algorithm.
(c) Write a program to achieve the purpose of the algorithm and check your answer to the first part of the task.
(d) Alter your algorithm so that the method will deal with a polynomial of any degree (say six).
(e) Write an algorithm that will transform a cubic polynomial

$$A(x - k)^3 + B(x - k)^2 + C(x - k) + D$$

into the form $ax^3 + bx^2 + cx + d$. (SMP)

Task M

For the whole of this task, we shall assume $-1 \leqslant x \leqslant 1$. Consider the recurrence relation

$$T_{n+1}(x) = 2xT_n(x) - T_{n-1}(x) \qquad n \geqslant 1$$

If $T_0(x) = 1$ and $T_1(x) = x$, use the recurrence relation to find the polynomials $T_2(x)$, $T_3(x)$, $T_4(x)$.

These are called the **Chebyshev polynomials**. It is required to express powers of x: x^0, x^1, x^2, x^3 and x^4 in terms of sums of Chebyshev polynomials. Clearly, $x^0 = 1 = T_0(x)$, $x^1 = x = T_1(x)$. x^2 is as given in the table below. Complete this table.

n	T	x
0	1	T_0
1	x	T_1
2	$2x^2 - 1$	$\frac{1}{2}T_0 + \frac{1}{2}T_2$
3	...	...
4	$8x^4 - 8x^2 + 1$	$\frac{3}{8}T_0 + \frac{1}{2}T_2 + \frac{1}{8}T_4$

Now write down p(x), the Taylor polynomial of degree four centred at the origin of the function f(x) = e^x. Write down also the error term E(x) associated with this polynomial. On the same axes, draw the graphs of $y = e^x$ and y = p(x) with $-1 \leqslant x \leqslant 1$. Use your graph to estimate the greatest error in $\left| e^x - \text{p}(x) \right|$.

Now consider the following technique.

Replace x^4 in your polynomial p(x) by the expression involving Chebyshev polynomials: so

replace x^4 by $\frac{3}{8}\text{T}_0 + \frac{1}{2}\text{T}_2 + \frac{1}{8}\text{T}_4$

Simplify the resulting expression to the form

$$\text{p}(x) = \text{polynomial of degree three} + \tfrac{1}{192}\text{T}_4(x)$$

Superimpose the graph of the polynomial of degree three on your graph of $y = e^x$ and y = p(x). Estimate the greatest error in using this cubic polynomial to approximate e^x.

Using any method that seems to you to be suitable, estimate the greatest value that can be taken by $\frac{1}{192}\text{T}_4(x)$.

(Hint: consider the maximum value of $\left| \text{T}_4(x) \right|$ for $-1 \leqslant x \leqslant 1$.)

The total error in using your cubic polynomial to approximate e^x is $\left| \text{E}(x) + \frac{1}{192}\text{T}_4(x) \right|$. Find the greatest value that can be taken by this error term.

What is happening here? What, if anything has been achieved?

Suppose that it is required to find a polynomial that approximates e^x with a maximum error of 0.05. How could the above method be used to find such a polynomial?

Task N

Consider the differential equation

$$\frac{\text{d}y}{\text{d}x} = x^2 + y^2 + 1 \qquad \text{y}(1) = 2$$

We want to know the solution of this differential equation when x = 1.1, 1.2 and 1.3. Write $x_0 = 1$, $x_1 = 1.1$, $x_2 = 1.2$, $x_3 = 1.3$, and $h = 0.1$. Use (a) Euler's method (b) a Runge-Kutta method, to obtain the solutions at these points.

From basic calculus $\displaystyle\int \frac{\text{d}y}{\text{d}x}\,\text{d}x = y$

Integrating both sides of the differential equation we get

$$\int_{x_0}^{x_1} \frac{\text{d}y}{\text{d}x}\,\text{d}x = \int_{x_0}^{x_1} (x^2 + y^2 + 1)\,\text{d}x$$

The left-hand side = $\Big[\, y \,\Big]_{x_0}^{x_1} = y_1 - y_0$

where $y_0 = \text{y}(x_0)$ and $y_1 = \text{y}(x_1)$

To evaluate (approximately) the right-hand side, observe that since y is a function of x (write $y = \text{y}(x)$) then $x^2 + [\text{y}(x)]^2 + 1$ is also a function of x. Hence, we may use the trapezium rule to approximate the integral.

If we imagine a graph

then $\displaystyle\int_{x_0}^{x_1} (x^2 + y^2 + 1)\,dx \approx (u_0 + u_1)\frac{h}{2}$

If $\quad x = x_1 \quad$ and $\quad y = y_1$

then $\quad u_1 = x_1^2 + y_1^2 + 1$

Write down an expression for u_0.

Putting the left-hand side and the right-hand side together

$$y_1 - y_0 \approx \frac{h}{2}(u_0 + u_1)$$

giving

$$y_1 - y_0 \approx \frac{0.1}{2}[u_0 + x_1^2 + y_1^2 + 1]$$

Use this equation to obtain a value for y_1.

Compare your answer with that achieved from Euler's method and a Runge-Kutta method.

Now use this method (called the **trapezium method**) to calculate (approximately) the values of y_2 and y_3.

Compare your answers with those achieved from Euler's method and a Runge-Kutta method of order two.

Use this method to obtain values for the solution of the differential equation

$$\frac{dy}{dx} = x\sin^2 y \qquad y(0) = 1$$

at $x = 0.1$, $x = 0.2$ and $x = 0.3$.

Comment on the advantages and disadvantages of the trapezium method and suggest a way in which it might be improved.

Task O

Consider the differential equation

$$\frac{dy}{dx} = f(x, y) \qquad y(a) = y_0$$

A difference equation that might be used to solve this differential equation is

$$w_{n+1} = w_n + \frac{h}{2}[f(x_n, w_n) + f(x_{n+1}, w_{n+1})]$$

where w_n represents an approximation to the solution at $x = x_n$ and h represents the step length. (This is the trapezium rule described in Task N.)

Show that for the differential equation

$$\frac{dy}{dx} = \frac{1}{x} + y^3 \qquad y(2) = 1 \qquad \ldots (1)$$

where the solution is required at $x = 2.2$, $x = 2.4$, $x = 2.6$, this difference equation becomes (with $n = 0$)

$$w_1 = w_0 + \frac{h}{2}\left(0.5 + 1 + \frac{1}{2.2} + w_1^3\right) \qquad \ldots (2)$$

Simplify this expression and explain how the value of w_1 could be obtained.

One way of overcoming the serious disadvantage inherent in this method is to use Euler's method applied to Equation (1) to obtain a first estimate of w_1, i.e. use Euler's method to estimate the value of y at $x = 2.2$.

The value obtained is called 'the predictor value'. Substitute this estimate of w_1 in the right-hand side of Equation (2).

The value taken by the right-hand side of Equation (2) then becomes the new estimate of w_1 [the corrected value].

The entire process is then repeated with w_2.

(a) Write down the difference equation associated with $n = 1$.
(b) Use Euler's method to obtain a first estimate of w_2.
(c) Substitute this value of w_2 in the right-hand side of the difference equation and obtain the new estimate of w_2, and so on.

The method is called the **predictor-corrector method**. Use the predictor-corrector method to calculate w_1, w_2 and w_3. Suggest a way in which the accuracy of the predictor-corrector method could be improved. Compare the solution obtained by the predictor-corrector method with the solution obtained by some other method. Give your opinion of this method, illustrating your argument with further examples.

Task P

The Redlich-Kwong equation is

$$p = \frac{Rt}{v - b} - \frac{A}{v(v + b)}$$

R is a constant and it is known that $R = 1.98$. On one occasion, $p = 54.1$, $t = 124.7$, $b = 3.11$ and $A = 0.0062$. Find the value of V. (Assume that all the values are in appropriate units.)

It is thought that each of p, t, b and A is correct to the number of digits written (using rounding). It may be assumed that the value of R is exact. Write the range of values that could be taken by p, t, b, and A.

Investigate how the value of V might vary under these circumstances.

Task Q

x	0	0.5	1	1.5
y	2	0.0625	–1	1.0625

It is required to find, as accurately as possible, the value of x at which $y = 0$.

Construct the cubic polynomial that interpolates these data and use a numerical technique to obtain the solutions of this polynomial. Could these solutions be used to approximate the values of x at which $y = 0$? Comment. Suggest an alternative method that might yield more accurate results.

We now consider a second method of finding a value of x at which $y = 0$. This technique is known as **inverse interpolation** and may be described as follows.

Write the data in the 'inverse' form

y	2	0.0625	–1	1.0625
x	0	0.5	1	1.5

and construct the cubic polynomial in y that interpolates the data. Using the fact that the required value of x occurs when $y = 0$, put $y = 0$ in your cubic polynomial and hence obtain an approximation to x. Why will this method, as described, fail to obtain both x-values? Suggest a modification that would enable this method to be used to approximate both values of x.

The data was generated by the equation $y = x^4 - 4x + 2$. Obtain the solutions of this equation in the interval [0, 1.5] correct to five decimal digits and comment on the accuracy of the different methods of solution.

Propose an equation of your own and use inverse interpolation to obtain (approximately) the solutions.

Task R

In a study of solar energy, L.L. Vant-Hall (*Solar energy* **18** 1976, page 33) derived the equation

$$C = \frac{\pi\left(\dfrac{h}{\cos A}\right)^2 F}{0.5\pi D^2(1 + \sin A - 0.5\cos A)}$$

If it is known that $h = 300$, $C = 1200$, $F = 0.8$ and $D = 14$ (in suitable units), find A and describe carefully your method of solution.

Task S

Let y be some function of x and assume that when x takes the value x_n, y takes the value y_n. Let $x_n - x_{n-1} = h$ and $x_{n+1} - x_n = h$.

Explain why $\dfrac{y_{n+1} - y_n}{h}$ could be taken to be an approximation to the value of $\dfrac{dy}{dx}$ when $x = x_n$.

A second approximation to $\dfrac{dy}{dx}$ (when $x = x_n$) is given by

$$\frac{dy}{dx} \approx \frac{y_{n+1} - y_{n-1}}{2h} \qquad \ldots (1)$$

and an approximation to $\dfrac{d^2y}{dx^2}$ (when $x = x_n$) is given by

$$\frac{d^2y}{dx^2} \approx \frac{y_{n+1} - 2y_n + y_{n-1}}{h^2} \qquad \ldots (2)$$

Consider the differential equation

$$\frac{d^2y}{dx^2} - 2\frac{dy}{dx} + y = 2e^x$$

where it is given that when $x = 3$, $y = 1$ and $\dfrac{dy}{dx} = 1$. We want to calculate (approximately) the value of the solution y(x) at $x = 3.1, 3.2, 3.3, 3.4$. Write $x = x_n$, $y = y_n$ and substitute expressions from Equations (1) and (2) for $\dfrac{dy}{dx}$ and $\dfrac{d^2y}{dx^2}$ to produce a difference equation for the differential equation. Why is the given information insufficient to obtain from the difference equation the values of y_1, y_2, y_3, and y_4?

(a) Write down the Taylor polynomial of degree one of y(x), centred at some suitable point on the x-axis, and use this to calculate (approximately) the value of y_1. Use your difference equation to calculate (approximately) the values y_2, y_3, and y_4.

(b) Now calculate y_1 using the Taylor polynomial of y(x) of degree two and calculate, as in (a), the values y_2, y_3, y_4.

(c) Finally, use the Taylor polynomial of degree five to calculate y_1 and use the difference equation to calculate y_2, y_3, y_4.

Plot all three solutions on a graph.

The exact solution of the differential equation is $y = [(9 + e^{-3}) - 6x + x^2]e^x$. Calculate the errors in your three sets of solutions and comment fully on the method.

Task T

Let $\quad y = e^x$

Use the approximation

$$\frac{dy}{dx} \approx \frac{f(a + h) - f(a)}{h} \qquad \ldots (1)$$

with $h = 0.1, 0.01, 0.001, 0.0001, 0.00001, 0.000001, 0.0000001$ to estimate the value of $\frac{dy}{dx}$ when $x = 1$.

Calculate the error in each approximation and draw up a table showing the error in each approximation against the associated value of h.

Repeat the exercise using the approximation

$$\frac{dy}{dx} \approx \frac{f(a+h) - f(a-h)}{2h} \qquad \ldots (2)$$

Comment on your results.

For the approximation in Equation (1), write down an expression for the error term.

Suppose now that $f(a + h)$ is subject to rounding error e_+. Write $f_+(a + h)$ to represent the computed value of $f(a + h)$.

Then $\quad f(a + h) = f_+(a + h) + e_+$

If $f(a)$ is subject to rounding error e_0 and $f_0(a)$ represents the computed value of $f(a)$, write down an expression linking $f(a)$, $f_0(a)$ and e_0.

Explain why we may now write

$$\frac{f(a+h) - f(a)}{h} = \frac{f_+(a+h) - f_0(a)}{h} + \frac{e_+ - e_0}{h} - \frac{1}{2}hf''(\eta)$$

Hence write down an expression for the total error in using $\frac{f_+(a+h) - f_0(a)}{h}$ to approximate $\frac{dy}{dx}$. Call this total error E. If rounding errors are bounded by 0.5×10^{-7} write down an expression in h for E. (Since the largest value to be taken by h is 0.1, assume that $0.9 \leqslant \eta \leqslant 1.1$.)

Find the value of h that gives a minimum value for E. Does this agree with your results obtained earlier? Comment.

For the approximation in Equation (2)

$$\frac{dy}{dx} = \frac{f(a+h) - f(a-h)}{2h} - \frac{h^2}{6}f'''(\eta) \qquad a - h < \eta < a + h$$

Write $f_-(a - h)$ to represent the computed value of $f(a - h)$.

Now repeat the second investigation using the approximation of Equation (2).

Task U

Let $y = f(x)$ be a function and suppose that the value of $f(x)$ is given at $n + 1$ points:

x	x_0	x_1	x_2	...	x_n
y	y_0	y_1	y_2		y_n

Let $p_{0,1,\dots,n}$ represent the polynomial of degree n that interpolates $f(x)$ at $x_0, x_1, x_2, \dots, x_n$. So, for example:

$p_{1,2}$ represents the polynomial (of degree one) that interpolates $f(x)$ at x_1, x_2
$p_{0,2,3}$ represents the polynomial (of degree two) that interpolates $f(x)$ at x_0, x_2, x_3
p_0 represents the polynomial (of degree zero) (a constant) that interpolates $f(x)$ at x_0.

Hence $p_0 = y_0$ and similarly, $p_1 = y_1, p_2 = y_2, \dots, p_n = y_n$.

If we have:

x	$x_0 = 1$	$x_1 = 3$	$x_2 = 5$	$x_3 = 6$	$x_4 = 7$
$y = f(x)$	$y_0 = 8$	$y_1 = 5$	$y_2 = 3$	$y_3 = 3$	$y_4 = 4.5$

Write down $p_{0,1}$ and $p_{1,2,3}$.

Now show that

$$p_{0,1} = \frac{(x - x_1)p_0 - (x - x_0)p_1}{x_0 - x_1}$$

(i.e. show that the right-hand side is a polynomial of degree one that interpolates $f(x)$ at x_0, x_1.)

In the same way, show that

(a) $$p_{0,1,4} = \frac{(x - x_0)p_{1,4} - (x - x_4)p_{0,1}}{x_4 - x_0}$$

(b) $$p_{0,1,2,3} = \frac{(x - x_2)p_{0,1,3} - (x - x_3)p_{0,1,2}}{x_3 - x_2}$$

In (b) we have seen a method of writing an interpolating polynomial of degree three ($p_{0,1,2,3}$) in terms of two interpolating polynomials of degree two ($p_{0,1,3}$ and $p_{0,1,2}$). This process may be expressed more generally in the form:

for any $0 < i, j < n$

$$p_{0,1,2,\dots,n} = \frac{(x - x_j)p_{0,1,\dots,j-1,j+1,\dots,n} - (x - x_i)p_{0,1,\dots,i-1,i+1,\dots,n}}{x_i - x_j} \qquad \dots (1)$$

This technique may be used to calculate, for a given set of data, interpolating polynomials of successively higher degrees. The technique is particularly useful if it is required to use polynomial interpolation to obtain ever more accurate approximations to $f(X)$ where X is not given in the data.

To illustrate: from the data in the last table it is required to calculate approximately the value of $f(2.5)$.

Using $p_{0,1}(x)$ we could write

$$p_{0,1}(2.5) = \frac{(2.5 - x_0)p_1 - (2.5 - x_1)p_0}{x_1 - x_0}$$

$$= \frac{(2.5 - 1)5 - (2.5 - 3)8}{3 - 1}$$

$$= 9.5$$

Also $$\mathrm{p}_{1,2}(2.5) = \frac{(2.5 - 3)3 - (2.5 - 5)5}{5 - 3}$$

$$= 5.5$$

Or, we could use $\mathrm{p}_{0,1,2}(x)$ and write

$$p_{0,1,2}(2.5) = \frac{(2.5 - 1)\mathrm{p}_{1,2} - (2.5 - 5)p_{0,1}}{5 - 1}$$

$$= \frac{1.5 \times 5.5 - (-2.5) \times 9.5}{4}$$

$$= 8$$

Write down three further approximations to f(2.5). Which of these approximations do you think will be the most accurate? Give reasons.

We could view this process as a table:

x_0	p_0				
x_1	p_1	p_{01}			
x_2	p_2	p_{12}	p_{012}		
x_3	p_3	p_{23}	p_{123}	p_{0123}	and so on

Table 1

Complete the table shown below.

1	8		
3	5	9.5	
5	3	5.5	8
6	3		
7	4.5		

This technique, of obtaining higher order polynomials to interpolate an increasing number of data points is known as Neville's algorithm. The increasing number of subscripts in the notation makes it difficult to implement this technique on a computer, but observe that the position of each p in the table may be specified using just two numbers:

(a) the row: which specifies the last x value on which the interpolation is to be performed,

(b) the column: which may be used to specify the degree of the interpolating polynomial.

So, for example, each item in the fourth row uses x_3 as the final interpolating point. Also, the third column consists of polynomials of degree one.
Describe the elements that would appear in (a) the tenth row (b) the eighth row (c) the ninth column (d) the sixth column.

Now write

Q_{ij} = polynomial of degree j that interpolates f(x) at the points $x_{i-j}, x_{i-j+1}, \ldots, x_i$

so that

Q_{20} = polynomial of degree zero that interpolates f(x) at x_2
Q_{31} = polynomial of degree one which interpolates f(x) at x_2, x_3
Q_{42} = polynomial of degree two that interpolates f(x) at x_2, x_3, x_4

Then Table 1 may be written:

$$\begin{array}{lllll} x_0 & Q_{00} & & & \\ x_1 & Q_{10} & Q_{11} & & \\ x_2 & Q_{20} & Q_{21} & Q_{22} & \\ x_3 & Q_{30} & Q_{31} & Q_{32} & Q_{33} \end{array}$$

Describe carefully Q_{11}, Q_{22}, Q_{33}.

To calculate approximately the value of f(2.5), we should calculate successively $Q_{11}(2.5), Q_{22}(2.5), Q_{33}(2.5), \ldots$ until satisfactory accuracy has been achieved. To assist in this, we use:

$$Q_{ij} = \frac{(x - x_{i-j})Q_{i,j-1} - (x - x_i)Q_{i-1,j-1}}{x_i - x_{i-j}}$$

Prove this result using Equation (1) on page 349.
Using this result we have Algorithm 2.1.
Use this algorithm to calculate approximately the value of f(2.5) using all the data from the table of x and f(x) on page 349.

x	1.0	1.3	1.6
y	0.765 197 7	0.620 086 0	0.455 402 2

For the data shown above write a program that will
(a) Calculate as accurately as possible the value of y when $x = 1.25$.
(b) Make a decision about whether sufficient accuracy has been achieved.
(c) Ask for further data and, if further data is available, calculate an improved approximation.

In fact, further data is available. It is known that when $x = 1.9$, $y = 0.281\,818\,6$ and when $x = 2.2$, $y = 0.110\,362\,3$.

Use your program to calculate as accurately as possible the value of y when $x = 1.25$.

Appendix 1

In everyday life, we use numbers in the decimal system. So when we write, for example, 216.95, we really mean

$$2 \times 10^2 + 1 \times 10^1 + 6 \times 10^0 + 9 \times 10^{-1} + 5 \times 10^{-2}$$

The system gets the name 'decimal' from the fact that each of the numbers 2, 1, 6, 9 and 5 multiplies a power of 10. We shall call 10 the base of the system and observe that to write a number in base 10, we use only the digits 0, 1, 2, ..., 9 (together with powers of 10).

But why choose 10 as a base? Many other choices of base are possible. In the binary system, we take the base to be 2. A system having a base of 2 would use only the digits 0 and 1 (together with powers of 2) and in a base 2 system, we would write, by analogy with the base 10 system

$$1001.101_2 \text{ to mean } 1 \times 2^3 + 0 \times 2^2 + 0 \times 2^1 + 1 \times 2^0 + 1 \times 2^{-1} + 0 \times 2^{-2} + 1 \times 2^{-3}$$

$$\text{so } 1001.101_2 \text{ represents } 8 + 0 + 0 + 1 + \tfrac{1}{2} + 0 + \tfrac{1}{8}$$

$$= 9\tfrac{5}{8}$$

$$= 9.625_{10} \quad \text{in base 10}$$

In the same way:

11011000.11_2 is shorthand for

$$1 \times 2^7 + 1 \times 2^6 + 0 \times 2^5 + 1 \times 2^4 + 1 \times 2^3 + 0 \times 2^2 + 0 \times 2^1 + 0 \times 2^0 + 1 \times 2^{-1} + 1 \times 2^{-2}$$

$$= 128 + 64 + 16 + 8 + \tfrac{1}{2} + \tfrac{1}{4}$$

$$= 216\tfrac{3}{4}$$

$$= 216.75_{10}$$

A number written in base 2 may be called a binary number[†].

[†] If the reader wishes to know more about binary numbers, Conte and de Boor 1981 *Elementary numerical analysis* 3rd edn. McGraw-Hill or a textbook on computer science should be consulted.

Appendix 2

Let $\quad y = 0.d_1 d_2 d_3 d_4 \ldots \times 10^e$

Theorem 1a

If y^* is the chopped three digit form of y then

$$|y - y^*| < 10^{e-3}$$

Further, if y^* is the chopped k digit form of y, then

$$|y - y^*| < 10^{e-k}$$

Proof

Write $\quad y = 0.d_1 d_2 d_3 d_4 d_5 \ldots \times 10^e$

Then $\quad y^* = 0.d_1 d_2 d_3 0 0 \times 10^e$

The roundoff error is $\quad |y - y^*|$

$$= 0.000 d_4 d_5 \ldots \times 10^e$$
$$= 0.d_4 d_5 \ldots \times 10^{e-3}$$

But $\quad 0.d_4 d_5 \ldots < 1$

Hence $\quad |y - y^*| = |0.d_4 d_5 \ldots \times 10^{e-3}| = |0.d_4 d_5 \ldots| \times 10^{e-3}$

$$< 10^{e-3}$$

The proof for the k digit form of y is similar.

Theorem 1b

If y^* is the rounded three digit form of y then

$$|y - y^*| < 0.5 \times 10^{e-3}$$

Further, if y^* is the rounded k digit form of y then

$$|y - y^*| < 0.5 \times 10^{e-k}$$

Proof

Write $y = 0.d_1 d_2 d_3 ... \times 10^e$

If y^* is the rounded three digit form of y then

$$y^* = 0.D_1 D_2 D_3 0 \times 10^e$$

where D_1, D_2, D_3 are the three digits that occur in the rounded form of y.

We consider two cases.

Case 1 $d_4 \geqslant 5$ (This is the situation that occurred in Example 1.6.)

In this case $D_3 = \begin{cases} d_3 + 1 & \text{if } d_4 = 5, 6, 7, 8 \\ 0 & \text{if } d_4 = 9 \end{cases}$

Now, $y^* > y$ and the roundoff error

$$|y - y^*| = y^* - y$$

We calculate $\begin{array}{l} 0.D_1 D_2 D_3 0 \quad \times 10^e \; - \\ 0.d_1 \, d_2 \, d_3 \, d_4 ... \times 10^e \end{array}$

A little thought will convince the reader that

$$\begin{aligned} y^* - y &\leqslant 0.0005 \times 10^e \\ &= 0.5 \times 10^{e-3} \end{aligned}$$

Hence $|y - y^*| \leqslant 0.5 \times 10^{e-3}$

Case 2 $d_4 < 5$

In this case $D_3 = d_3$ and $y^* \leqslant y$

The roundoff error

$$\begin{aligned} |y - y^*| = y - y^* = \; & 0.d_1 \, d_2 \, d_3 \, d_4 ... \times 10^e \\ & \underline{0.D_1 D_2 D_3 0 \quad\;\; \times 10^e} \\ & 0.0 \; 0 \; 0 \; d_4 ... \times 10^e \end{aligned}$$

Hence $\begin{aligned} y - y^* &< 0.0005 \times 10^e \\ &= 0.5 \times 10^{e-3} \end{aligned}$

giving $|y - y^*| < 0.5 \times 10^{e-3}$

In both cases $|y - y^*| \leqslant 0.5 \times 10^{e-3}$

and this gives the required result.

The proof for the k digit form of y is similar.

Appendix 3

To establish the Runge-Kutta methods of order two for the solution of the differential equation $\frac{dy}{dx} = f(x, y)$.

We begin with a Taylor method of order two and write

$$y(x_0 + h) \approx y(x_0) + hy'(x_0) + \frac{h^2}{2!}y''(x_0) \qquad \ldots (1)$$

To simplify expressions, write:

$$y_0 = y(x_0) \qquad y_0' = y'(x_0) \qquad y_0'' = y''(x_0)$$

Then $\quad y_0' = f(x_0, y_0)$

We may write $y''(x) = \frac{d}{dx}y'(x)$ and from Chapter 7 we have

$$y_0'' \approx \frac{y'(x_0 + h^*) - y'(x_0)}{h^*}$$

where h^* is not necessarily equal to h.

Substituting for $y'(x_0 + h^*)$ and $y'(x_0)$ gives

$$y_0'' \approx \frac{f(x_0 + h^*, y(x_0 + h^*)) - f(x_0, y_0)}{h^*} \qquad \ldots (2)$$

But $\quad y(x_0 + h^*) \approx y_0 + h^*y_0'$

Substituting this approximation for $y(x_0 + h^*)$ in Equation (2) gives

$$y_0'' \approx \frac{f(x_0 + h^*, y_0 + h^*f(x_0, y_0)) - f(x_0, y_0)}{h^*}$$

Substituting the expressions for y_0' and y_0'' in Equation (1) gives

$$y(x_0 + h) \approx y(x_0) + hf(x_0, y_0) + \frac{h^2}{2!}\left[\frac{f(x_0 + h^*, y_0 + h^*f(x_0, y_0)) - f(x_0, y_0)}{h^*}\right]$$

Now write $h^* = \lambda h$ and rewrite this equation in terms of w_0, w_1

$$w_1 = w_0 + h\mathrm{f}(x_0, w_0) + \frac{h^2}{2}\left[\frac{\mathrm{f}(x_0 + \lambda h, w_0 + \lambda h\mathrm{f}(x_0, w_0)) - \mathrm{f}(x_0, w_0)}{\lambda h}\right]$$

$$= w_0 + h\left(1 - \frac{1}{2\lambda}\right)\mathrm{f}(x_0, w_0) + \frac{h}{2\lambda}\mathrm{f}(x_0 + \lambda h, w_0 + \lambda h\mathrm{f}(x_0, w_0))$$

Write $\quad A = \left(1 - \frac{1}{2\lambda}\right) \qquad B = \frac{1}{2\lambda} \qquad a = \lambda \qquad b = \lambda$

and we have

$$w_1 = w_0 + Ah\mathrm{f}(x_0, w_0) + Bh\mathrm{f}(x_0 + ah, w_0 + bh\mathrm{f}(x_0, w_0))$$

or $\quad w_1 = w_0 + Ak_1 + Bk_2$

which is the required expression.

Observe that $A + B = \left(1 - \frac{1}{2\lambda}\right) + \frac{1}{2\lambda} = 1$ and that $Ba = Bb = \frac{1}{2}$. These conditions on A, B, a, b may then be used to define a Runge-Kutta method of order two.

Bibliography

Burden and Faires 1985 *Numerical analysis* 3rd edn. Prindle, Weber and Schmidt

Conte and de Boor 1981 *Elementary numerical analysis* 3rd edn. (international student edn.) McGraw-Hill

Golub (ed.) 1984 *Studies in numerical analysis* Mathematical Association of America

Hall and Knight 1891 (latest impression 1964) *Higher algebra* 4th edn. Macmillan

Ralston and Rabinowitz 1978 *A first course in numerical analysis* 2nd edn. (international student edn.) McGraw-Hill

Yakowitz and Szidarovsky 1989 *An introduction to numerical computation* 2nd edn. Macmillan

Answers

Answers will be given to five decimal digit (rounded) accuracy unless there are good reasons for doing otherwise.

Exercise 1

1 (a) $0.543\,291 \times 10^3$ (b) $0.678\,91 \times 10^{-2}$ (c) $0.290\,03 \times 10^2$
(d) -0.6839×10^{-1} (e) $0.516\,321 \times 10^0$ (f) -0.1×10^{-4}

2 (b), (c), (e), (f)

3 (a) 0.478×10^0 (b) 0.152×10^{-3} (c) 0.999×10^{-1}
(d) 0.181×10^4 (e) 0.615×10^7

4 (a) 0.6815×10^{-1} (b) 0.7592×10^4 (c) 0.8178×10^0
(d) 0.9522×10^{-2} (e) 0.5316×10^{-3}

5 (a) 538 (b) 0.78 (c) 0.000 541 (d) 2.236
(e) −0.000 015 (f) 25.700 (g) 6.882

6	Rounding	Chopping	Exponent
(a)	$0.888\,888\,89 \times 10^1$	$0.888\,888\,88 \times 10^1$	1
(b)	$0.108\,166\,54 \times 10^2$	$0.108\,166\,53 \times 10^2$	2
(c)	$0.928\,173\,64 \times 10^{-2}$	$0.928\,173\,63 \times 10^{-2}$	−2
(d)	$0.170\,997\,59 \times 10^1$	$0.170\,997\,59 \times 10^1$	1
(e)	$0.194\,931\,77 \times 10^{-1}$	$0.194\,931\,77 \times 10^{-1}$	−1
(f)	$0.967\,741\,94 \times 10^1$	$0.967\,741\,93 \times 10^1$	1
(g)	$0.110\,633\,16 \times 10^2$	$0.110\,633\,16 \times 10^2$	2

7 0.514×10^{-1}

8 -0.372×10^{-3}

9 4.932, 0.98×10^{-3}

11 (a) 3257 (b) $0.469\,81 \times 10^{-2}$

12 (a) 0.2×10^{-5} (b) $0.333\,44 \times 10^{-3}$

13 675.9 (a) 0.073 (b) $0.107\,99 \times 10^{-3}$

14	**Roundoff error**	**Absolute error**	**Relative error**
(a)	-0.8×10^{-2}	0.8×10^{-2}	$0.117\,91 \times 10^{-4}$
(b)	0.1432×10^{-3}	0.1432×10^{-3}	$0.246\,29 \times 10^{-2}$
(c)	0.17×10^{-2}	0.17×10^{-2}	$0.212\,99 \times 10^{-3}$
(d)	0.2×10^{-2}	0.2×10^{-2}	$0.100\,54 \times 10^{-3}$
(e)	-0.2×10^{-8}	0.2×10^{-8}	$0.111\,23 \times 10^{-2}$
(f)	0.52×10^{-4}	0.52×10^{-4}	$0.556\,03 \times 10^{-2}$
(g)	2	2	$0.228\,23 \times 10^{-4}$

15 $\frac{9}{35}$ (a) 0.214×10^{-2} (b) 0.833×10^{-2}

16 (a) 0.168 802 659 (b) 0.17 (c) 0.11973×10^{-2}, 0.70931×10^{-2}

17 –552 797.7092, 25, 552 822.7092, –1

18 (a) 1649.069 25 (b) 1670 (c) 20.931 (d) –0.012 69

19 (a) 0.964 285 714 (b) 0.962, $0.228\,57 \times 10^{-2}$, $0.237\,03 \times 10^{-2}$

20 (a) 2.458 333 33, 2.4 (b) 25, 29

21 (a) –3.199 636 364 (b) –3

22 (a) $x = -0.299\,999\,97$, $y = -2.566\,666\,7$
(b) $x = -0.312$, $y = -2.56$

(c)	**Absolute error**	**Relative error**
x	0.012 000 03	–0.04
y	$0.666\,67 \times 10^{-2}$	$-0.259\,74 \times 10^{-2}$

23 (a) $x = -173\,334$, $y = 130\,000$ (b) $x = -1\,730$, $y = 1\,300$

24 (a) $x = 186\,113.2$, $y = 167\,500$ (b) no solution

25 (a) $x_1 = 2$, $x_2 = -1.422\,65$ (b) $x_1 = 2$, $x_2 = -1.4$

(c)	**Absolute error**	**Relative error**
x_1	0	0
x_2	0.022 65	–0.015 921

26 (a) $x_1 = 3$, $x_2 = -5.425\,297\,03$ (b) $x_1 = 2.7$, $x_2 = -5$

(c)	**Absolute error**	**Relative error**
x_1	0.3	0.1
x_2	0.425 30	–0.078 391

27 [0.005 475, 0.005 49]

28 [16.416,16.747]

29 [0.0351, 0.0358]

30 [57.3, 59.4]

31 [0.316 85, 0.320 17], 0.166×10^{-2}

32 [0.367 89, 0.368 47], $0.896 40 \times 10^{-3}$

33 [–151, –49.667]

34 [0.238 88, 0.252 10], 0.027 554

35 (a) [0.73, 0.84] (b) [1.9481, 2.1053]
(c) [42.76, 47.17] (d) [2.25, 2.56]
(e) [16.173, 17.194] (f) [0.176 12, 0.200 60]
(g) [–0.029 167, –0.027 636] (h) 2575.9, 3065.3]
(i) $[0.169 62 \times 10^{-2}, 0.186 12 \times 10^{-2}]$

36 (a) [0.961 26, 0.970 30] (b) [–12.706, –7.5958]
(c) [0.978 15, 0.984 81] (d) [1.2479, 1.2746]

37 $A \in [30.190, 32.179]$, $C \in [19.477, 20.112]$

38 [1271.5, 1463.3]

39 [1.8794, 2.9344]

40 [52, 52.2]

41 [1.7337, 1.8557]

42 [1.3157, 1.3732], [–0.428 43, –0.406 39]

43 (a) $x \approx 0°$, $\sin^2 x$
(b) $x \approx 45°$, $\cos x$
(c) $x \approx 0$, $\ln\left(1 + \frac{x}{10}\right)$
(d) $x \approx 4$, $\frac{4 - x}{2 + \sqrt{x}}$
(e) x is large, $\frac{0.1}{x + \sqrt{x^2 - 0.1}}$

44 $\sqrt{x + 1} + \sqrt{x} = 0.223 61 \times 10^{-2}$

45 –0.4, –0.0002

46 (a) –8.000 074 999, $0.749 992 968 9 \times 10^{-4}$
(b) 4.333 325 642, $0.769 232 134 5 \times 10^{-5}$
(c) 9.000 236 105, $-0.236 104 917 2 \times 10^{-3}$

47 x 'large', $4x + 4\sqrt{x^2 - 1}$, 79 999.9998

48 $0.229 65 \times 10^{-3}$

49 $0.161 42 \times 10^{-2}$

50 $0.108 25 \times 10^{-1}$

51 $0.392\,16 \times 10^{-1}$

52 $0.497\,52 \times 10^{0}$

53 $1.4137, 0.651\,75 \times 10^{-2}$

54 $0.805\,58 \times 10^{-2}$

55 $0.305\,55 \times 10^{-1}$

56 (a) 0.0505 (b) 0.0505 (c) 0.2025

57 (a) 0.3 (b) 0.45 (c) 0.6

58 (a) 0.038 (b) 0.076 (c) 0.256 (d) 0.178

59 To three decimal digits:

relative error	(a) 0.0190	(b) 0.0106	(c) 0.0127	(d) 0.0192
absolute error	(a) 0.159×10^{-2}	(b) 0.235×10^{-2}	(c) 0.827×10^{-5}	(d) 35100

60 2.59, 5.90, –3.31 max. errors 0.016, 0.049, 0.65

61 $5 \times 10^{-3}(n_1 + n_2)$, 34

62 1.3288 max. error = 0.257 66

63 $[0.809\,02 \times 10^{10}, 0.817\,71 \times 10^{10}]$

64 [34.930, 35.779]

65 [1.9596, 2.4563]

66 $\frac{1}{2}b + \frac{1}{2}c + h\left(\frac{a+d}{h+r}\right)$

67 $\delta = \dfrac{ex^{12}}{1 + x^6}$

68 (i) -0.658×10^{-2} (ii) –0.01 (ii) is better

69 (a) 85 970 (b) 85 950

71 $P_0 e^{ct}\left(e^{\delta t} + \dfrac{Ee^{\delta t}}{P_0} - 1\right)$

Exercise 2

1 (a) $5x + 5$ (b) $4x^2 - 7x + 14$ (c) $3x^2 - 18x - 1$
(d) $2x + 7$ (e) $2x^3 - 6x^2 - 5x + 3$ (f) $\frac{11}{4}x^2 + x + 3$
(g) $16x^3 - 38x^2 + x + 12$ (h) $-x^3 + 3x^2 + x - 40$

2 (a) $x^2 + 4x + 3$ (b) $x^3 - 5x^2 + 7x - 3$ (c) $x^3 - 6x^2 + 9x - 2$
(d) $x^3 - 7x + 6$ (e) $x^3 + x^2 - 5x + 3$ (f) $x^4 - 2x^3 + 2x^2 - 2x + 1$
(g) $6x^3 + 5x^2 - 6x + 1$ (h) $4x^3 - 20x^2 + 13x + 12$
(i) $12x^4 + 2x^3 + x^2 + 8x - 3$ (j) $64x^4 + 32x^3 - 12x^2 - 4x + 1$
(k) $\frac{4}{3}x^3 + \frac{5}{3}x^2 - \frac{10}{3}x + 1$ (l) $\frac{1}{5}x^4 + 2x^3 - \frac{9}{10}x^2 + x - \frac{1}{2}$

3 $y = -\frac{4}{3}x + \frac{13}{3}$

4 $y = \frac{3}{5}x + \frac{1}{5}$

5 13

6 $y_1 = a_1x_1 + a_0 \qquad a_1 = \dfrac{y_0 - y_1}{x_0 - x_1} \qquad a_0 = \dfrac{(y_1x_0 - y_0x_1)}{x_0 - x_1}$

7 $x = 3\frac{5}{7}$

8 (a) 0.9741 (b) 0.973 92 (c) 0.974 16

9 1.2817

10 1.0365

11 $y = -1.5x^2 + 10.1x - 7.7$ (a) 8.82 (b) 9.02 (c) 6.22

12 $y = 28.4x^2 - 36.6x + 25$ (a) –36.6 (b) –65 (c) 48.6 (d) 133.8

13 $f(x) = 12\frac{4}{9}x^2 - 124x + 360$, 190

14 $p = -0.375x^2 + 2.8x + 0.075$ (a) 4.6302, 2.8364
(b) Year = 3.7333 Greatest profit = 5.3017

15 2.7, 5.1 Former more reliable

16 $-3\frac{1}{5}(x-2)(x-3)(x-4) + (x-1.5)(x-3)(x-4) - \frac{4}{3}(x-1.5)(x-2)(x-4) + \frac{4}{5}(x-1.5)(x-3)(x-4)$

(a) 6.3417 (b) 9.542 (c) 6.98

17 $-1.154\,166\,7x^3 - 3.95x^2 + 3.704\,166\,67x + 4.5$
(a) (i) $f(x) = -9.4750$ (ii) 1.8047 (iii) 5.2203
(b) (i) $f'(x) = 5.6542$ (ii) 6.7885 (iii) –1.1115

18 $f(x) \approx 0.0209x^3 - 0.064\,588x^2 - 0.027\,529x + 0.167\,749$, 0.036 763

19 $-0.011\,933x^3 + 0.136\,049\,99x^2 - 0.584\,516\,647x + 1.1675$

	f(2.5)	f(3.5)
Degree 1	0.381 70	0.279 35
Degree 2	0.374 54	0.272 19
Degree 3	0.370 07	0.276 68

20 $y = -0.164\,58x^2 + 1.186\,89x - 1.022\,31$

23 $x^2 - 3.5x + 1$

24 $y = 2.7869x - 5.8873$ Max. error = 0.1

25 $y = -0.232\,54x + 0.600\,42$ Max. error = 0.045 985

26 $-0.048\,15x^2 + 0.558\,65x + 0.4895$ Max. error = 0.024 056

27 $-0.4375x^2 + 1.3803x - 0.101\,36$ Max. error = 0.038 625

28 (b) 0.25

29 $-0.0323x^2 + 0.449\,20x + 0.329\,40$ Max. error = $0.475\,19 \times 10^{-2}$

30 1.6839

31 (a) By fitting a cubic polynomial, f(4.72) = 0.701 36
(b) $x^3 - 3x^2 - 2x + 5$

32 $x^3 - 6x^2 + 14x - 11$ Solution = 1.5466 22.28

34 $h < 0.686\,83$

35 Error = $\frac{f'''(\eta)}{3!}(x-1)(x-1.5)(x-2)$

36 (a) $y = 1$ (b) $y = -x^2 + 2$ (c) $y = 0 \times x^3 - 0.9x^2 + 1.9$
(e) $y = 1.3333$ (f) $y = -1.1428x^2 + 2$
(g) $y = 0 \times x^3 - 0.941\,15x^2 + 1.8823$

39 $x^3 - 2.5x^2 - 35.5x + 114$ 3.6318

40 degree 1 $-3.450\,56x + 10.775\,04$ 3.1227
degree 2 $-1.027\,66x^2 + 3.743\,06x - 1.556\,88$ 3.1634
degree 3 $0.623\,08x^3 - 6.635\,38x^2 + 19.9431x - 16.5108$ 3.1334

41 $2.4897x^5 - 9.7828x^4 + 1.8319x^3 + 19.833x^2 + 0.064\,628x$

Exercise 3

3 (a) 1.876 (b) 0.5874 (c) 0.1416 (d) 2.456

4 (a) 1.5377 (b) −0.879 39, 1.3473, 2.5321
(c) −0.903 21, 3.7093 (d) −1.9530, −0.301 98, 1.3110

5 (a) $x_{13} = 1.854\,553\,22$ (b) $x_{13} = 1.854\,553\,22$ (c) $x_{12} = 1.854\,614\,26$

6 (a) $x_7 = 2.277\,343\,75$ (b) $x_7 = 2.277\,343\,75$ (c) $x_6 = 2.273\,437\,5$

7 (a) 1.8241 (b) −1.3500, 0.806 47
(c) −0.735 79, 0.735 79 (d) −2.1149, 0.254 10, 1.8608

8 (a) −1.790, −0.1204, 1.160 (b) 0.8029
(c) 147.4 (d) 5.248, 7.329

9 (a) $x_4 = -0.912\,719\,213$ $x_5 = 1.118\,331\,21$
(b) $x_5 = -0.912\,765\,379$ $x_6 = 1.118\,325\,59$
$x_5 = -0.912\,765\,379$ $x_5 = 1.118\,331\,21$

12 1.054

13 0.433

14 (a) 0.7391 (b) −4.996 (c) 0.5302 (d) 1.141

15 (a) −0.62 (b) 1.2 (c) 0.24

16 10.980 692 5 (a) $x_8 = 10.978\,716\,7$ (b) $x_{10} = 10.980\,282\,8$
(c) $x_7 = 10.976\,354\,2$

17 −1.82, 0.823

18 −2.489, −0.2892, 2.778

19 –0.1987, 1.286, 3.912

22 (a) 0.2427 (b) 1.855 (c) 5.760

23 (a) 0.5 (b) 0.25 (c) 0.125 (d) 2^{-6} (e) 2^{-11}

24 (a) 2^{-4} (b) 2^{-7} (c) 2^{-13} (d) $2^{-(n+1)}$

25 Assuming the initial intervals to be [1,2] or [18,19], 10 iterations are necessary

26 13 iterations

27 (a) $|\sin x| < 0.25$ (b) $x^2 > 1$, –1.6180 (c) $x < 2\ln(2\sqrt{3})$, –0.564 66
(d) $-11.871 < x < -11.323$ $-8.6771 < x < -8.1292$

28 (b) (ii) and (iii) (c) (ii)

29 (b) (i)

30 (a) 4.5616, 0.438 45
(b) $-0.485\,06 < k < 0$ $0 < k < 0.485\,07$
(c) $k = -0.242\,53$ $k = 0.242\,53$

31 (b) Take solution ≈ 0.7 (i) $k \approx 0.143\,44$ (ii) 0.655 430 839

33 (a) 0.547 155 848 (b) –0.386 597 457
(c) 0.586 626 746 (d) 5.76
(e) –2.315 940 43, 0.467 090 379, 1.848 850 05

35 –1.7895, –0.120 43, 1.1600

36 0.66

37 –3.096, 1.306

38 0.5671

39 –0.607 01

40 1.926

41 –2.292, 1.538

42 (c) 2.094 551 48

43 (a) 0.5 (b) 0.25 (c) 0.125 (d) $2^{-(n+1)}$ $n = 9$, 0.198 242 188

45 $h^3 - 12h^2 + 64 = 0$ 2.6108

46 1.02

47 0.5961

Exercise 4

1 (i) and (ii)

2 $x = 2, y = -3, z = 1$

3 $x = -2, y = -1, z = 4$

4 $x = -3, y = 2, z = -1$

5 $a = -1, b = -2, c = -3$

6 $x_1 = 2, x_2 = -2, x_3 = -1$

8 $x = 4, y = 1, z = 2$

9 $x_1 = -3, x_2 = 4, x_3 = -2$

10 $a = 1.5, \quad b = 4, c = -0.5$

11 $p = 1, q = -1, r = -3$

12 $x = 3, y = -1, z = 1, w = -2$

13 $a = 3.5, b = 2.5, c = 1.5, d = 0.5$

14 $x = 3, y = -2, z = 4$

15 $x_1 = -2, x_2 = -3, x_3 = -4$

16 $x = 4, y = -1, z = 2, w = 0$

17 $a = 4, b = 0, c = -2, d = 1$

18 $x_1 = 3, x_2 = 1, x_3 = -1, x_4 = 1$

19 (a) $x = 10, y = 0.3997$ without pivoting
(b) $x = 8, y = 0.4$ with pivoting
(c) $x = 8, y = 0.4$

20 (a) $a = 7.5, b = 1.501$ without pivoting
(b) $a = 10, b = 1.5$ with pivoting

21 $x_1 = 7.338, x_2 = 0.024\,95$ without pivoting
$x_1 = 7.333, x_2 = 0.025\,13$ with pivoting

22 $9.1179 \leqslant x \leqslant 48.333 \qquad -66.667 \leqslant y \leqslant -11.765$

23 $72.429 \leqslant x \leqslant 84.333 \qquad 142.86 \leqslant y \leqslant 166.67$

24 (a) $x = 359\,998, y = -299\,999$ (b) No solution

25 Ill-conditioned: (a), (b) and (d)

26 Ill-conditioned

27 Very ill-conditioned

28 $x = -0.2609 \qquad y = 1.491 \qquad z = -0.3652$

29 $x_1 = 7.692 \times 10^{-2} \qquad x_2 = 0.115 \qquad x_3 = 0.577$

30 $x_1 = 0.080 \qquad x_2 = 0.106 \qquad x_3 = 0.310$

31 $x = 1.4588 \qquad y = -1.0832 \qquad z = -0.186\,12$

32 $x_1 = 0.897 \qquad x_2 = 0.764 \qquad x_3 = 0.615$

33 $x = 1.103 \qquad y = -0.433 \qquad z = 0.778$

34 $x_1 = -0.396 \quad x_2 = 0.117 \quad x_3 = -0.338$

35 Determinant $= -0.44 \times 10^{-2}$
$x = -0.55 \quad y = 1.59 \quad z = 0.55$

36 $x_1 = 0.15873b_1 + 0.074977b_2 - 0.00038273b_3$
$x_2 = 0.14761b_2 + 0.063084b_3$
$x_3 = 0.20428b_3$

$$\begin{pmatrix} 0.15873 & 0.074977 & -0.00038273 \\ 0 & 0.14761 & 0.063084 \\ 0 & 0 & 0.20428 \end{pmatrix}$$

37 (a) $x = 3.2 \quad y = 0 \quad z = 0.65$
(b) $x = 2.58 \quad y = 0.4 \quad z = 0.652$
(c) $x = 2.5949367 \quad y = 0.386075955 \quad z = 0.651898735$

Exercise 5

1 $p_3(x) = x - \frac{1}{6}x^3$
Calculator (a) 0.099833 (b) 0.47943 (c) 0.84147
Polynomial (a) 0.099833 (b) 0.47917 (c) 0.83333

2 $e^x \approx 1 + x + \frac{x^2}{2} + \frac{x^3}{6}$
(a) 1.2213 (b) 1.4907 (c) 0.33333 (d) 23.667

3 $\ln(1+x) \approx x - \frac{1}{2}x^2 + \frac{1}{3}x^3 \qquad \ln(1-x) \approx -x - \frac{1}{2}x^2 - \frac{1}{3}x^3$

4 $p_3(x) = 1 - \frac{1}{2}x + \frac{3}{8}x^2 - \frac{5}{16}x^3$

5 $x - x^2$

6 $1.6 + 0.52x + 0.175x^2$
(a) 1.6819 (b) 1.7409 (c) 1.8034 (a)

7 $5.3 + 1.6x + 0.2x^2$
(a) (i) 5.628 (ii) 5.7125 (b) (i) 1.66 (ii) 1.68

8 (a) $x - \frac{1}{6}x^3 + \frac{1}{120}x^5$ (b) $1 - \frac{1}{2}x^2 + \frac{1}{24}x^4 + 0x^5$
(c) $x + \frac{1}{3}x^3 + \frac{2}{15}x^5$
(a) 1 (b) $-\frac{1}{2}$ (c) 2

9 $x - \frac{1}{2}x^2 + \frac{1}{6}x^3 - \frac{1}{12}x^4 \qquad -\frac{1}{2}$

10 (a) $1 - x^2 + 0x^3$ (b) $\frac{1}{2} - \frac{1}{2}(x-1) + \frac{1}{4}(x-1)^2 + 0(x-1)^3$

11 (a) $4.1132504 + 1.4542536(x-2)$
(b) $4.1132504 + 1.4542536(x-2) + 0.075296445(x-2)^2$
(c) $4.1132504 + 1.4542536(x-2) + 0.075296445(x-2)^2 + 0.01147283883(x-2)^3$

12 $0.99833417 - 0.0333000117(x-0.1) - 0.16616696(x-0.1)^2$

13 $20.5 + 1.67(x-2) + 0.39(x-2)^2$
(a) 20.671 (b) 20.8496 (c) 20.4503

14 (a) $1 + 0.8(x - 1) + 0.05(x - 1)^2 + 0.011\,666\,7(x - 1)^3$
(b) 0.841 907 (b) 1.080 51 (c) 0.805 087 5

15 (a) $41.15 + 4.05(x - 10) - 0.0025(x - 10)^2 + 0.000\,166\,67(x - 10)^3$
(i) 43.1744, 4.0476 (ii) 35.0688, 4.0586 (b) (i)

16 1.5 $0.799\,64 + 0.183\,50(x - 1.5) - 0.094\,841(x - 1.5)^2$ 0.791 74

17 $4.4817 - 4.4817(x + 1.5) + 2.2408(x + 1.5)^2$ 0.153 94

18 $\frac{1}{3} - \frac{1}{9}(x - 2) + \frac{1}{27}(x - 2)^2 - \frac{1}{81}(x - 2)^3$ 0.031 25

19 (a) 0 (b) 4

20 (a) −10 (b) −6 (c) 23.25

21 (a) −31 (b) −1

22 (a) 1301 (b) 309.125 (c) 127.625 (d) 1385.096

24 (a) −2.25, −9.5 (b) −24.25, −33.5

25 (a) 18 (b) 2 (c) 366

26 (a) −4 (b) 148.25 (c) 2

27 (a) $-1\,\mathrm{m\,s^{-1}}$ (b) $-20\,\mathrm{m\,s^{-1}}$ (c) $-57\,\mathrm{m\,s^{-1}}$

28 (a) −11 (b) 0

31 (a) 8 (b) −8 (c) −2

32 (a) 6 (b) 2 (c) −34

33 (a) 10, 14 (b) 61.375, 61.25 (c) −40.944, 48.56

34 (a) 101 (b) 346

37 $\alpha + \frac{1}{3}\alpha^3$

38 $f'(0) = 0$ $f''(0) = -3$ $f'''(0) = 15$

39 (a) $0.367\,88 - 0.367\,88(x - 1) + 0.183\,94(x - 1)^2$

$$- 0.061\,313(x - 1)^3 + \frac{1}{4!}e^{-\eta}(x - 1)^4$$

(b) $0.5 - 0.0625(x - 3) + 0.011\,718\,8(x - 3)^2$

$$- 0.002\,441\,4(x - 3)^3 + \frac{105}{16 \times 4!}(\eta + 1)^{-\frac{9}{2}}(x - 3)^4$$

(c) $1 - \left(x - \frac{\pi}{2}\right)^2 - \frac{8\cos(2\eta)}{4!}\left(x - \frac{\pi}{2}\right)^4$

40 (a) $p_2(x) = \frac{x(x-1)}{-0.25} \times 0.303\,27 + \frac{x(x-0.5)}{0.5} \times 0.367\,88$

(b) $T_2(x) = x - x^2$

41 (a) $0.333\,333 - 0.055\,555\,6(x - 5) + 0.009\,259\,25(x - 5)^2$
$- 0.001\,543\,21(x - 5)^3$

(b) $f(5.1) \approx 0.327\,87$ $f(5.5) \approx 0.307\,68$

(c) $\dfrac{0.006\,172\,84}{4!}(x-5)^4$

(d) Error in f(5.1) $\leqslant 0.257\,20 \times 10^{-7}$
Error in f(5.5) $\leqslant 0.160\,75 \times 10^{-4}$

43 (a) $1 + x + \dfrac{x^2}{2} + \dfrac{x^3}{6} + \dfrac{x^4}{24}$ $\quad r(x) = \dfrac{e^{\eta}x^5}{5!}$ $\quad 0 < \eta < x$ (b) 8

44 $x + \frac{1}{6}x^3 + \frac{1}{120}x^5$ $\quad y_1 = 0.100167$ $\quad y_2 = 0.2010$ $\quad 0.2013$

46 $y_0'' = 0$ $\quad y_0''' = 2$ $\quad y_0^{(4)} = -6$ $\quad y_0^{(5)} = 0$
$y_1 \approx 1.7003$ $\quad y_1'' \approx 0.170\,03$
$y_2 \approx 1.4017$ $\quad y_3 \approx 1.1046$

Exercise 6

1 (a) 1.218 (b) –2.211, 1.241 (c) –4.874, –0.8503, 0.7240

2 (a) $x_0 = 0$; 0.248 044 589 $\quad x_0 = 2$; 1.941 897 76
(b) $x_0 = -3$; –2.278 863 94 $\quad x_0 = 2$; 2.278 862 87

3 (a) $x_0 = 1$; 1.854 909 3 (b) $x_0 = -5$; –4.627 365 08
(c) $x_0 = 1$; 0.806 465 994

4 $x_0 = -2$, $x_1 = -2.375$, $x_2 = -2.306$, $x_3 = -2.303$ (4 d.d.)

5 $x_0 = 1$, $x_1 = 0.2$, $x_2 = -3.738$, $x_3 = -2.686$
$x_0 = -1$, $x_1 = -3$, $x_2 = -2.238$, $x_3 = -1.840$

6 98.950 062 9

7 2.086

8 0.135 902 635, 1.075 376 46

9 –0.563 57

10 Solution is 1.267 527 8 (8 d.d.)

12 (a) –1.2361, 2.0000, 3.2361
(b) –1.4476, 0.30135, 1.1462
(c) –2.3894, 0.091 046, 2.2984
(d) –1.2273, –0.679 70, 0.180 77, 1.3263
(e) –0.471 78, –1.0000, 1.2718
(f) –1.3317, 0.115 25, 0.930 75

13 (a) –1.6696, 0.194 71, 0.237 42 ± 0.984 39i
(b) –0.459 47, –1.0000, 0.729 73 ± 0.581 43i
(c) 0.823 61, –0.411 81 ± 1.5029i
(d) 0.205 57, 1 , –2.1028 ± 0.665 46i
(e) –0.328 14, 2.1661, 0.206 03 ± 0.81302i
(f) –0.329 04, 0.789 52 ± 0.946 69i

14 10.2, 0.0123

15 1.5

16 0.818

17 (a) $x_{n+1} = x_n \left(\frac{21 - x_n^2}{14} \right)$ 2.65

18 2.17 (b) 2.18

19 −0.2034, −0.7446, 1.8892

21 $e_{n+1} \approx \frac{3}{5}\lambda^2 e_n$

22 0.46

23 (a) −2.4321, 0.579 81, 1.5208, 2.3315

(b) $x = -1.5$ $x = 1$ $x = 2$
$y = -22.4375$ $y = 1$ $y = -1$

24 If $a = -3.7913$ and $b = 1.6180$ $Q(x) = x^2 - 0.173\,253\,855x - 0.489\,042\,778$
$x = 0.791\,29$, $x = -0.618\,03$

25 4.3 $-3 \pm 4i$

26 (a) [5.81, 5.82] (b) 5.816 19 (c) $p = \alpha$ $q = -\frac{1}{\alpha}$
(d) 0.029 41, −5.846 (e) $0.294\,125 \times 10^{-1}$, −5.845 60

Exercise 7

1 (a) 0.65 (b) −0.7

2 −1.3, −1.3, 0.1, 1.4

use backward difference $\frac{dy}{dx} \approx \frac{y_6 - y_7}{x_6 - x_7}$

3 1.5, 2.5, 4.5, 8

4 1.35, 0.45, −0.8, −0.85, −0.55

5	**(a)**	**(b)**	**(c)**	**(d)**
Approx	0.414	0.3564	0.3179	0.2897
Exact	0.408 25	0.353 55	0.316 23	0.288 68
Error	−0.005 75	−0.002 85	−0.001 67	−0.001 02

6 x	**2.25**	**2.75**	**3.25**	**3.5**	**3.75**
(a)	0.8428	0.696	0.5928	0.552	0.5164
(b)	0.8926	0.7292	0.6166	0.5724	0.5342
Exact	0.8889	0.7273	0.6154	0.5714	0.5333

7 (a) (i) 0.59 (ii) 1.655
(b) (i) −0.38 (ii) 0.105
(c) (i) −0.41 (ii) −0.395

8

x	10	10.5	11	11.5
$\frac{dy}{dx}$	−2.5	−1.8	0.6	2.3
$\frac{d^2y}{dx^2}$	−0.4	3.2	6.4	0.4

9 (a) 0.1043 (b) 0.0346 (c) 0.0151 (d) 0.0079
(a) 0.089 44 (b) 0.031 62 (c) 0.014 27 (d) 0.007 543
Error −0.014 86 −0.002 98 −0.000 83 −0.000 357

10 (a) 1.05, 0.35, −0.5, −0.55
(b) −0.3, −1.1, −0.6, 0.5
At $t = 3$ 1.05, −0.3 At $t = 6$ −0.55, 0.5

11 (b) (i) −1 (ii) 5 (iii) 17
(c) (i) 2 (ii) −4 (iii) −7
(d) (i) 3 (ii) 6 (iii) 9

12 (b) (i) 2 (ii) 2 (iii) 11
(c) (i) −1 (ii) −1 (iii) −1
(d) (i) −1 (ii) −1 (iii) −1

13 (a) 3.5 (b) 0.5 (c) 5.145

14 (a) 2.25 (b) 5 (c) 1.125 (d) −1.5 (e) 2

15 (a) 22.804 (b) 0.25 (c) 1.4979 (d) 4 (e) (d) is exact

16 9.275

17 (a) 1.755 (b) 6.525 (c) 4.46 (d) 0.14 (e) 49.215

18 0.399 64

19 3.6246

20 Cubic polynomial, f(2.6) ≈ 64.4 62.96

21 (a) 0.556 19, 0.002 604 2 (b) 0.558 08, 0.001 89

22 $0.673\,66 \times 10^{-4}$ $0.174\,72 \times 10^{-5}$ $0.659\,97 \times 10^{-4}$

23 $0.999\,08 \times 10^{-2}$ $0.114\,31 \times 10^{-2}$ $0.967\,90 \times 10^{-2}$

24 (a) 105.43 (b) 75.122 (c) 66.573
(a) 150.64 (b) 37.660 (c) 9.4151 63.618

25 Error considerations give $n = 6$ (but $n = 5$ gives a solution of required accuracy)
Exact solution = 0.708 073

26 (a) 8.6667 (b) 5.3 (c) −2.7

27 (a) 7.7333 (b) 4.6667 (c) −10.1 (d) −0.15

28 (a) 9.4708 (b) 0.269 26 (c) 2.4281 (d) 12.125 (e) 4.4490
(f) 2.1434

29 (a) 6.2742 (b) 0.864 96 (c) 20.833 (d) 0.296 34
(e) 0.292 89 (f) 11.943 (g) −0.706 41 (c) is exact

30 (a) 10.15 (b) 2.7033 (c) 102.43 (d) −4.62 (e) 3.4167
(f) −0.410 72

31 1.451 726 95 1.444 565 98 Error = 0.007 161 0

32 49.318 329 4 Error = 0.3×10^{-3}

33 (a) 1.347 690 (b) 1.347 676 85 (c) 1.347 678 12 1.348

34 (a) 2.083 333 (b) 2.079 761 9 (c) 2.079 509 38 2.079 44

35 6 45.2235

36 (a) 1.0432 (b) 1.037 774 66

37 (a) 0.706 096 (b) 0.707 107 94

38 0.100 51 $-\frac{1}{6}(x+4)^{-3}$ $-0.228\,62 \times 10^{-3}$ 0.100 335 35

39 1.218 865 51 $\frac{1}{3072}(x+3)^{-\frac{7}{2}}$ $0.287\,72 \times 10^{-4}$ 1.218 951 42

40 Error considerations give (a) 148 (b) 14

41 Error considerations give (a) 30 (b) 6
Errors (a) $-0.423\,25 \times 10^{-4}$ (b) -0.9835×10^{-5}

42 J_1' 0.499 0.4975 0.492 0.4825 0.47
J_1'' −0.03 −0.08 −0.11 −0.14
Using central difference formula except for $J_1'(0)$

43 3.5, 6.75, 3 15, −2.5

44 (a) −0.184 04 (b) $-\frac{1}{24} f'''(\eta)$ $1.5 < \eta < 2.5$ (c) $\frac{5}{64}$

45 −1.4 −1.9

46 1.991 531 4

47 (a) 25.35 (b) 24.966 667

48 (a) 0.785 (b) 0.334 0.9614

49 2.004, 0.2%

50 (a) 1.553 Error = $0.48k$ (b) 0.569 Error = $0.12k$

51 1.776

52 (a) 0.8821 (b) 0.0040 3.14

53 $2\varepsilon + \dfrac{2}{3n}$ $n > 167$

54 0.855 65, 0.000 07

Exercise 8

2 $y = x^2 + 5x - 3$

3 $y = x\tan x$

5 $w_{n+1} = (1 - 2h)w_n + 3hx_n$ 1.1, 1.21, 1.328, 1.4524, 1.58192

6 11.808, 13.795, 15.967, 18.333, 20.900

7 (a) 0.65, 0.825517092, 1.03020908, 1.26821587
(b) 0.656313555, 0.83969888, 1.05409293, 1.30395842

8 (a) 2.6, 3.381, 4.3993, 5.72809001
(b) 2.645125, 3.49987781, 4.63401341, 6.13978273

9 (a) 8.04, 8.07843137, 8.11538462, 8.1509434, 8.18518519
(b) 8.03960396, 8.0776699, 8.11428572, 8.14953271, 8.18348624

10 With $h = 0.5$ 11.25, 21.125, 35.688, 57.281, 89.422

11 This question should be attempted on a computer. Take $h \leqslant 0.001$.

12 Exact values (to 5 d.d.) (a) 4.7474 (b) 10.426

13

x	**1.2**	**1.4**	**1.6**	**1.8**	**2**
(a)	−0.81654	−0.68480	−0.58511	−0.50669	−0.44314
(b)	−0.82492	−0.69949	−0.60495	−0.53092	−0.47122

14 0.8, 0.64, 0.512 d.e. $= \dfrac{y_{n+1} - y_n}{h} + 2y_n$

Since $y_{n+1} = y_n + hy'_n + \dfrac{h^2}{2}y''(\eta)$

$$\text{d.e} = \frac{hy'_n + \frac{1}{2}h^2y''(\eta)}{h} + 2y_n = y'_n + \frac{h}{2}y''(\eta) + 2y_n$$

$$= \frac{h}{2}y''(\eta) = \frac{h}{2}\,4y(\eta) = 2hy(\eta)$$

Since $0.81873 \leqslant y(\eta) \leqslant 1$ $\quad 0.16375 \leqslant \text{d.e.} \leqslant 0.2$

$0.81638 \leqslant y(1.1) \leqslant 0.82$

15 $w_{n+1} = w_{n-1} + \dfrac{8he^{x_n}}{w_n}$ 2.0873

(a) 2.1514 (b) 2.2346 (c) 2.3136 (d) 2.4022

16 w_1 (a) 2 (b) 1.96 (c) 1.957333
w_2 (a) 1.84 (b) 1.824 (c) 1.8229332
w_3 (a) 1.616 (b) 1.5696 (c) 1.56650628

17 2.98, 4.7261, 8.7501

18 x	**1**	**1.1**	**1.2**	**1.3**	**1.4**	**1.5**
y	0	0.11	0.241 99	0.397 94	0.579 47	0.787 38

19 0.076, 0.424 41, 1.3736, 3.6475, 8.8433

21 (a) 2.46, 3.9222, 5.3877
(b) 2.4608, 3.9243, 5.3916
(c) 2.4610, 3.9247, 5.3922

22 (a) 2.1, 2.199, 2.294 92, 2.3856, 2.4687
(b) 2.1, 2.1980, 2.2916, 2.3788, 2.4571

23 (a) 2.1789 (b) 2.0732 (c) 2.0880

25 (a) 4.608, 7.6235, 13.564, 25.912, 53.068
(b) 4.042, 4.1750, 4.4152, 4.7863, 5.3214
(c) 0.545 38, 0.732 75, 0.912 62, 1.0855, 1.2518
(d) 2.8715, 3.1902, 3.4720, 3.7301

26 (7) (a) 0.662 76, 0.854 20, 1.0786, 1.3406
(b) 0.663 01, 0.854 78, 1.0795, 1.3421
(8) (a) 2.6905, 3.6214, 4.8778, 6.5746
(b) 2.6974, 3.6401, 4.9158, 6.6431
(9) (a) 8.04, 8.0784, 8.1154, 8.1509, 8.1852
(b) 8.0392, 8.0769, 8.1132, 8.1481, 8.1818
(10) With $h = 0.5$ 12.813, 26.008, 47.138, 81.161, 136.14
(11) Again, take a very small step size

(13) x	**1.2**	**1.4**	**1.6**	**1.8**	**2**
(a)	−0.831 87	−0.711 83	−0.621 79	−0.551 70	−0.495 56
(b)	−0.832 95	−0.713 65	−0.624 17	−0.554 56	−0.498 86

27 $h = 0.1$: 2.5168, 3.4848, 5.8010, 15.237

29 1.2576, 1.5569, 1.9018, 2.2956, 2.7414

30 $h = 0.2$ 10.022, 10.048, 10.078, 10.111, 10.149

31 $h = 0.2$ 2.3175, 3.6262, 5.9949, 10.271

32 (a) 4.4, 5.0970, 6.3699, 8.9668
(b) 4.5485, 5.6073, 8.0127, 16.018
(c) 4.5453, 5.6021, 8.0287, 16.430

33 (a) −2.9989, −4.5345, −6.8690, −10.392, −15.685
(b) −3.0064, −4.5569, −6.9189, −10.492, −15.870

34 (a) 8.1984, 6.7214, 5.5104, 4.5177, 3.7037
(b) 13.239, 27.077, 43.554, 63.322, 87.172

35 −3.0975, −3.1903, −3.2785, −3.3624, −3.4421

36 1.41 (Use a very small step size)

37 2.9623, 2.9259, 2.8908, 2.8570, 2.8243

38 4.5524, 5.6391, 8.2614, 20.087

39 $h = 0.2$ −0.6, −0.375 55, −0.244 40, −0.164 72, −0.114 49
$h = 0.1$ −0.687 77, −0.501 81, −0.382 15, −0.300 64, −0.242 63
$h = 0.05$ −0.694 48, −0.510 24, −0.390 66, −0.308 68, −0.250 03

40 (a) −19, 361, −6859
(b) 181, 32 761, 5 929 700
(c) 5514.3, 30 408 000, 0.168×10^{10}
All wildly inaccurate

41 (a) $\frac{du}{dx} = v \qquad \frac{dv}{dx} = 4v - 2u + x \qquad u(3) = -1,\ v(3) = 0$
(b) $\frac{du}{dx} = v \qquad \frac{dv}{dx} = -2v + 7u + \ln(x) + 1 \qquad u(-1) = 2,\ v(-1) = 3$
(c) $\frac{du}{dx} = v \qquad \frac{dv}{dx} = \frac{1 - 3u^2}{x} \qquad u(50) = 0.2,\ v(50) = 0.4$
(d) $\frac{du}{dx} = v \qquad \frac{dv}{dx} = \frac{5u^2 - xv}{\sqrt{x - 1}} \qquad u(10) = 1.4,\ v(10) = -0.2$

42 (a) 2.9, 3.91, 5.055, 6.3647, 7.8745
(b) 0.2, 0.48, 0.868 28, 1.3988, 2.1143
(c) 0.3, 0.87, 1.7515, 2.9784, 4.5789
(d) 0.681 61, 0.816 44, 0.957 43, 1.1020, 1.2472
(e) 0.45, 0.405, 0.364 89, 0.329 35, 0.297 96

43 (a) 2.955, 4.0462, 5.3063, 6.7752, 8.5010
(b) 0.24, 0.587 24, 1.0767, 1.7540, 2.6781
(c) 0.435, 1.1839, 2.2845, 3.7679, 5.6606
(d) 0.685 90, 0.822 44, 0.962 02, 1.1015, 1.2378
(e) 0.4525, 0.409 88, 0.371 83, 0.337 96, 0.307 85

44 (a) 2.9594, 4.0567, 5.3255, 6.8062, 8.5482
(b) 0.244 23, 0.596 63, 1.0931, 1.7802, 2.7180
(c) 0.443 34, 1.2031, 2.3169, 3.8159, 5.7265
(d) 0.685 43, 0.821 32, 0.960 00, 1.0985, 1.2334
(e) 0.452 49, 0.409 84, 0.371 75, 0.337 84, 0.307 69

46 (a) 1.2, 3.0468, 7.5470, 16.123, 30.754
(b) 2.0234, 5.7844, 13.955, 28.925, 54.062
(c) 2.2019, 6.2840, 14.993, 30.827, 57.304

50 2.3, 2.6695, 3.1358, 3.7425

51 1.1125, 1.2607, 1.4635, 1.7562, 2.2126 Exact 1.653 842 (7 d.d.)

52 Exact 2.317 911 7, 5.843 519 9, 16.086 815 (8 d.d.)
(a) 4.9960 (b) 12.393

53 $h = 0.1$ 2.1938, 5.1596, 13.028

54 0.42667×10^{-1}, 0.18229

55 $y = \dfrac{1}{1-x}$ 1, 1.25, 1.6406, 2.3135, 3.6517, 6.9853

56 y(1.1) = 1 y(1.2) = 1.03

x	1.4	1.6	1.8	2
y	1.232 636	1.461 909	1.891 159	2.337 651

Exact solution: $y = \dfrac{1}{4}x^3 + \dfrac{3}{4x}$ y(2) = 2.375

59 1.9958, 2.3302, 2.9615

60 (a) 0.91, 0.834 73, 0.770 87, 0.716 02, 0.668 43

(b) $\dfrac{h^3}{6}y'''(\eta)$ $1 < \eta < 1.5$ (c) −0.001 $1.1 < \eta < 1.2$

61 (a) 0.496 993 (b) 0.496 920 0.496 896

62 1, 0.980, 0.920 77, 0.824 47, 0.694 47, 0.535 19, 0.352 06, 0.151 49, −0.059 16

64 1.1657, 1.2763, 1.3417, 1.3674, 1.3592

65 As a guide, when $n = 1.5$, y(2) ≈ 1.286 67

Index

Pages on which definitions are to be found are shown in **bold** type.

absolute error **30**–34, 38
  in addition and subtraction 48–9
  error in 48, 53
  of polynomials 65
Aitken's process 130–31
algebraic solution to equation **94**
algorithms 13–16, 23
analytic solution to differential equation **284**
augmented matrix of system of equations
  149–52, **150**

backward substitution **143**
binary numbers 4–5, 352
bisection (binary search) method 97–100
  algorithm 100, 102
  convergence 104
  error analysis 120–21
bounds of error 35, 122
byte **5**

calculus in error analysis 46–7
central difference formula 235–6
  error term 237–41
Chebyshev polynomials **342**
chopping 27–8
  roundoff error 31, 37, 353
coefficients of polynomial **65**
computer arithmetic, errors in 9, 38
computer storage of numbers 4–9
constant term of polynomial **65**
continuous functions **16–17**
  approximation by polynomials 67
  see also polynomial interpolation
convergence 213
  bisection method 104
  fixed point iterative method 122, 123,
    124–31, 207–8, 209, 215
  Newton-Raphson method 212–13
  secant method 109, 122, 123
cubic equation, solving 223

data sets 10–11, 68
difference equation **288**
differential equations 13, 282–320, **283**
differentiation *see* numerical differentiation
discontinuous functions 17
division, error in 51–2

elementary operations **144**
elementary row operations **151**
equivalent systems of equations **142**
errors 3, **25**–53
  analysis 38, 39–53; *see also under*
    bisection method; fixed point
    iterative method; secant method
  in computer arithmetic 9, 38
  in Euler's method 293–5
  in use of Gaussian elimination 157–60
  in use of Lagrange interpolating polynomial
    80–85, 187–8
  measurement 30–38
  roundoff **29,** 31–2, 36–7, 163, 353–4
Euler's method for solving differential equations
  288–91, 300, 318
  algorithm 290, 313
  compared with Runge-Kutta methods 306–7, 310
  error 293–5
  geometrical interpretation 291–2
  modified *see* Runge-Kutta methods,
    simple
exponent **5, 27**

first derivative, calculation 233–6, 295–8
first order differential equations **284**, 286, 288–90
fixed point iterative method 109–16
  algorithm 112
  convergence 122, 123, 124–31, 207–8, 209,
    215
  error analysis 122–4
  geometrical interpretation 116–20
  Newton-Raphson method as special case 214
floating point form 6, **27**
forward difference formula 233–4
  error term 237, 239–40

Gaussian elimination 143–9
  algorithm 152–3
  errors using 157–60
  in tridiagonal systems 155–6;
    algorithm 156–7
  using augmented matrix 149–52
  using pivoting strategies 160–63;
    algorithm 162
  with sum check 153–5
general solution to differential equation **285**–6

Heun's method 307–8
  algorithm 307
Horner's method of polynomial
  evaluation 188–96;
    algorithm 195, 217

ill-conditioned systems 163–7, **164**
infinite decimals 26
  computer storage 8

integration *see* numerical integration
intermediate value theorem 19–20
interval arithmetic 39–44
inverse interpolation **346**
iterative function **112**
  choice of 114, 116, 120
iterative methods **97**

Lagrange interpolating polynomial 71–80
  algorithm for calculation 77
  errors 80–85, 187–8
linear interpolation *see* secant method
local discretisation error **293**

machine number **6–7**
Maclaurin polynomials (series) 175, **187**
mantissa **5, 27**
matrix of coefficients of system of equations
  149, 166–7
multiplication
  error in 50–51
  of polynomials 66

Neville's algorithm 77
Newton-Raphson method 204–10, 213,
  214–16, 222
  algorithm 205
  convergence 215
  graphical representation 210–13
  use in evaluating polynomials 216–25
numerical analysis **3–4**
numerical differentiation **231**, 232–41
numerical integration **232**, 241–68
numerical methods **94**
numerical solutions 11–12, **94**

partial pivoting 161
particular solution to differential equation **286**
pivotal equation **145**
pivoting strategies **160**–63
polynomial interpolation 67–80
  errors 80–85
polynomials **64**–85, 174–96
  evaluation (Horner's method) 188–96;
    algorithm 195
  finding real solutions 216–25;
    algorithm 219; program 220–21
predictor-corrector method **345**
proper integrals 232

quadratic equations 44–6
Quasimodo function **81**

relative error **34**–5, 38
  in addition and subtraction 53
  in multiplication and division 50–52
Rolle's theorem 18–19
  applications 82, 184
Romberg integration **341**
rounding 28
  roundoff error 32, 36, 353–4
roundoff error **29,** 31–2, 36–7, 163, 353–4
Runge-Kutta methods 304–20
  of order four 309–10, 320;
    algorithms 309, 319–20
  or order two 355–6
  simple 305–7, 314–17, 318–19;
    algorithms 305, 314

secant method 105, 106–9
  algorithm 109
  convergence 109, 122, 123, 207–8, 209, 215
  error analysis 122
  Newton-Raphson method as special case 216
second derivative, calculation 236–7, 240–41
second order differential equations **284**, 286,
  311–17
Simpson's rule 250–55, **252**, 264–5
  algorithm 258
  composite 255–7, **256**, 266–8, **267**
simultaneous equations 12, 141
solution to differential equation **284**–6
solution to equation **94**–5
starting value **97**
step size **287,** 290, 294–5
stopping rules (iterative methods) 101–4, 107–9
subtraction
  error in 49–50, 53
  of polynomials 66
subtractive cancellation 42–4, **43**
systems of linear equations 141–67
  ill-conditioned 163–7
solution *see* Gaussian elimination

Taylor methods 298–303
Taylor polynomials (series) 174–83, **179,** 187
  errors in use *see* Taylor's theorem
Taylor's theorem 183–8
trapezium method **344**
trapezium rule 242–8, **244**, 257, 259–61
  algorithm 250
  composite **248**–50, 261–3, **262**
triangular systems of equations **142**–3
tridiagonal systems of equations 155–7